326

Topics in Current Chemistry

Topics in Current Chemistry

Recently Published and Forthcoming Volumes

NMR of Proteins and Small Biomolecules
Volume Editor: Guang Zhu
Vol. 326, 2012

Molecular Imprinting
Volume Editor: Karsten Haupt
Vol. 325, 2012

Activity-Based Protein Profiling
Volume Editor: Stephan A. Sieber
Vol. 324, 2012

Beauty in Chemistry
Volume Editor: Luigi Fabbrizzi
Vol. 323, 2012

Constitutional Dynamic Chemistry
Volume Editor: Mihail Barboiu
Vol. 322, 2012

EPR Spectroscopy
Volume Editors: Malte Drescher, Gunnar Jeschke
Vol. 321, 2012

Radicals in Synthesis III
Volume Editors: Markus R. Heinrich, Andreas Gansäuer
Vol. 320, 2012

Chemistry of Nanocontainers
Volume Editors: Markus Albrecht, F. Ekkehardt Hahn
Vol. 319, 2012

Liquid Crystals: Materials Design and Self-Assembly
Volume Editor: Carsten Tschierske
Vol. 318, 2012

Fragment-Based Drug Discovery and X-Ray Crystallography
Volume Editors: Thomas G. Davies, Marko Hyvönen
Vol. 317, 2012

Novel Sampling Approaches in Higher Dimensional NMR
Volume Editors: Martin Billeter, Vladislav Orekhov
Vol. 316, 2012

Advanced X-Ray Crystallography
Volume Editor: Kari Rissanen
Vol. 315, 2012

Pyrethroids: From Chrysanthemum to Modern Industrial Insecticide
Volume Editors: Noritada Matsuo, Tatsuya Mori
Vol. 314, 2012

Unimolecular and Supramolecular Electronics II
Volume Editor: Robert M. Metzger
Vol. 313, 2012

Unimolecular and Supramolecular Electronics I
Volume Editor: Robert M. Metzger
Vol. 312, 2012

Bismuth-Mediated Organic Reactions
Volume Editor: Thierry Ollevier
Vol. 311, 2012

Peptide-Based Materials
Volume Editor: Timothy Deming
Vol. 310, 2012

Alkaloid Synthesis
Volume Editor: Hans-Joachim Knölker
Vol. 309, 2012

Fluorous Chemistry
Volume Editor: István T. Horváth
Vol. 308, 2012

NMR of Proteins and Small Biomolecules

Volume Editor: Guang Zhu

With Contributions by

K. Chen · N.K. Goto · R. Ishima · H. Li · R. Powers · J. Qin ·
T. Qureshi · J.L. Stark · H. Sun · K.H. Sze · N. Tjandra · H.S. Tse ·
O. Vinogradova · Q. Wu · X. Zhao · G. Zhu

Springer

Editor
Prof. Dr. Guang Zhu
Department of Biochemistry
The Hong Kong University of Science and Technology
Clear Water Bay
Kowloon, Hong Kong
China

ISSN 0340-1022 e-ISSN 1436-5049
ISBN 978-3-642-28916-3 e-ISBN 978-3-642-28917-0
DOI 10.1007/978-3-642-28917-0
Springer Heidelberg Dordrecht London New York

Library of Congress Control Number: 2012935239

Printed on acid-free paper

Springer is part of Springer Science+Business Media (www.springer.com)

Volume Editor

Prof. Dr. Guang Zhu

Department of Biochemistry
The Hong Kong University of Science and Technology
Clear Water Bay
Kowloon, Hong Kong
China

Editorial Board

Topics in Current Chemistry Also Available Electronically

Topics in Current Chemistry is included in Springer's eBook package *Chemistry and Materials Science*. If a library does not opt for the whole package the book series may be bought on a subscription basis. Also, all back volumes are available electronically.

For all customers with a print standing order we offer free access to the electronic volumes of the series published in the current year.

If you do not have access, you can still view the table of contents of each volume and the abstract of each article by going to the SpringerLink homepage, clicking on "Chemistry and Materials Science," under Subject Collection, then "Book Series," under Content Type and finally by selecting *Topics in Current Chemistry*.

You will find information about the

- Editorial Board
- Aims and Scope
- Instructions for Authors
- Sample Contribution

at springer.com using the search function by typing in *Topics in Current Chemistry*.

Color figures are published in full color in the electronic version on SpringerLink.

Aims and Scope

The series *Topics in Current Chemistry* presents critical reviews of the present and future trends in modern chemical research. The scope includes all areas of chemical science, including the interfaces with related disciplines such as biology, medicine, and materials science.

The objective of each thematic volume is to give the non-specialist reader, whether at the university or in industry, a comprehensive overview of an area where new insights of interest to a larger scientific audience are emerging.

Thus each review within the volume critically surveys one aspect of that topic and places it within the context of the volume as a whole. The most significant developments of the last 5–10 years are presented, using selected examples to illustrate the principles discussed. A description of the laboratory procedures involved is often useful to the reader. The coverage is not exhaustive in data, but rather conceptual, concentrating on the methodological thinking that will allow the non-specialist reader to understand the information presented.

Discussion of possible future research directions in the area is welcome.

Review articles for the individual volumes are invited by the volume editors.

In references *Topics in Current Chemistry* is abbreviated *Top Curr Chem* and is cited as a journal.

Impact Factor 2010: 2.067; Section "Chemistry, Multidisciplinary": Rank 44 of 144

Preface

Nuclear Magnetic Resonance (NMR) is one of only a few scientific techniques that have been widely applied in many different areas such as physics, chemistry, biology and medicine. Today, NMR plays a crucial role in structure-function studies of organic and inorganic compounds and of large biomolecules, particularly proteins and DNA. In medicine it is one of the most important imaging techniques available to physicians and it also has extensive applications in pharmaceutical research. In this special issue of NMR of Proteins and Small Biomolecules, review chapters cover a wide-range of topics on some of the latest developments in NMR techniques and their applications.

A review of recent developments in structure-based drug discovery begins this volume by showing how combining 1D and 2D NMR techniques with molecular docking can efficiently screen and identify novel "druggable" leads. The success of such an approach relies on the selection of worthwhile therapeutic targets; presumably proteins that are critical in particular diseases. In that context, studying protein-ligand interactions with NMR techniques that have unique powers for the study of weak protein-ligand interactions, as described in chapter 2, followed by the use of residual dipolar coupling approaches, reviewed in chapter 3, may be essential. Chapter 4 is dedicated to the study paramagnetic metalloproteins, giving an overview of paramagnetic NMR and ^{13}C directly detected protonless NMR spectroscopy. Since NMR is especially powerful for the study of proteins dynamics, more recent developments in transverse relaxation dispersion experiments, which have extended the range of NMR relaxation studies to the milli-micro second timescale, are illustrated in chapter 5. Solid and liquid state NMR techniques for studying membrane proteins represent another "hot" area of contemporary research. Chapters 6 and 7 systematically describe these approaches. This volume concludes with chapter 8, a review of the considerable sensitivity enhancement attainable in magnetic resonance spectroscopic imaging (MRSI) through the use of dynamic nuclear polarization transfer. This non-invasive technique can be applied to the measurement of metabolites *in vivo* to allow early diagnosis and assessment of diseases in personalized medicine.

I hope that the reviews of these unique topics in NMR techniques and their applications as presented in this volume NMR of Proteins and Small Biomolecules are informative and fun to read; and I thank all the authors who made this project possible.

Hong Kong Guang Zhu

Contents

Application of NMR and Molecular Docking in Structure-Based Drug Discovery 1
Jaime L. Stark and Robert Powers

NMR as a Unique Tool in Assessment and Complex Determination of Weak Protein–Protein Interactions 35
Olga Vinogradova and Jun Qin

The Use of Residual Dipolar Coupling in Studying Proteins by NMR 47
Kang Chen and Nico Tjandra

NMR Studies of Metalloproteins 69
Hongyan Li and Hongzhe Sun

Recent Developments in ^{15}N NMR Relaxation Studies that Probe Protein Backbone Dynamics 99
Rieko Ishima

Contemporary Methods in Structure Determination of Membrane Proteins by Solution NMR 123
Tabussom Qureshi and Natalie K. Goto

Protein Structure Determination by Solid-State NMR 187
Xin Zhao

Dynamic Nuclear Polarization: New Methodology and Applications 215
Kong Hung Sze, Qinglin Wu, Ho Sum Tse, and Guang Zhu

Index 243

Top Curr Chem (2012) 326: 1–34
DOI: 10.1007/128_2011_213

Published online: 14 September 2011

Application of NMR and Molecular Docking in Structure-Based Drug Discovery

Jaime L. Stark and Robert Powers

Abstract Drug discovery is a complex and costly endeavor, where few drugs that reach the clinical testing phase make it to market. High-throughput screening (HTS) is the primary method used by the pharmaceutical industry to identify initial lead compounds. Unfortunately, HTS has a high failure rate and is not particularly efficient at identifying viable drug leads. These shortcomings have encouraged the development of alternative methods to drive the drug discovery process. Specifically, nuclear magnetic resonance (NMR) spectroscopy and molecular docking are routinely being employed as important components of drug discovery research. Molecular docking provides an extremely rapid way to evaluate likely binders from a large chemical library with minimal cost. NMR ligand-affinity screens can directly detect a protein-ligand interaction, can measure a corresponding dissociation constant, and can reliably identify the ligand binding site and generate a co-structure. Furthermore, NMR ligand affinity screens and molecular docking are perfectly complementary techniques, where the combination of the two has the potential to improve the efficiency and success rate of drug discovery. This review will highlight the use of NMR ligand affinity screens and molecular docking in drug discovery and describe recent examples where the two techniques were combined to identify new and effective therapeutic drugs.

Keywords Drug discovery, FAST-NMR, *In silico* screening, Ligand affinity screens, Molecular docking, Nuclear magnetic resonance, Virtual screening

J.L. Stark and R. Powers (✉)
Department of Chemistry, University of Nebraska – Lincoln, 722 Hamilton Hall, Lincoln, NE 68588-0304, USA
e-mail: rpowers3@unl.edu

Contents

1 Introduction 2
2 NMR Ligand Affinity Screens 4
 2.1 Ligand-Based NMR Screens 4
 2.2 Target-Based NMR Screens 6
3 Molecular Docking 7
 3.1 Docking 8
 3.2 Scoring 10
 3.3 Protein Flexibility 13
 3.4 Virtual Screening and Assessment 14
4 Combining Molecular Docking with NMR Ligand Affinity Screens 17
 4.1 Identification of New Therapeutic Targets 18
 4.2 Rapid Protein–Ligand Structure Determination 20
 4.3 Lead Identification 23
5 Concluding Remarks 26
References 26

1 Introduction

The completion of the human genome project [1] coupled with an increase in R&D investments was widely anticipated to be the cornerstone of personalized medicine with a corresponding explosion in new pharmaceutical drugs targeting a range of diseases. Nearly a decade later, the rate at which new drugs enter clinical development and reach the market has declined dramatically despite the influx of novel therapeutic targets and R&D investments. In the past 5 years the number of new molecular entities (NMEs) receiving FDA approval has decreased by 50% from the previous 5 years [2]. There are several reasons for this decline, but most stem from the fact that drug discovery is a complex and costly endeavor. Approximately 80–90% of drugs that reach the clinical testing phase fail to make it to market [3, 4]. Efforts to reduce costs often lead pharmaceutical companies to invest their time and money in proven therapies, like "best-in-class" drugs, instead of "first-in-class" drugs that target new mechanisms of action or diseases. As a result, many diseases are "orphaned" and lack any therapeutic compounds in the discovery pipeline. Addressing these issues will require fundamental changes to create a more efficient drug discovery process.

The enormous costs and high failure rates inherent to the pharmaceutical industry are clearly contributing factors to the declining number and diversity of new therapeutics. Efforts that minimize costs without restricting research endeavors will evidently benefit the development of drugs for various human diseases. The availability of hundreds of whole-genome sequences for numerous organisms provides an invaluable data set for drug research [1, 5, 6]. Identifying a novel "druggable" protein target is a critical first step for a successful and efficient drug discovery effort. Unfortunately, bioinformatics analysis alone does not generally provide enough information to justify embarking upon an expensive drug discovery program [7, 8]. Instead, knowing the three dimensional structure of a protein greatly enhances the value of the bioinformatics analysis. Protein structures often provide insights into the molecular basis of the protein's biological function and its relationship to a particular disease. A protein structure also provides detailed information on the sequence and structural characteristics that govern ligand binding interactions. Building a drug discovery effort based on structural information promises to help in the identification of novel therapeutic targets, in the discovery of new lead compounds, and in the optimization of drug-like properties to improve efficacy and safety. Currently, the drug discovery process within the pharmaceutical industry employs high-throughput screening (HTS) as the primary method for identifying lead compounds. However, the high false positive rate [9–12] combined with a significant cost in time and money has encouraged the development of alternative methods to drive the drug discovery process [13, 14].

Nuclear magnetic resonance (NMR) spectroscopy is uniquely qualified to assist in making the drug discovery process more efficient [15, 16]. NMR is useful for several reasons: (1) it directly detects the interaction between the ligand and protein using a variety of techniques, (2) samples are typically analyzed under native conditions, (3) hundreds of samples can be analyzed per day, and (4) information on the binding site and binding affinity can be readily obtained. These features allow NMR to be an effective tool at multiple steps in the drug discovery pathway, which includes verifying HTS and virtual screening hits [15, 17–19], screening fragment-based libraries [15, 20–22], optimizing lead compounds [15, 17, 23, 24], evaluating ADME-toxicology [25–27], and identifying and validating therapeutic targets [28, 29]. Nevertheless, there are still intrinsic costs to maintaining an NMR instrument, screening a compound library, and producing significant quantities of a protein. One way to significantly reduce experimental costs is to utilize *in silico* methodologies to supplement the lead identification and optimization steps of the drug discovery process [30].

Molecular docking is a computational tool that predicts the binding site location and conformation of a compound when bound to a protein [30–32]. This approach has been found to be fairly successful in redocking compounds into previously solved protein–ligand co-structures [33], where more than 70% of the redocked ligands reside within 2 Å root mean squared deviation (RMSD) of the actual ligand pose. During the prediction of protein–ligand co-structures, molecular docking programs calculate a binding score that allows for the selection of the best ligand pose. The binding score is typically based on a combination of geometric and energetic functions (bond lengths, dihedral angles, van der Waals forces, Lennard-Jones and electrostatic interactions, etc.) in conjunction with empirical functions unique to each specific docking program [34–39]. A large variety of docking programs are available that include AutoDock [40], DOCK [41], FlexX [42], Glide [43], HADDOCK [44], and LUDI [45, 46].

Binding energies are also routinely used to rank different ligands from a compound library after being docked to a protein target. The virtual or *in silico* screening of a library composed of thousands of theoretical compounds can be accomplished in a day with minimal cost [47–49]. Thus, a virtual screen can significantly accelerate the hit identification and optimization process while reducing the amount of experimental effort. However, a virtual screen does have significant limitations that prevent it from completely replacing traditional HTS [50–52]. These limitations include inaccurate scoring functions, use of rigid proteins, and simplified solvation models. In essence, a virtual screen only increases the likelihood that a predicted ligand actually binds the protein target, experimental verification is essential. Despite the individual drawbacks, NMR ligand affinity screens and molecular docking are complementary techniques. This review will highlight the use of NMR ligand affinity screens and molecular docking in drug discovery and describe recent examples where the combination of the two techniques provides a powerful approach to identify new and effective therapeutic drugs.

2 NMR Ligand Affinity Screens

NMR ligand affinity screening is a versatile technique that is useful for multiple stages of the drug discovery process [15, 17, 22, 53]. This versatility arises from the ability of NMR to directly detect protein–ligand binding based on changes in several NMR parameters. A binding event is detected by the relative differences between the protein or ligand NMR spectrum in the bound and unbound states. However, the specific type of information obtained about the binding process depends on whether a ligand-based or target-based NMR experiment is used.

2.1 Ligand-Based NMR Screens

Ligand-based NMR screens typically monitor the NMR spectrum of a ligand under free and bound conditions. Distinguishing between a free ligand and a protein–ligand complex is generally based on the large molecular weight difference that affects several NMR parameters. Small molecular weight molecules have slow relaxation rates (R_2), negative NOE cross-peaks, and large translational diffusion coefficients (D_t). If a protein–ligand binding event occurs, the ligand adopts the properties of the larger molecular-weight protein, increasing R_2, producing positive NOE cross-peaks, and decreasing D_t, all of which can be observed by NMR [54]. Most ligand-based NMR screens use one-dimensional (1D) ^{1}H-NMR experiments to monitor these changes, which provide significant benefits for a high-throughput screen. 1D NMR experiments are typically fast (2–5 min) and routinely use mixtures without the need to deconvolute [55]. The deconvolution of mixtures is avoided by ensuring that NMR ligand peaks do not overlap in the NMR spectrum (Fig. 1). The application of mixtures allows for hundreds to thousands of compounds to be screened in a single day. Another advantage of ligand-based NMR methods is the minimal amount of protein required (<10 μM) for each experiment. Additionally, isotopically labeled proteins are not needed for the

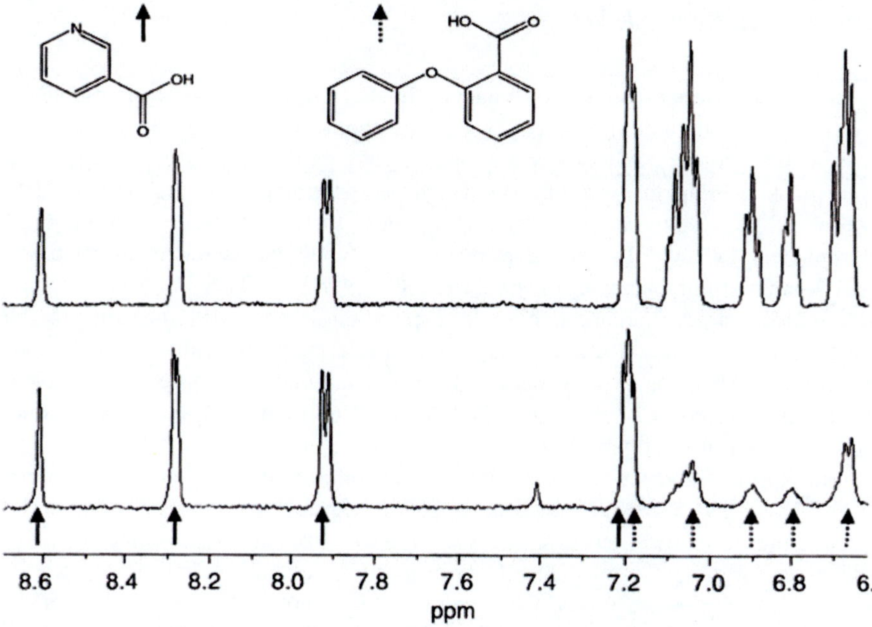

Fig. 1 An example of the use of a ligand-detect NMR experiment to observe the line broadening (increase R_2) that occurs when one compound, in a mixture of two compounds, binds a protein target. The ^{1}H-NOESY spectra of nicotinic acid (*left* structure) and 2-phenoxybenzoic acid (*right* structure) in a mixture without protein (*top* spectrum) and with the protein, p38 MAP kinase, added (*bottom* spectrum). The *solid* and *dashed arrows* represent the resonances of nicotinic acid and 2-phenoxybenzoic acid, respectively. In this case, the resonances corresponding to 2-phenoxybenzoic acid are broadened, indicating binding of this compound to the protein. (Reprinted with permission from [178], copyright 2001 by Academic Press)

NMR ligand affinity screen and protein molecular weight is not a limiting factor [21]. In fact, higher molecular-weight proteins enhance the observation of a binding event in a ligand-based NMR screen. All of these characteristics make ligand-based NMR screens a routinely used drug discovery technique.

There are several screening techniques created from ligand-based NMR experiments: line broadening [56], STD NMR [57], WaterLOGSY [58], SLAPSTIC [59], TINS [60], transferred NOEs [61], FAXS [62, 63], FABS [64, 65], and diffusion measurements [66, 67]. Each of these methods utilizes a specific NMR parameter that indicates ligand-binding, such as a change in ligand NMR peak width or diffusion, a saturation transfer from the protein or solvent to the ligand, an NOE transfer between the free and bound ligand, a spin-label induced paramagnetic relaxation, or fluorine chemical shift anisotropy. The choice of which method to use typically depends upon the protein target and the compound library being screened. In addition, line broadening and STD, among other techniques, can be used to measure dissociation constants (K_D) [68, 69]. Conversely, ligand-based NMR screens don't provide any structural information about the protein–ligand complex.

2.2 Target-Based NMR Screens

A target based screen focuses on changes in the protein (or other target) NMR spectrum to identify a binding event. Typically, chemical shift perturbations (CSPs) occur in the protein NMR spectrum upon ligand binding. The complexity and severe peak overlap in a protein 1D ^{1}H NMR spectrum makes it impractical to observe subtle CSPs for weak binding ligands. Instead, two-dimensional (2D) heteronuclear NMR [70–72] experiments are typically used for target-based NMR ligand affinity screens [73]. 2D^{1}H-^{13}C/^{15}N HSQC/TROSY NMR experiments require a significant increase in experiment time (>10 min) due to the additional dimension and the need to collect a reference spectrum for the ligand-free protein. Also, the protein needs to be ^{15}N and/or ^{13}C isotopically labeled. Importantly, 2D^{1}H-^{13}C/^{15}N HSQC/TROSY NMR experiments provide additional information about the ligand binding site.

A binding ligand often results in the observation of CSPs of the resonances in a 2D^{1}H-^{15}N- or ^{1}H-^{13}C-HSQC spectrum (Fig. 2a). These CSPs are usually caused by a change in the chemical environment for residues proximal to the bound ligand or residues undergoing ligand-induced conformational changes. The availability of the protein structure and the NMR sequence assignments (correlation of an NMR resonance with a specific amino acid residue) allows for the CSPs to be mapped onto a three-dimensional (3D) representation of the protein's surface. A cluster of residues on the protein surface with observed CSPs often identifies the ligand-binding site.

The ligand binding affinity or K_{D} is also routinely determined from CSPs measured from a series of 2D ^{1}H-^{13}C/^{15}N HSQC/TROSY NMR experiments. The magnitude of the CSPs at varying ligand concentrations is correlated to the K_{D} for the protein–ligand complex using the following equation [74, 75]:

$$\mathrm{CSP}_{\mathrm{obs}} = \mathrm{CSP}_{\mathrm{max}} \frac{(K_{\mathrm{D}} + [L] + [P]) - \sqrt{(K_{\mathrm{D}} + [L] + [P])^2 - (4[L][P])}}{2[P]}, \quad (1)$$

where $[P]$ is the protein concentration, $[L]$ is the ligand concentration, $\mathrm{CSP}_{\mathrm{max}}$ is the maximum CSP observed for a fully bound protein, and $\mathrm{CSP}_{\mathrm{obs}}$ is the observed CSP at a particular ligand concentration. A least squares fit of (1) to the experimental CSP data is used to calculate a K_{D} (Fig. 2b).

As previously mentioned, since target-based screens require the use of multidimensional NMR experiments, data collection is significantly longer relative to ligand-based NMR screens. Also target-based screens require higher protein concentrations (>50 μM compared to <10 μM). This severely limits the utility of target-based NMR screens for the high-throughput analysis of large compound libraries. Instead, the approach is typically used to validate hits from a high-throughput screen or the analysis of relatively small fragment-based libraries [76–78]. A fragment-based library consists of low molecular-weight compounds (<250–350 Da) that are fragments of known drugs or have drug-like properties

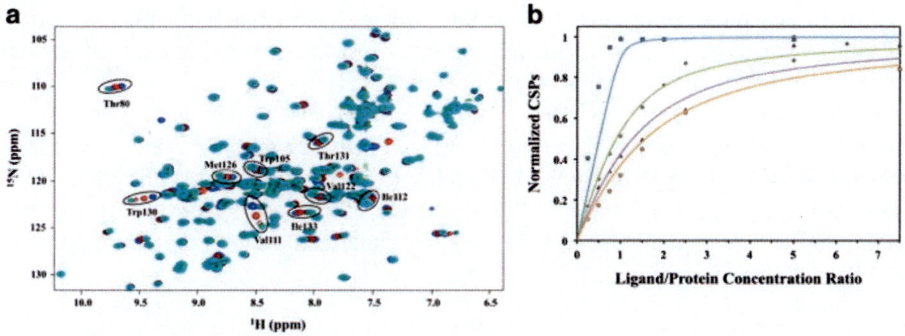

Fig. 2 (**a**) An overlay of the 2D ^{1}H-^{15}N HSQC spectra for the protein YndB titrated with increasing amounts of chalcone. The perturbed residues can be used to identify a consensus binding site. (**b**) NMR titration data for YndB bound to chalcone (*blue*), flavanone (*green*), flavone (*purple*), and flavanol (*orange*). The magnitude of the chemical shift perturbation can be used to calculate the dissociation constants for each compound. (Reprinted with permission from [112], copyright 2010 by John Wiley and Sons)

[79]. Recent advances like the SOFAST-HMQC experiment [80, 81] and the Fast-HSQC experiment [82] have decreased the time and amount of protein necessary for a target-based screen. Nevertheless, NMR ligand affinity screens are still very resource intensive, requiring a significant amount of time and material. Also, since any high-throughput screen produces a significant amount of negative data (most ligands don't bind or inhibit a protein), a more efficient approach is to screen a library of compounds with a higher probability of binding the protein target. In effect, a virtual or *in silico* screen can be used to enrich a library with likely binders.

3 Molecular Docking

An accurate prediction of the interactions between two molecules requires an in-depth understanding of the energetics that led to a stable biomolecular complex. Unfortunately, a model that correctly accounts for all the factors involved in a productive protein–ligand interaction is currently unknown. Further, the problem is exponentially more complex than just modeling the specifics of a protein–ligand interaction. A protein contains thousands of atoms that have specific interactions with each other, with the solvent, and with other ions; in addition to the bound ligand. Because of this complexity, computational efforts that attempt to model protein–ligand interactions require significant amounts of processing power and time. Many efforts that utilize molecular dynamics and distributed computing [83, 84] are generally limited to a detailed analysis of a *single* system. These methods are generally not practical for the majority of researchers interested in conducting a virtual screen of a library containing upwards of millions of compounds. To make molecular docking computationally feasible and easily accessible, many simplifications and trade-offs in the process are necessary.

Many computer programs are available to perform or assist with molecular docking. The vast number of docking programs makes it impractical to describe them all in detail within a single review (for other reviews please see [85–89]). Each docking program does have some unique features that make them particularly useful for a given situation or problem. However, nearly all the docking programs consist of two primary components: docking (or searching) and scoring [30, 31]. Docking refers to the sampling of the ligand's conformation space and its orientation relative to a receptor. Scoring is used to evaluate and rank the current pose of the ligand.

3.1 Docking

The docking process requires, at a minimum, two inputs: the three-dimensional structures of the receptor (protein) and the ligand. The most common simplification to the docking process is to keep the structure of the receptor rigid and stationary. Only the ligand is typically allowed to be flexible as it is docked to the protein. Keeping the protein rigid significantly minimizes the complexity of the calculation. Sampling the conformations and orientations of the ligand is done using systematic or stochastic methods [30, 31].

Systematic search methods attempt to sample all of the possible conformations of a ligand by incrementing the torsional angles of each rotatable bond. Unfortunately, this technique is computationally expensive due to the exponential increase in the number of possible conformations (N_{conf}) as the number of rotatable bonds increases:

$$N_{\text{conf}} = \prod_{i=1}^{N} \prod_{j=1}^{n_{\text{inc}}} \frac{360}{\theta_{i,j}}, \tag{2}$$

where N represents the number of rotatable bonds, n_{inc} is the number of incremental rotations for each rotatable bond, and $\theta_{i,j}$ is the size of the incremental rotation for each rotatable bond. As a result, purely brute force systematic approaches are generally not used. Instead, most systematic searches require the use of efficient shortcuts. As an illustration, MOLSDOCK [90] uses mutually orthogonal Latin squares (MOLS) to identify optimal ligand conformations. Latin squares are an $N \times N$ matrix, where each parameter (torsion angle value) occurs only once in each row and column. Orthogonal Latin squares are two or more superimposed $N \times N$ matrices, where each parameter still only occurs once in each row and column. MOLS are used to identify the N^2 subset of ligand conformations used to calculate binding energies. Simply, only a small subset of the possible ligand conformations is sampled to construct the potential surface and identify the minima.

Perhaps the most commonly utilized systematic search method is incremental construction, which is used by DOCK [41], FlexX [42], E-Novo [91], LUDI [45, 46], ADAM [92], and TrixX [93]. In this particular method, the ligand is split into fragments. The most rigid fragments are often used as the core or anchor and are docked first into the receptor binding pocket. The remaining fragments are incrementally added back onto the core fragment, where each addition is systematically rotated to evaluate the most optimal conformation. Thus, incremental construction drastically reduces the number of possible conformations that need to be searched in order to identify the optimal pose.

Another systematic approach uses rigid docking in combination with a predefined library of ligand conformations, which is implemented in OMEGA [94], FLOG [95], Glide [43], and the TrixX Conformer Generator [96]. This technique generates several low energy conformers for a ligand that are clustered by RMSD. A representative conformer from each cluster is then docked into the receptor. The approach is very fast because the docking process keeps the ligand rigid, eliminating the need to spend computation time on searching torsional space. A tradeoff for this increase in speed is a potential loss in accuracy, since the binding potential for all possible conformers may not be explored. Conversely, a major benefit of the technique is the fact that the library of structural conformers only needs to be generated once. This is a significant savings in time for the pharmaceutical industry, where screening libraries may consist of millions of compounds.

Unlike systematic approaches that attempt to sample all possible ligand conformations, stochastic searches explore conformational space by making random torsional changes to a single ligand or a population of ligands. The structural changes are then evaluated using a probability function. There are three types of stochastic searches: Monte Carlo algorithms [97], genetic algorithms [98], and tabu search algorithms [99]. The most basic stochastic method is the Monte Carlo algorithm, which utilizes a Boltzmann probability function to determine whether to accept a particular ligand pose:

$$P \sim \exp\left[\frac{-(E_1 - E_0)}{K_B T}\right], \tag{3}$$

where P is the probability the conformation is accepted, E_0 and E_1 are the ligand's energy before and after the conformational change, K_B is the Boltzmann constant, and T is the temperature. The simple scoring function used by the Monte Carlo algorithms is more effective than molecular dynamics in avoiding local minima and finding the global minimum. Alternatively, genetic algorithms utilize the theory of evolution and natural selection to search ligand conformation space. In this case, the conformations, orientations, and coordinates of a ligand are encoded into variables representing a "genetic code." A population of ligands with random genetic codes is allowed to evolve using mutations, crossovers, and migrations. The new population is evaluated using a fitness function that eliminates unfavorable ligand poses. Eventually, a final population converges to ligands with the most favorable "genes" or conformations (Fig. 3). Tabu searches, like other stochastic methods, randomly modify the conformation and coordinates of a ligand, score the conformer, and then repeat the process for a new conformation. Tabu searches

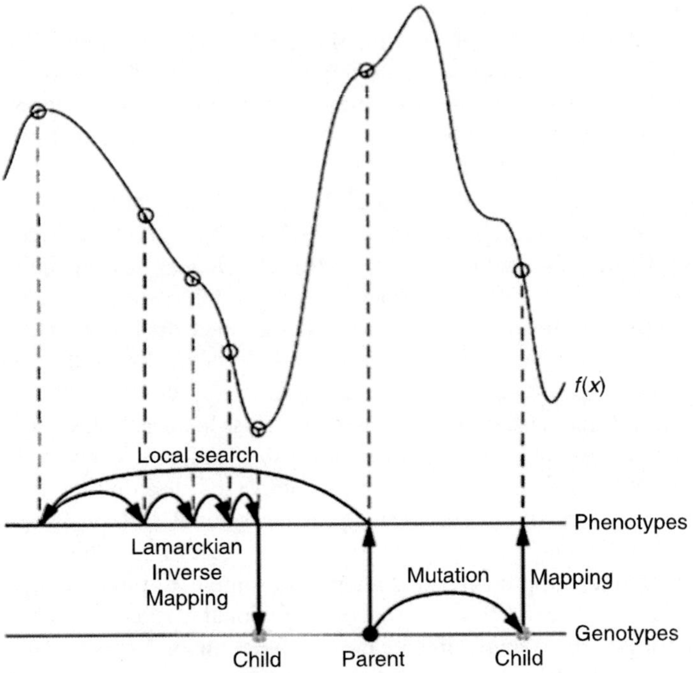

Fig. 3 An illustration of the genetic algorithm approach, where the states of the ligand (translation, orientation, and conformation relative to the protein) are interpreted as the ligand genotype and the atomic coordinates represent the phenotype. A plot of the change in the fitness function (*f(x)*) as the ligand population is allowed to mutate, crossover, and migrate. The genetic evolution of the ligand effectively samples conformational space where the best conformer is identified by a minimum in the fitness function (Reprinted with permission from [179], copyright 1998 by John Wiley and Sons)

utilize a tabu list to remember previous ligand states. A pose is immediately rejected if it is close to a prior conformation. The tabu list encourages the search to progress to unexplored regions of conformational space.

3.2 Scoring

While docking algorithms are generally efficient at generating the correct ligand pose, it is important for the docking program to actually select the correct ligand conformation from an ensemble of similar conformers. In essence, the scoring function should be able to distinguish between the true or optimal binding conformation and all the other poses. The scoring function is also used to rank the relative binding affinities for each compound in the library. Ideally, the scoring function should be able to calculate the free energy ($\Delta G_{\text{binding}}$) of the protein–ligand binding interaction, which is directly related to the K_D:

$$\Delta G_{\mathrm{binding}} = -RT\ln\frac{1}{K_{\mathrm{D}}}. \quad (4)$$

Unfortunately, accurately calculating the binding free energy is very challenging due to the many forces that influence binding. In molecular docking, there are five primary types of scoring functions: force field-based, empirical, knowledge-based, shape-based, and consensus [100–102].

Force field-based scoring functions [30, 31] are used to calculate the free energy of binding by combining the receptor–ligand interaction energy and the change in internal energies of the ligand based on its bound conformation (Fig. 4). The internal energy of the receptor is usually ignored since the receptor is kept rigid in most docking programs. The protein–ligand binding energies are typically defined by van der Waal forces, hydrogen bonding energies, and electrostatic energy terms. The van der Waals and hydrogen bonding terms often utilize a Lennard-Jones potential function, while the electrostatic terms are described by a coulombic function. Unfortunately, these interaction energies were originally derived from measuring enthalpic interactions in the gas phase. Of course, receptor–ligand binding interactions actually occur in an aqueous solution, which introduces additional interactions between the solvent molecules, the receptor, and the ligand. Protein–ligand binding energies are also dependent on the entropic changes that occur upon binding, which include torsional, vibrational, rotational, and translational entropies. Most entropy and solvation-based energy terms can't be calculated using force field-based scoring functions. As a result, force field-based scoring functions are incomplete and inaccurate.

Empirical scoring functions [103–106] are similar to force field-based scoring functions since they use a summation of individual energy terms. But empirical scoring functions also attempt to include solvation and entropic terms. This is typically achieved by using experimentally determined binding energies of known ligand–receptor interactions to train the scoring system using regression analysis. Empirical scoring functions are fast, but the accuracy is completely dependent upon the experimental data set used to train the scoring function. In general, empirical scoring functions are reliable for ligand–receptor complexes that are similar to the training set.

Knowledge-based scoring functions [107–109] are fundamentally different from force field-based and empirical scoring functions. Knowledge-based scoring functions don't attempt to calculate the free energy of binding. Instead, these scoring functions utilize a sum of protein–ligand atom pair interaction potentials to calculate a binding affinity. The atom pair interaction potentials are generated based upon a probability distribution of interatomic distances found in known protein–ligand structures. The probability distributions are then converted into distance-dependent interaction energies. In this manner, knowledge-based scoring functions allow for the modeling of binding interactions that are not well understood. The approach is also very simple, which is useful for screening large compound libraries. Unfortunately, knowledge-based scoring functions are designed

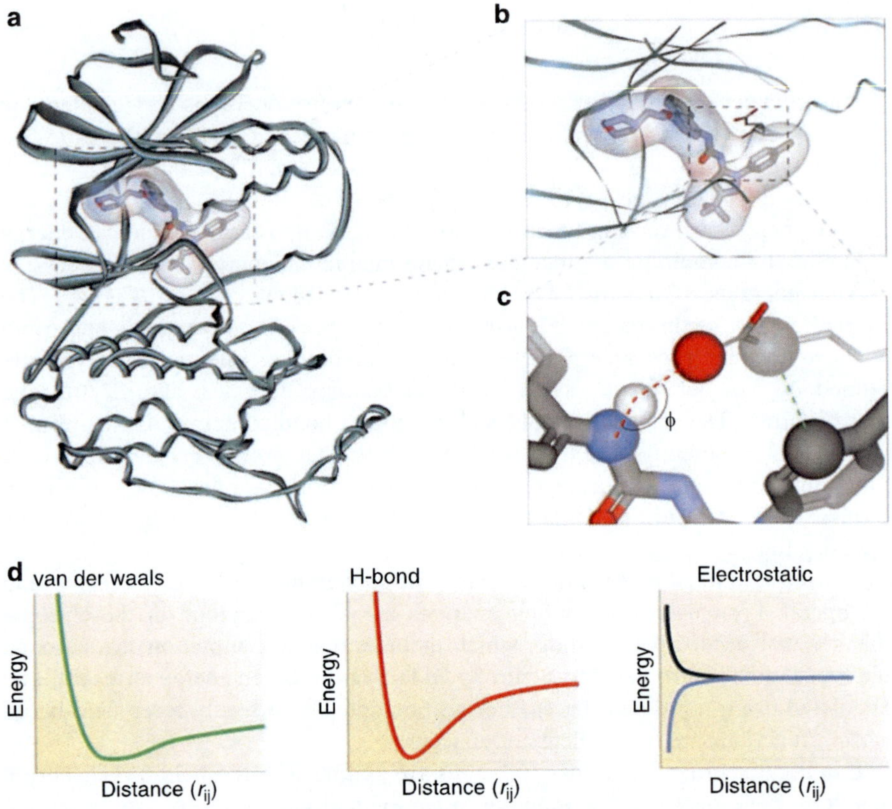

Fig. 4 (**a**) A representation of p38 mitogen-activated protein kinase structure bound to BIRB796 and (**b**) an expanded view of the binding site. (**c**) A representation of the hydrogen-bonding (*red*) and electrostatic interactions (*green*) between the atoms of the protein and the atoms of the ligand. (**d**) A representation of three force-field energy terms (van der Waals, hydrogen-bonding, and electrostatic) as distance between the interacting atom pairs change. (Reprinted with permission from [30], copyright 2004 by the Nature Publishing Group)

to reproduce known experimental structures, and the binding score generated has little relevance to an actual binding affinity. This is an issue similar to empirical scoring functions; the accuracy of the scoring function is strongly dependent on the similarity of the protein–ligand complex to the training data set.

As implied, shape-based scoring functions are based on a shape match between the ligand and the ligand binding site [110]. These scoring functions are typically used as prefilters to eliminate compounds that are unable to fit into the ligand binding site [111, 112]. Shape-based scoring functions are very fast, but are limited relative to more accurate scoring functions that calculate binding affinities. Shape-based scoring functions typically generate smooth energy surfaces using Gaussian functions [111], which are more tolerant to atomic variations and make protein clash interactions "softer." This essentially helps minimize the effect of small structural variations that may occur during ligand binding.

While the above scoring methods are generally useful in describing protein–ligand interactions, the simplifications used in each approach limits the overall accuracy in predicting the correct docked ligand pose [113, 114]. The major weakness of most docking programs has been shown to be the scoring function. One approach to compensate for this deficiency is to use a consensus score from a combination of scoring functions to rescore a docked pose. Consensus scoring [31, 115] has been shown in several examples to improve docking results compared to a single scoring function. However, like individual scoring functions, the improvement is not consistent and the proper choice of scoring functions to calculate a consensus score is typically based on trial and error.

3.3 Protein Flexibility

Proteins are inherently flexible and undergo a range of motions over different time scales, and thus the use of rigid protein structures by molecular docking is problematic [116, 117]. This is especially troublesome for therapeutic targets where only an apo-structure is available. Conformational changes upon ligand-binding may range from small perturbations in side chain conformation at the site of ligand binding to large rearrangements of the entire protein structure. Not accounting for such structural changes during ligand docking can drastically alter the ability to identify reliable protein–ligand models correctly [118–122]. Conversely, attempting to dock a large library of flexible ligands to a completely flexible protein structure using molecular dynamics is too computationally expensive to be practical.

Several approaches to "solve" the protein flexibility problem have been explored. The first generally applicable approach utilized soft docking in the scoring function, which reduces the van der Waals repulsion terms in the empirical scoring function [123, 124]. This allows for some overlap between ligand and protein atoms. While this approach is simple and fast, it can only accommodate very small changes in side chain conformations. Other approaches attempt to implement protein structural changes into the docking process. For example, a library of side chain rotamers for residues only in the ligand binding site is routinely used [40, 125]. This dramatically reduces the number of active rotatable bonds during the docking process and has a lower computational cost compared to molecular dynamics. However, the inclusion of a library of rotamers in the docking protocol is significantly slower than rigid protein docking. Furthermore, the approach is limited to local side chain conformational changes.

The most common docking technique that attempts to account for protein flexibility uses multiple protein structures. The ensemble of structures is expected to represent the range of conformations sampled by the protein and has the benefit of being able to evaluate both small and large conformational changes. The molecular docking is repeated for each individual protein conformation, which results in a proportional increase in computational time. Also, the results may be ambiguous, since there may be several equally valid ligand poses for each different protein conformation. This is especially apparent in virtual screening approaches where enrichment factors suffer when docking to multiple structures (please see Sect. 3.4). This is likely due to an increase in the number of false positives among the top hits [126]. Ensemble docking is an alternative to docking multiple structures that removes the ambiguity [118]. All the protein structures from the ensemble are superimposed in order to generate an average structure or an average receptor grid. The docking is then performed against the average structure or average receptor grid (Fig. 5). The ensemble docking approach allows for a single docking at a significantly lower computational cost; however, it may suffer from accuracy problems if the ensemble is biased towards the unbound form of the protein. Effectively, a biased ensemble may negate the goal of incorporating protein flexibility if it represents a single conformation.

3.4 Virtual Screening and Assessment

Using molecular docking to identify lead candidates is an attractive approach for the pharmaceutical industry; it allows for the rapid evaluation of millions of chemical compounds while using minimal resources compared to traditional HTS. The process by which molecular docking is used to rank compounds within a library based on a predicted binding affinity is known as virtual screening [127, 128]. The potential benefit to drug discovery has inspired the development and evaluation of numerous virtual screening methodologies. A virtual screen requires a balance between optimizing speed and maximizing accuracy. Specifically, the goal of a drug discovery virtual screen is the rapid and efficient separation of a small subset of active compounds from a relatively large random library of inactive compounds. Unfortunately, determining the effectiveness of a specific virtual screening process is challenging, where independent evaluators routinely generate inconsistent results [87, 129–131].

The ambiguous nature of the results from a virtual screen requires additional methods to evaluate its success. Typically, a virtual screening process is evaluated against a protein target with a set of known binders. Assessing the performance of a virtual screen is primarily based on the accuracy of the predicted ligand pose and binding affinity. The correct binding pose is often evaluated by calculating the RMSD between the docked and experimental ligand structures. The evaluation of binding affinity is typically based on the accurate ranking of known binders instead of the absolute scores because of the known limitations with calculating a binding energy. Other modes of performance assessment involve evaluating enrichment and generating diverse hit lists.

In a virtual screening protocol, every compound in a library (N_{tot}) is docked to the protein and a corresponding binding score is calculated. The binding score for the ligand's best docked pose is used to rank the ligand relative to the entire library.

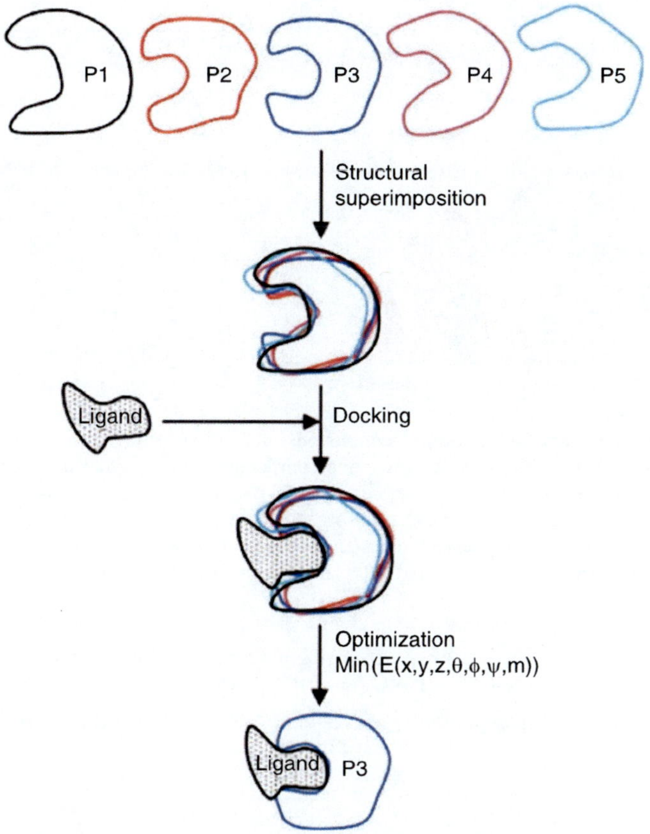

Fig. 5 A cartoon illustration of ensemble docking, where five individual protein structures are superimposed to create a single scoring parameter for the docked ligand. Ensemble docking minimizes the computational effort since a single docking occurs to select the best conformer instead of five separate molecular docking simulations. (Reprinted with permission from [118], copyright 2007 by John Wiley and Sons)

A virtual screen never results in all the truly active compounds being top ranked. Instead, most virtual screening protocols set a binding score or ranking threshold to identify the predicted active compounds or "hits." In general, top ranked compounds are expected to be enriched with active compounds compared to a random selection (Fig. 6a). A high enrichment factor ($EF > 10$) is considered the benchmark of success for a virtual screening [132]. Enrichment is dependent on sensitivity (Se) and specificity (Sp). Sensitivity represents the true positive rate, which is the ratio of true positives (TP) found by the virtual screening vs the total number of actives (A) in the library. The number of actives corresponds to both true positive (TP) and false negative (FN):

$$Se = \frac{TP}{TP + FN}. \tag{5}$$

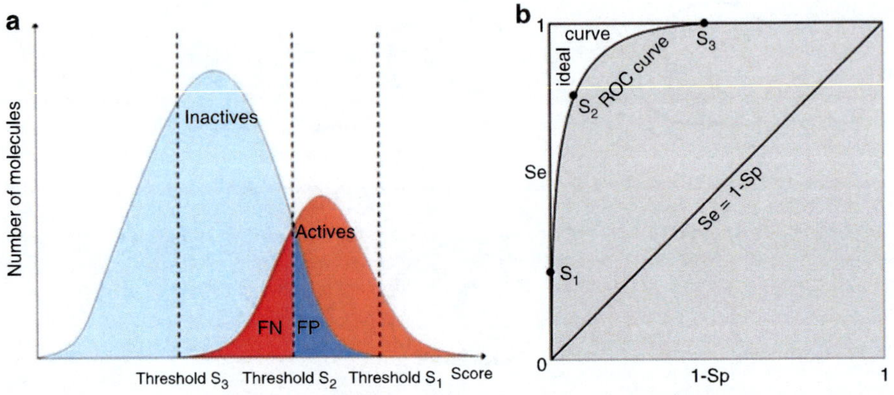

Fig. 6 (**a**) A theoretical distribution of compounds in a virtual screen based upon the docking score. The overlap between active and inactive compounds indicates that the scoring threshold used to identify a hit by virtual screening is critical. (**b**) A ROC curve is used to evaluate the enrichment of a virtual screen and select a scoring threshold. A ROC curve that approaches $Se = 1$ and $1\text{-}Sp = 0$ represents perfect enrichment. The area under the ROC curve (AUC) represents the probability that a true active is identified. (Reprinted with permission from [131], copyright 2008 by Springer)

Specificity is the measure of the true negative rate, which represents the ratio of true negatives (*TN*) to the total number of inactive compounds. The number of inactive compounds corresponds to both true negatives (*TN*) and false positives (*FP*):

$$Sp = \frac{TN}{TN + FP}. \tag{6}$$

The enrichment factor is a common method for evaluating the enrichment capabilities of a virtual screen:

$$EF = \frac{\left(\frac{TP}{TP+FP}\right)}{\left(\frac{TP+FN}{N_{\text{tot}}}\right)}. \tag{7}$$

The enrichment factor is dependent upon the ratio of active compounds to the total number of compounds in the library. As a result, enrichment scores are difficult to compare between virtual screens with different libraries. Also, the enrichment factor does not distinguish between high and low ranking compounds.

Perhaps the more popular approach for evaluating enrichment is to generate a receiver operating characteristic (ROC) curve [133]. The ROC curve is a plot of the true positive rate (*Se*) against the false positive rate ($1-Sp$) at varying thresholds for determining a hit. A ROC curve allows for the evaluation of a virtual screening method without using an arbitrary scoring threshold. Enrichment occurs when the resulting data point at a particular threshold resides above the diagonal ($Se = 1-Sp$), which corresponds to a random selection of compounds. In a perfect virtual screen where every active compound is identified as a hit and every inactive compound falls below the threshold, the ROC curve approaches the top left corner ($Se = 1$ and $1-Sp = 0$) (Fig. 6b).

Hit list diversity is also an important consideration for the success of a virtual screen since there is more value in identifying a few unique compounds instead of many compounds all based on the same chemical scaffold. One way that diversity can be determined is by comparing the structural similarities of hits from a virtual screen by using the Tanimoto index [134] and then clustering the results. Basically, a Tanimoto index is calculated based on the fraction of similar chemical substructures present in two structures. Generally, 1,365 chemical substructures are used to describe a structure. The substructures include individual elements, two-atom substructures, single rings, condensed rings, aromatic rings, other rings, chains, branches, and functional groups:

$$TI = \frac{C}{A + B + C}, \tag{8}$$

where A represents the substructural features present in the first structure, B represents the substructural features present in the second structure, and C represents the substructural features common to both structures. Identical structures have a TI score of 1, where completely dissimilar structures have a TI value of 0.

4 Combining Molecular Docking with NMR Ligand Affinity Screens

The vast majority of initial leads in drug discovery are identified from HTS [13, 135, 136]. Pharmaceutical companies have invested heavily in developing and maintaining large chemical libraries (>1,000,000 compounds), which are screened using automated, biological assays intended to monitor a specific response or biological effect [136]. Unfortunately, HTS is extremely inefficient due to the high cost of developing, maintaining, and screening such large libraries of compounds. Furthermore, the random search for an effective drug in the vastness of chemical space (~10^{60} compounds) [137] is almost guaranteed to fail. Thus, HTS hit rates are typically very low, where <0.5% of compounds exhibit any inhibitor activity in an assay [138]. Correspondingly, HTS assays are highly inefficient since most of the screening effort is spent on the analysis of negative data. Additionally, HTS assays, by nature, are mechanistic "black boxes," and a response does not provide any information on the mechanism of inhibition. This often leads to numerous false positives from undesirable interactions [11, 12, 139] that may lead the drug discovery project astray. Improving the efficiency of drug discovery requires the implementation of advanced techniques that better guide the selection of lead candidates without sacrificing speed.

Ideally, an entirely *in silico* approach to screening a large compound library would significantly improve efficiency and reduce costs [140, 141]. However, several assessments of virtual screens have concluded that, without prior in-depth analysis of the protein's ligand binding site, only a marginal improvement in finding successful leads is observed relative to standard HTS [32]. NMR can complement a virtual screen by providing rapid experimental validation of lead compounds. NMR allows for a ligand-binding event to be directly observed instead of relying on false-positive prone activity assays. Also, NMR provides detailed structural information about the ligand binding site and the orientation of the bound ligand. An NMR ligand affinity screen can be used to validate upwards of thousands of predicted hits from a virtual screen [142]. Thus, combining NMR with virtual screens may provide a more efficient approach to lead identification and drug discovery.

4.1 Identification of New Therapeutic Targets

The functional assignment of unannotated proteins is essential to the drug discovery process. Greater than 40% of protein sequences encoded in eukaryotic genomes consist of proteins of unknown function and represent an important opportunity to identify new therapeutic targets [143]. Assigning a function to an uncharacterized protein is an arduous and time-consuming task. The process often requires detailed biochemical studies that may include analyzing cell phenotypes through knockout libraries, monitoring of gene expression levels, or utilizing pull-down assays [144–147].

Since the interactions of proteins with other biomolecules or small molecules is the basis of a functional definition or classification, identifying the functional ligand, the functional epitope or ligand binding site, and the 3D structure of the protein–ligand complex are invaluable for a functional annotation. A functional epitope or ligand binding site is evolutionarily conserved relative to the rest of the protein structure in order for the protein to maintain its biological function. Therefore, proteins that share similar binding site structures are expected to be functional homologs and bind a similar set of ligands [28, 29]. Correspondingly, numerous *in silico* approaches attempt to infer a function for an uncharacterized protein by predicting ligand binding sites using geometry-based, information-based, and energy-based algorithms [148–150]. Unfortunately, unambiguously identifying the ligand binding site on a protein can be challenging without experimental evidence, especially for proteins with no known function.

Functional Annotation Screening Technology using NMR (FAST-NMR) [28, 29] is one approach that combines HTS by NMR with molecular docking and bioinformatics analysis in order to assign a function to a protein (Fig. 7). In this process, a compound library that contains approximately 430 biologically relevant compounds [151] is screened by NMR using a multistep approach [152]. First, a ligand-based screen using 1D NMR ^{1}H line-broadening experiments identifies

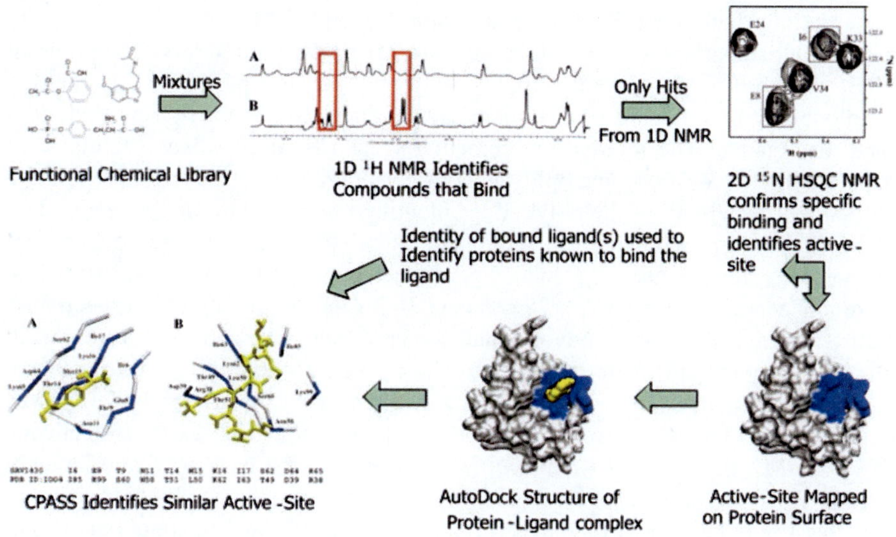

Fig. 7 A flow diagram of the FAST-NMR process. Mixtures of biologically active compounds are first assayed in a ligand-based 1D line broadening screen against the protein of interest. Compounds that are identified as hits are then verified using CSPs from a 2D ^{1}H-^{15}N HSQC experiment that define a binding site on the protein surface. The CSPs are used to guide and filter an AutoDock molecular docking calculation to generate a protein–ligand co-structure. The ligand binding site defined by the co-structure is then compared to other experimental binding sites in the PDB using CPASS. (Reprinted with permission from [28], copyright 2008 by Elsevier)

potential binders. These hits are then verified in a target-based screen using a 2D ^{1}H-^{15}N HSQC experiment, where the occurrence of CSPs allows for the identification of the ligand binding site. Molecular docking is used to generate a rapid protein–ligand co-structure [121] that serves as input for the Comparison of Protein Active-Site Structures (CPASS) program [153]. CPASS compares the sequence and structure of this NMR modeled ligand binding site to ~36,000 unique experimental ligand binding sites from the RCSB Protein Databank [143]. Thus, a protein of unknown function can be annotated from a protein with a known function that shares a similar ligand binding site [154]. The FAST-NMR and CPASS approach has been used for the successful annotation of two hypothetical proteins, SAV1430 from *S. aureus* [29] and PA1324 from *P. aeruginosa* [155]. It has also been used to identify a structural and functional similarity between the bacterial type III secretion system and eukaryotic apoptosis [156].

The FAST-NMR approach was recently applied to protein YndB from *Bacillus subtilis* to generate a functional annotation [112]. FAST-NMR was augmented by the inclusion of a virtual screen using the Nature Lipidomics Gateway library that contains ~22,000 lipids. Eight major categories of lipids are represented in the library (fatty acyls, glycerolipids, glycerophospholipids, sphingolipids, sterol lipids, prenol lipids, saccharolipids, and polyketides), which are further divided into a total of 538 distinct subclasses. The initial goal was to identify lipid scaffolds that preferentially bound YndB to infer the natural ligand. OMEGA [94] was used to generate a database of ~10,000,000 conformers from the lipid library. The program FRED was then used to dock the lipid conformer library to YndB. FRED [111] used rigid docking based on shape complementarity and a consensus scoring system to rank the ligands. The relative enrichment for each lipid class was calculated at different thresholds. Only one lipid category, the polyketides, had a positive relative enrichment, where all of the polyketides identified belonged to the flavonoid class of lipids. Within the flavonoids, three subclasses emerged as favorable hits from the virtual screen, where chalcones/hydroxychalcones, flavanones, and flavones/flavonols accounted for 44.9%, 28.6%, and 14.3% of the top 50 hits, respectively. *trans*-Chalcone, flavanone, flavone, and flavonol were selected to represent each class. The compounds were titrated into YndB to confirm binding and to measure K_D. The titrations were followed using a series of 2D ^{1}H-^{15}N HSQC NMR experiments, where CSPs were measured to calculate K_Ds (Fig. 2). *trans*-Chalcone (K_D <1 μM), flavanone (K_D 32 ±3 μM), flavone (K_D 62 ± 9 μM), and flavonol (K_D 86 ± 16 μM) were all shown to bind YndB in the same ligand binding site with K_Ds that mimicked the virtual screen ranking. Chalcones and flavonoids have not been identified among the natural products of *Bacillus* organisms, but are important precursors to plant antibiotics. The screening results are consistent with the symbiotic relationship between *B. subtilis* and plants. *B. subtilis* YndB is proposed to be part of a stress-response network that senses chalcone-like molecules during a plant's response to a pathogen infection. The stress-response may induce *B. subtilis* sporulation or the production of antibiotics to assist in combating the plant pathogens.

4.2 *Rapid Protein–Ligand Structure Determination*

A protein–ligand complex is instrumental to a structure-based approach to drug discovery. A new protein–ligand structure is required for each iteration of the lead modification process, until the compound has been evolved into a drug candidate. As a result, rapid protein–ligand structure determination benefits the drug discovery process. There are several methods that utilize NMR CSPs from a protein–ligand binding interaction with molecular docking to generate a corresponding co-structure. Some recent techniques include the McCoy and Wyss method [157], LIGDOCK [158], NMRScore [159], AutoDockFilter [121], QCSP-Steered Docking [160], and HADDOCK [44]. Basically, the CSPs are used to guide the docking process qualitatively and then to steer or filter the docking quantitatively. The docked model is validated by an agreement with the experimental CSPs.

AutoDockFilter (ADF) utilizes a post-filtering approach for rapidly (~35–45 min) generating a co-structure. First, CSPs from the 2D ^{1}H-^{15}N HSQC spectrum are mapped onto the protein surface to define the AutoDock 4.0 3D search grid. A 100 docked ligand poses are generated within the CSP defined search grid. Second, the CSPs are used to filter the ligand conformers and select the best pose with the AutoDockFilter (ADF) program. ADF calculates a pseudodistance (d_{CSP}) based on the magnitude of the CSPs and compares it to the shortest distance (d_{S}) between any atom in the residue that incurred the CSP with any atom in the docked ligand pose. A violation energy is attributed to each protein residue that is further from the docked ligand pose then predicted by the CSP pseudodistance. The sum of these violation energies generates an overall NMR energy (E_{NMR}) for the docked ligand conformer:

$$E_{\mathrm{NMR}} = k\sum_{i=1}^{n}(\Delta_{\mathrm{Dist}})^2 \Delta_{\mathrm{Dist}} = \begin{cases} d_{\mathrm{CSP}} - d_{\mathrm{S}} \; d_{\mathrm{CSP}} < d_{\mathrm{S}} \\ 0 \quad d_{\mathrm{S}} \leq d_{\mathrm{CSP}} \end{cases}. \tag{9}$$

The conformer with the lowest NMR energy corresponds to the best protein-ligand co-structure based on a consistency with the experimental CSPs. The NMR energy also provides a qualitative way to evaluate the reliability of the co-structure, with high NMR energies correlating to unreliable co-structures (Fig. 8).

NMRScore [159] is very similar to ADF. NMRScore uses poses generated by AutoDock and seven other docking programs. CSPs are calculated for each pose using DivCon, where a CSP RMSD is determined between the calculated and experimental CSPs. The best pose corresponds to the conformer with the lowest CSP RMSD. The McCoy and Wyss method [157] also uses simulated chemical shift changes. But, unlike the NMRScore approach, the docked ligand is replaced by a number of randomly placed amino-acid probes within the ligand binding site. Proton chemical shifts, primarily from ring-current effects, are calculated for the protein with and without the docked amino-acid probes. The proton chemical shifts are calculated using the SHIFTS program [161], where CSPs are determined based on the difference between the two sets of calculated proton chemical shifts. The best pose for the amino-acid probe is chosen based on a minimal difference between the experimental and calculated proton CSPs. The ligand is then docked to the protein by aligning the ligand with the amino-acid probes.

Instead of simulated chemical shifts, the HADDOCK [44] and LIGDOCK [158] programs use CSPs to define ambiguous interaction restraints (AIRs) [162]. AIRs are an intermolecular distance restraint between all atoms of the residue with the CSP and all atoms of the ligand. Importantly, other experimental information (STDs, mutational data, etc.) can also be used to define AIRS. HADDOCK and LIGDOCK employ a three-tiered approach to refining the protein–ligand complex. First, the ligand is docked to a rigid protein structure. Next, the protein–ligand structure is refined with simulated annealing in torsional space [163]. Finally, the structure is optimized with explicit solvent to remove any remaining structural problems. HADDOCK and LIGDOCK are particularly beneficial since the protein–ligand co-structure is directly refined against the experimental CSPs. The methods do suffer from long computation times and potential difficulties with proper parameterization of the ligand. HADDOCK was initially developed to dock protein–protein interactions and was later modified to accommodate ligands, whereas LIGDOCK was specifically designed to generate protein–ligand co-structures.

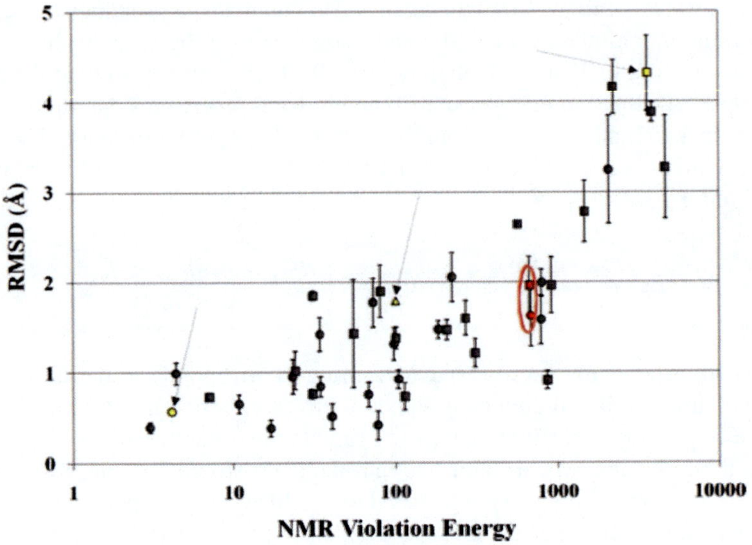

Fig. 8 A comparison of the NMR docking energy from AutoDockFilter to the rmsds between the best docked ligand conformers and the experimental protein–ligand co-structure. An improved correlation is observed for the docking of ligands to the bound form of the protein (*circles*) compared to the apo-protein structure (*squares*). The *red* data points correspond to AutoDockFilter docking results using experimental CSPs for staphylococcal nuclease (PDB-ID: 1EY0, 1SNC) [180–182]. The *yellow* data points correspond to a docking to the apo-structure of acetylcholinesterase (PDB-ID:1ACJ, 1QIF) that resulted in a high rmsd. However, the inclusion of side chain flexibility for residues in the ligand binding site resulted in an improved docking and lower rmsd. (Reprinted with permission from [121], copyright 2008 by the American Chemical Society)

Gonzalez-Ruiz and Gohlke describe a conceptual hybrid (QCSP-Steered Docking) of the AutoDockFilter and the HADDOCK/LIGDOCK procedures, effectively combining the best features of both methods [160]. AutoDock 3.0.5 was modified to incorporate a new hybrid scoring scheme utilizing the DrugScore target function [164] with an amended CSP energy function. Basically, AutoDock is used to generate poses similar to AutoDockFilter, but when an energetically acceptable pose is obtained, CSPs are calculated for the pose. The calculated CSPs are based only on ring current effects [165] from aromatic rings in the ligand. A comparison between the calculated and experimental CSPs is used to calculate an energy violation. Instead of an absolute difference, a Kendall's rank correlation coefficient is used to account for magnitude differences between the experimental and calculated CSP values. The pose with the lowest DrugScore and CSP energy is chosen. Thus, QCSP-Steered Docking is as fast as AutoDockFilter, but allows for direct refinement against the experimental CSPs like HADDOCK/LIGDOCK.

4.3 Lead Identification

Several recent approaches have investigated the combination of NMR and molecular docking for identifying inhibitors for specific proteins. Typically, these approaches apply one of two methodologies: (1) a virtual screen of a large compound library followed by validation of potential binders by NMR or (2) a fragment-based screen using NMR followed by the use of molecular docking to generate a protein–ligand co-structure for optimization.

Virtual screening followed by NMR validation is perhaps the most commonly used combination of these two techniques. Several recent studies have highlighted the use of this approach [166–169]. Branson et al. [166] used a virtual screen with NMR to identify inhibitors of lupindiadenosine $5',5'''$-P^1,P^4-tetraphosphate (Ap_4A) hydrolase. These proteins are found in eukaryotes, prokaryotes, and archaea and have been proposed to be involved in several biological functions, ranging from apoptosis, DNA repair, to gene expression. In bacteria, it has also been shown to be involved in pathogenesis, which makes this a potential target for developing antimicrobial agents. There is also a significant difference in sequence between the bacterial and animal forms of the protein, which makes this even more attractive as a drug target. In this study, a virtual screen using DOCK 4 [41] was performed on Ap_4A hydrolase from *Lupinusangustifolius* with a database of ~120,000 compounds. The docked poses from DOCK were reranked according to consensus scoring using six different scoring functions, where the top 100 ranked ligands were selected and then filtered again to remove all compounds with a logP of 3 or greater in order to select for compounds likely to be water soluble. The result was seven compounds, of which six were commercially available. These six compounds were then subjected to isothermal titration calorimetry to identify any inhibition of hydrolase activity. From that analysis, one compound (NSC51531), which contains a 1,4-diaminoanthracene-9,10-dione core, showed significant binding affinity (~1 μM K_D) and was chosen for analysis by 2D ^1H-^{15}N HSQC. The NMR analysis showed CSPs consistent with the ATP binding site of the protein. In addition, introducing NSC51531 to the human Ap_4A hydrolase showed non-specific binding and had no apparent toxic effects against human fibroblasts. This is likely due to structural differences between the binding sites of the lupin and human forms of Ap_4A hydrolase. Potentially, a scaffold based upon NSC51531 could result in an inhibitor with specificity towards the bacterial form of the protein leading to an effective microbial agent (Fig. 9a).

Veldcamp and coworkers [169] utilized a similar method that targeted the chemokine CXCL12, which activates the CXCR4 receptor shown to be involved with cancer progression. In this approach, nearly 1.5 million compounds from the ZINC database [170] were screened using DOCK 3.5 [171] against the region of CXCL12 that interacts with CXCR4. Specifically, a sulfotyrosine (sY21) was targeted since it was anticipated to be an important residue for the CXCL12-CXCR4 interaction. The top 1,000 hits were manually inspected to identify five compounds with a favorable interaction with sY21. These five compounds were

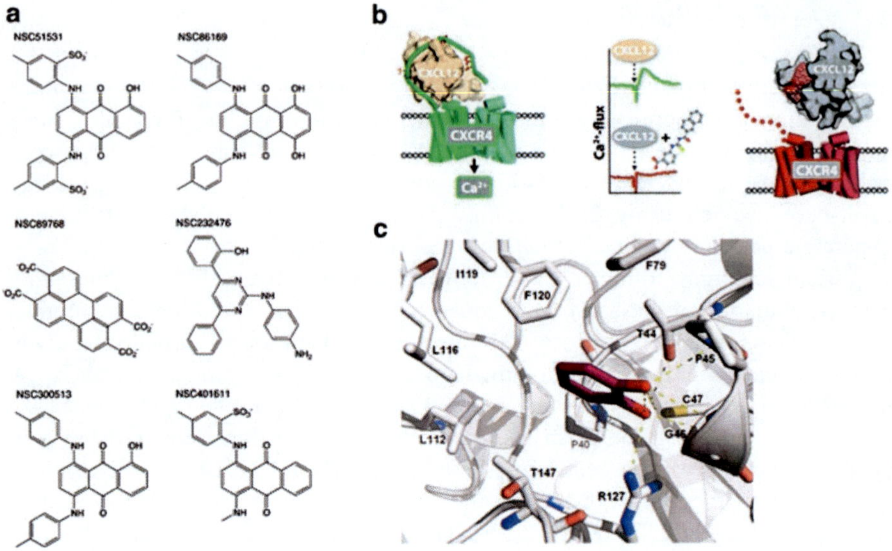

Fig. 9 (**a**) The inhibitors to lupin Ap_4A hydrolase, where NSC51531, NSC232476, and NSC89768 were identified by the virtual screen and NSC86169, NSC300513, and NSC401611 were structural analogs of NSC51531. (Reprinted with permission from [166], copyright 2009 by the American Chemical Society). (**b**) A representation of the interaction between the three sulfotyrosine groups of chemokine CXCL12 and the N-terminal region of the G-protein-coupled receptor CXCR4. Virtual screening and NMR identified 3-(naphthalene-2-carbonylthiocarbomoylamino)benzoic acid (ZINC 310454) as a possible inhibitor of the binding between CXCL12 and CXCR4, which was verified with a calcium flux assay. (Reprinted with permission from [169], copyright 2010 by the American Chemical Society). (**c**) The docked pose of fragment F152 (*magenta*) in the active site of human peroxiredoxin 5 with the hydroxyl groups oriented towards catalytic cysteine (C47). (Reprinted with permission from [174], copyright 2010 by PLoS)

then screened using 2D ^1H-^{15}N HSQCs, which showed that four of the compounds bound weakly, but specifically, to CXCL12 in the region of interest. The strongest binder, ZINC 310454, had a K_D of ~64 μM. Additional NMR screens with analogs to ZINC 310454 showed the importance of the carboxylic acid and naphthyl group, since analogs lacking these features showed no binding in the 2D ^1H-^{15}N HSQC experiments. Furthermore, a calcium flux assay demonstrated that 100 μM ZINC 310454 inhibited CXCL12-mediated signaling. Correspondingly, ZINC 310454 may be a useful scaffold for drug development (Fig. 9b). The results also reinforced the validity of chemokines as a target for drug discovery.

Using molecular docking to screen a large compound library does reduce the time and resources relative to an HTS assay, but it still suffers from an unfocused approach. In general, virtual screens or HTS assays don't efficiently sample chemical space or improve the diversity of hits. Molecular modeling also requires a priori knowledge of the binding site to guide the virtual screen, which may be difficult when dealing with new potential therapeutic targets. One approach to these problems may be to utilize NMR as the primary screening tool and molecular docking to generate protein–ligand co-structures. Since it is not practical to use NMR to screen the large library of compounds typically utilized by HTS or virtual screening, a more focused approach with a smaller compound library is employed.

Fragment-based screening utilizes a significantly smaller library consisting of simple, low molecular-weight (<250–350 Da) molecules [15, 20–22]. These fragment-like molecules typically have weaker binding affinities (millimolar range) compared to hits found in high-throughput screens (micromolar range), but NMR is sensitive enough to detect these weak protein–ligand interactions. Importantly, fragment-based libraries are more efficient in covering chemical space. Simply, the number of possible compounds decreases drastically as the number of atoms is reduced. Thus, a smaller chemical library actually covers a larger percentage of chemical space. An even greater structural diversity can be achieved by chemically linking multiple fragments. This also results in an additive improvement in binding affinity. Evolving a drug from smaller fragments in this manner has the added benefit of improving ligand efficiency, which typically results in a more bioavailable compound that minimizes non-specific and unfavorable interactions [172, 173].

A recent study [174] by Barelier and colleagues utilized fragment-based screening by NMR and molecular docking in the investigation of the human peroxiredoxin 5 (PRDX5) ligands. Peroxiredoxins are important enzymes that catalyze the reduction of hydroperoxides through a conserved cysteine. However, very few ligands have been identified that bind these proteins despite the availability of crystal structures for PRDX5 bound with benzoate (PDB ID: 1HD2, 1H40) [175]. A compound library of 200 fragment compounds was screened by NMR using STD and WaterLOGSY experiments, where six fragments were identified as binders. STD experiments were also used to calculate the binding affinities for the six fragment molecules, which were in the 1–5 mM range. Since the 1D experiments did not provide information about the location of the binding site, AutoDock 4 [40] was used to dock the fragments to the PRDX5 protein structure. The docking was done against the entire protein structure; a grid search focusing on the benzoate ligand binding site was not used. Not surprisingly, ambiguous results were obtained. The molecular fragments bound to several locations on the PRDX5 structure that were indistinguishable based on binding energies.

Of necessity, the NMR backbone assignments for PRDX5 were obtained to enable the identification of the ligand binding site by monitoring CSPs in 2D ^{1}H-^{15}N HSQC experiments. All the fragments were shown to generate a similar set of CSPs consistent with a binding site that included the proposed catalytically active cysteine. The docked binding conformation was also further confirmed from CSPs for derivatives of these fragments. Analysis of the PRDX5 structure with the docked fragments identified the presence of a potentially important hydroxyl functional group that was pointed towards the catalytic cysteine (Fig. 9c). Interestingly, the benzoate compound found in the PRDX5 crystal structure did not show binding by NMR. But, derivatives of benzoate that included a hydroxyl functional group showed improved affinity, further indicating the importance of this hydroxyl group in ligand binding to PRDX5. These results provide further validation of the value of combining fragment-based NMR screens with molecular docking to generate chemical leads.

While fragment-based screens have been shown to be an effective approach to drug discovery, NMR ligand affinity screens require more time and material than a virtual screen. However, fragment-based screens are extremely helpful for new therapeutic targets with unknown binding sites. Also, the approach has the added benefit of providing information about the druggability of the protein target. There is a correlation between the hit rate of a fragment-based NMR screen and the ability of the protein target to bind drug-like compounds with high affinity [176, 177].

5 Concluding Remarks

Significant advances continue to be made in the fields of molecular docking and NMR ligand affinity screens that are benefiting drug discovery. Molecular docking provides an extremely rapid way to evaluate likely binders from a large chemical library with minimal cost. Unfortunately, limitations in the accurate ranking of true binders by molecular docking programs require further experimental validation. Conversely, NMR ligand-affinity screens can directly detect a protein–ligand interaction, can measure a corresponding K_D, and can reliably identify the ligand binding site. However, NMR-ligand affinity screens are resource intensive and are generally limited to relatively small chemical libraries. Thus, the strengths and weakness of virtual screens and NMR ligand affinity screens are perfectly complementary. Combining the two screening techniques has the potential of significantly improving the efficiency of drug discovery. The combination of NMR and molecular modeling techniques has been shown to enable the rapid determination of reliable protein–ligand co-structures, the identification of new therapeutic targets, and the successful discovery of new drug leads.

References

1. Venter JC et al (2001) The sequence of the human genome. Science 291(5507):1304–1351
2. Paul SM et al (2010) How to improve R&D productivity: the pharmaceutical industry's grand challenge. Nat Rev Drug Discov 9(3):203–214
3. Kola I, Landis J (2004) Can the pharmaceutical industry reduce attrition rates? Nat Rev Drug Discov 3(8):711–715
4. Cuatrecasas P (2006) Drug discovery in jeopardy. J Clin Invest 116(11):2837–2842
5. Bernal A, Ear U, Kyrpides N (2001) Genomes OnLine Database (GOLD): a monitor of genome projects world-wide. Nucleic Acids Res 29(1):126–127
6. Frishman D et al (2003) The PEDANT genome database. Nucleic Acids Res 31(1):207–211
7. White RH (2006) The difficult road from sequence to function. J Bacteriol 188(10): 3431–3432
8. Gerlt JA, Babbitt PC (2000) Can sequence determine function? Genome Biol 1(5): REVIEWS0005

9. Rishton GM (1997) Reactive compounds and in vitro false positives in HTS. Drug Discov Today 2(9):382–384
10. Seidler J et al (2003) Identification and prediction of promiscuous aggregating inhibitors among known drugs. J Med Chem 46(21):4477–4486
11. McGovern SL et al (2003) A specific mechanism of nonspecific inhibition. J Med Chem 46(20):4265–4272
12. McGovern SL et al (2002) A common mechanism underlying promiscuous inhibitors from virtual and high-throughput screening. J Med Chem 45(8):1712–1722
13. Kenny BA et al (1998) The application of high-throughput screening to novel lead discovery. Prog Drug Res 51:245–269
14. Macarron R (2006) Critical review of the role of HTS in drug discovery. Drug Discov Today 11(7–8):277–279
15. Powers R (2009) Advances in nuclear magnetic resonance for drug discovery. Expert Opin Drug Discov 4(10):1077–1098
16. Pellecchia M et al (2008) Perspectives on NMR in drug discovery: a technique comes of age. Nat Rev Drug Discov 7(9):738–745
17. Roberts GCK (2000) Applications of NMR in drug discovery. Drug Discov Today 5(6): 230–240
18. Huth JR et al (2004) ALARM NMR: a rapid and robust experimental method to detect reactive false positives in biochemical screens. J Am Chem Soc 127(1):217–224
19. Dalvit C et al (2006) NMR-based quality control approach for the identification of false positives and false negatives in high throughput screening. Curr Drug Discov Technol 3(2): 115–124
20. Schade M (2007) Fragment-based lead discovery by NMR. Front Drug Des Discov 3:105–119
21. Zartler ER, Mo H (2007) Practical aspects of NMR-based fragment discovery. Curr Top Med Chem 7(16):1592–1599
22. Dalvit C (2009) NMR methods in fragment screening: theory and a comparison with other biophysical techniques. Drug Discov Today 14(21/22):1051–1057
23. Fesik SW (1993) NMR structure-based drug design. J Biomol NMR 3(3):261–269
24. Kubinyi H (1998) Structure-based design of enzyme inhibitors and receptor ligands. Curr Opin Drug Discov Devel 1(1):4–15
25. Ishihara K et al (2009) Identification of urinary biomarkers useful for distinguishing a difference in mechanism of toxicity in rat model of cholestasis. Basic Clin Pharmacol Toxicol 105(3):156–166
26. Ott K-H, Aranibar N (2007) Nuclear magnetic resonance metabonomics: methods for drug discovery and development. Methods Mol Biol 358:247–271
27. Powers R (2009) NMR metabolomics and drug discovery. Magn Reson Chem 47(S1): S2–S11
28. Powers R, Mercier KA, Copeland JC (2008) The application of FAST-NMR for the identification of novel drug discovery targets. Drug Discov Today 13(3–4):172–179
29. Mercier KA et al (2006) FAST-NMR: functional annotation screening technology using NMR spectroscopy. J Am Chem Soc 128(47):15292–15299
30. Kitchen DB et al (2004) Docking and scoring in virtual screening for drug discovery: methods and applications. Nat Rev Drug Discov 3(11):935–949
31. Halperin I et al (2002) Principles of docking: an overview of search algorithms and a guide to scoring functions. Proteins 47(4):409–443
32. Warren GL et al (2006) A critical assessment of docking programs and scoring functions. J Med Chem 49(20):5912–5931
33. Hartshorn MJ et al (2007) Diverse, high-quality test set for the validation of protein-ligand docking performance. J Med Chem 50(4):726–741
34. Jacobsson M et al (2003) Improving structure-based virtual screening by multivariate analysis of scoring data. J Med Chem 46(26):5781–5789

35. Loving K, Salam NK, Sherman W (2009) Energetic analysis of fragment docking and application to structure-based pharmacophore hypothesis generation. J Comput Aided Mol Des 23(8):541–554
36. Salaniwal S et al (2007) Critical evaluation of methods to incorporate entropy loss upon binding in high-throughput docking. Proteins 66(2):422–435
37. Vasilyev V, Bliznyuk A (2004) Application of semiempirical quantum chemical methods as a scoring function in docking. Theor Chem Acc 112(4):313–317
38. Wei D et al (2010) Binding energy landscape analysis helps to discriminate true hits from high-scoring decoys in virtual screening. J Chem Inf Model 50(10):1855–1864
39. Zavodszky MI et al (2009) Scoring ligand similarity in structure-based virtual screening. J Mol Recognit 22(4):280–292
40. Morris GM et al (2009) AutoDock4 and AutoDockTools4: automated docking with selective receptor flexibility. J Comput Chem 30(16):2785–2791
41. Ewing TJ et al (2001) DOCK 4.0: search strategies for automated molecular docking of flexible molecule databases. J Comput Aided Mol Des 15(5):411–428
42. Rarey M et al (1996) A fast flexible docking method using an incremental construction algorithm. J Mol Biol 261(3):470–489
43. Friesner RA et al (2004) Glide: a new approach for rapid, accurate docking and scoring. 1. Method and assessment of docking accuracy. J Med Chem 47(7):1739–1749
44. Dominguez C, Boelens R, Bonvin AM (2003) HADDOCK: a protein-protein docking approach based on biochemical or biophysical information. J Am Chem Soc 125(7): 1731–1737
45. Bohm HJ (1992) LUDI: rule-based automatic design of new substituents for enzyme inhibitor leads. J Comput Aided Mol Des 6(6):593–606
46. Bohm HJ (1992) The computer program LUDI: a new method for the de novo design of enzyme inhibitors. J Comput Aided Mol Des 6(1):61–78
47. Cerqueira NMFSA et al. (2010) Virtual screening of compound libraries. Methods Mol Biol 572:57–70 (Ligand-Macromolecular Interactions in Drug Discovery)
48. Ripphausen P et al (2010) Quo vadis, virtual screening? A comprehensive survey of prospective applications. J Med Chem 53(24):8461–8467
49. Sousa SF et al (2010) Virtual screening in drug design and development. Comb Chem High Throughput Screen 13(5):442–453
50. Eckert H, Bajorath J (2007) Molecular similarity analysis in virtual screening: foundations, limitations and novel approaches. Drug Discov Today 12(5&6):225–233
51. Merz KM Jr (2010) Limits of free energy computation for protein-ligand interactions. J Chem Theory Comput 6(5):1769–1776
52. Proschak E et al (2007) Shapelets: possibilities and limitations of shape-based virtual screening. J Comput Chem 29(1):108–114
53. Wyss DF, McCoy MA, Senior MM (2002) NMR-based approaches for lead discovery. Curr Opin Drug Discov Devel 5(4):630–647
54. Lepre CA, Moore JM, Peng JW (2004) Theory and applications of NMR-based screening in pharmaceutical research. Chem Rev 104(8):3641–3676
55. Mercier KA, Powers R (2005) Determining the optimal size of small molecule mixtures for high throughput NMR screening. J Biomol NMR 31(3):243–258
56. Hajduk PJ, Olejniczak ET, Fesik SW (1997) One-dimensional relaxation- and diffusion-edited NMR methods for screening compounds that bind to macromolecules. J Am Chem Soc 119:12257–12261
57. Mayer M, Meyer B (1999) Characterization of ligand binding by saturation transfer difference NMR spectroscopy. Angew Chem Int Ed 38(12):1784–1788
58. Dalvit C et al (2000) Identification of compounds with binding affinity to proteins via magnetization transfer from bulk water. J Biomol NMR 18(1):65–68
59. Jahnke W, Rudisser S, Zurini M (2001) Spin label enhanced NMR screening. J Am Chem Soc 123(13):3149–3150

60. Vanwetswinkel S et al (2005) TINS, target immobilized NMR screening: an efficient and sensitive method for ligand discovery. Chem Biol 12(2):207–216
61. Fejzo J et al (1999) The SHAPES strategy: an NMR-based approach for lead generation in drug discovery. Chem Biol 6(10):755–769
62. Dalvit C et al (2003) Fluorine-NMR experiments for high-throughput screening: theoretical aspects, practical considerations, and range of applicability. J Am Chem Soc 125(25): 7696–7703
63. Dalvit C et al (2002) Fluorine-NMR competition binding experiments for high-throughput screening of large compound mixtures. Comb Chem High Throughput Screen 5(8):605–611
64. Dalvit C et al (2003) A general NMR method for rapid, efficient, and reliable biochemical screening. J Am Chem Soc 125(47):14620–14625
65. Dalvit C et al (2004) Reliable high-throughput functional screening with 3-FABS. Drug Discov Today 9(14):595–602
66. Price SW (1997) Pulsed-field gradient nuclear magnetic resonance as a tool for studying translational diffusion: part 1. Basic theory. Concepts Magn Reson 9:299–336
67. Price SW (1998) Pulsed-field gradient nuclear magnetic resonance as a tool for studying translational diffusion: part II. Experimental aspects. Concepts Magn Reson 10:197–237
68. Shortridge MD et al (2008) Estimating protein-ligand binding affinity using high-throughput screening by NMR. J Comb Chem 10(6):948–958
69. Ji Z, Yao Z, Liu M (2009) Saturation transfer difference nuclear magnetic resonance study on the specific binding of ligand to protein. Anal Biochem 385(2):380–382
70. Muhandiram DR et al (1993) A gradient 13C NOESY-HSQC experiment for recording NOESY spectra of 13C-labeled proteins dissolved in H2O. J Magn Reson B 102(3):317–321
71. Sklenar V et al (1993) Gradient-tailored water suppression for proton-nitrogen-15 HSQC experiments optimized to retain full sensitivity. J Magn Reson A 102(2):241–245
72. Per VK et al (1997) Attenuated T2 relaxation by mutual cancellation of dipole-dipole coupling and chemical shift anisotropy indicates an avenue to NMR structures of very large biological macromolecules in solution. Proc Natl Acad Sci USA 94(23):12366–12371
73. Shuker SB et al (1996) Discovering high-affinity ligands for proteins: SAR by NMR. Science 274(5292):1531–1534
74. Fielding L (2007) NMR methods for the determination of protein-ligand dissociation constants. Prog Nucl Magn Reson Spectrosc 51:219–242
75. Morton CJ et al (1996) Solution structure and peptide binding of the SH3 domain from human Fyn. Structure 4(6):705–714
76. Stoll F (2003) Library design. Chimia 57(5):224–228
77. Erlanson DA, McDowell RS, O'Brien T (2004) Fragment-based drug discovery. J Med Chem 47(14):3463–3482
78. Siegal G, Ab E, Schultz J (2007) Integration of fragment screening and library design. Drug Discov Today 12(23&24):1032–1039
79. Lipinski CA (2004) Lead- and drug-like compounds: the rule-of-five revolution. Drug Discov Today Technol 1(4):337–341
80. Schanda P, Kupce E, Brutscher B (2005) SOFAST-HMQC experiments for recording two-dimensional heteronuclear correlation spectra of proteins within a few seconds. J Biomol NMR 33(4):199–211
81. Schanda P, Brutscher B (2006) Hadamard frequency-encoded SOFAST-HMQC for ultrafast two-dimensional protein NMR. J Magn Reson 178(2):334–339
82. Mori S et al (1995) Improved sensitivity of HSQC spectra of exchanging protons at short interscan delays using a new fast HSQC (FHSQC) detection scheme that avoids water saturation. J Magn Reson B 108(1):94–98
83. Taufer M et al (2005) Study of an accurate and fast protein-ligand docking algorithm based on molecular dynamics. Concurr Comput 17(14):1627–1641
84. Garcia-Sosa AT, Sild S, Maran U (2009) Docking and virtual screening using distributed grid technology. QSAR Comb Sci 28:815–821

85. Kuntz ID, Meng EC, Shoichet BK (1994) Structure-based molecular design. Acc Chem Res 27(5):117–123
86. Krovat EM, Steindl T, Langer T (2005) Recent advances in docking and scoring. Curr Comput Aided Drug Des 1(1):93–102
87. Cole JC et al (2005) Comparing protein-ligand docking programs is difficult. Proteins 60 (3):325–332
88. Wandzik I (2006) Current molecular docking tools and comparisons thereof. MATCH 55 (2):271–278
89. Dias R, de Azevedo WF Jr (2008) Molecular docking algorithms. Curr Drug Targets 9 (12):1040–1047
90. Viji SN, Prasad PA, Gautham N (2009) Protein-ligand docking using mutually orthogonal Latin squares (MOLSDOCK). J Chem Inf Model 49(12):2687–2694
91. Pearce BC et al (2009) E-novo: an automated workflow for efficient structure-based lead optimization. J Chem Inf Model 49(7):1797–1809
92. Mizutani MY, Tomioka N, Itai A (1994) Rational automatic search method for stable docking models of protein and ligand. J Mol Biol 243(2):310–326
93. Schlosser J, Rarey M (2009) Beyond the virtual screening paradigm: structure-based searching for new lead compounds. J Chem Inf Model 49(4):800–809
94. Bostrom J, Greenwood JR, Gottfries J (2003) Assessing the performance of OMEGA with respect to retrieving bioactive conformations. J Mol Graph Model 21(5):449–462
95. Miller MD et al (1994) FLOG: a system to select quasi-flexible ligands complementary to a receptor of known three-dimensional structure. J Comput Aided Mol Des 8(2):153–174
96. Griewel A et al (2009) Conformational sampling for large-scale virtual screening: accuracy versus ensemble size. J Chem Inf Model 49(10):2303–2311
97. Hart TN, Read RJ (1994) Multiple-start Monte Carlo docking of flexible ligands. Birkhaeuser, Boston
98. Fuhrmann J et al (2010) A new Lamarckian genetic algorithm for flexible ligand-receptor docking. J Comput Chem 31(9):1911–1918
99. Cao T, Li T (2004) A combination of numeric genetic algorithm and tabu search can be applied to molecular docking. Comput Biol Chem 28(4):303–312
100. Huang S-Y, Zou X (2010) Advances and challenges in protein-ligand docking. Int J Mol Sci 11:3016–3034
101. Huang S-Y, Grinter SZ, Zou X (2010) Scoring functions and their evaluation methods for protein-ligand docking: recent advances and future directions. Phys Chem Chem Phys 12(40): 12899–12908
102. Huang S-Y, Zou X (2010) Mean-force scoring functions for protein-ligand binding. Annu Rep Comput Chem 6:281–296
103. Bohme A et al (1998) Piperacillin/tazobactam versus cefepime as initial empirical antimicrobial therapy in febrile neutropenic patients: a prospective randomized pilot study. Eur J Med Res 3(7):324–330
104. Eldridge MD et al (1997) Empirical scoring functions: I. The development of a fast empirical scoring function to estimate the binding affinity of ligands in receptor complexes. J Comput Aided Mol Des 11(5):425–445
105. Tao P, Lai L (2001) Protein ligand docking based on empirical method for binding affinity estimation. J Comput Aided Mol Des 15(5):429–446
106. Wang R, Lai L, Wang S (2002) Further development and validation of empirical scoring functions for structure-based binding affinity prediction. J Comput Aided Mol Des 16(1): 11–26
107. Muegge I, Martin YC (1999) A general and fast scoring function for protein-ligand interactions: a simplified potential approach. J Med Chem 42(5):791–804
108. Gohlke H, Hendlich M, Klebe G (2000) Knowledge-based scoring function to predict protein-ligand interactions. J Mol Biol 295(2):337–356

109. Velec HF, Gohlke H, Klebe G (2005) DrugScore(CSD)-knowledge-based scoring function derived from small molecule crystal data with superior recognition rate of near-native ligand poses and better affinity prediction. J Med Chem 48(20):6296–6303
110. Kortagere S, Krasowski MD, Ekins S (2009) The importance of discerning shape in molecular pharmacology. Trends Pharmacol Sci 30(3):138–147
111. McGann MR et al (2003) Gaussian docking functions. Biopolymers 68(1):76–90
112. Stark JL et al (2010) Solution structure and function of YndB, an AHSA1 protein from Bacillus subtilis. Proteins 78(16):3328–3340
113. Merlitz H, Herges T, Wenzel W (2004) Fluctuation analysis and accuracy of a large-scale in silico screen. J Comput Chem 25(13):1568–1575
114. Tirado-Rives J, Jorgensen WL (2006) Contribution of conformer focusing to the uncertainty in predicting free energies for protein-ligand binding. J Med Chem 49(20):5880–5884
115. Charifson PS et al (1999) Consensus scoring: a method for obtaining improved hit rates from docking databases of three-dimensional structures into proteins. J Med Chem 42(25): 5100–5109
116. Peng JW (2009) Communication breakdown: protein dynamics and drug design. Structure 17(3):319–320
117. Hayward S, de Groot BL (2008) Normal modes and essential dynamics. Methods Mol Biol 443:89–106 (Molecular Modeling of Proteins)
118. Huang SY, Zou X (2007) Ensemble docking of multiple protein structures: considering protein structural variations in molecular docking. Proteins 66(2):399–421
119. Erickson JA et al (2004) Lessons in molecular recognition: the effects of ligand and protein flexibility on molecular docking accuracy. J Med Chem 47(1):45–55
120. Sherman W et al (2006) Novel procedure for modeling ligand/receptor induced fit effects. J Med Chem 49(2):534–553
121. Stark J, Powers R (2008) Rapid protein-ligand costructures using chemical shift perturbations. J Am Chem Soc 130(2):535–545
122. B-Rao C, Subramanian J, Sharma SD (2009) Managing protein flexibility in docking and its applications. Drug Discov Today 14(7–8):394–400
123. Jiang F, Kim SH (1991) "Soft docking": matching of molecular surface cubes. J Mol Biol 219 (1):79–102
124. Claussen H et al (2001) FlexE: efficient molecular docking considering protein structure variations. J Mol Biol 308(2):377–395
125. Alberts IL, Todorov NP, Dean PM (2005) Receptor flexibility in de novo ligand design and docking. J Med Chem 48(21):6585–6596
126. Barril X, Morley SD (2005) Unveiling the full potential of flexible receptor docking using multiple crystallographic structures. J Med Chem 48(13):4432–4443
127. Klebe G (2006) Virtual ligand screening: strategies, perspectives and limitations. Drug Discov Today 11(13–14):580–594
128. Schneider G, Bohm HJ (2002) Virtual screening and fast automated docking methods. Drug Discov Today 7(1):64–70
129. Chen H et al (2006) On evaluating molecular-docking methods for pose prediction and enrichment factors. J Chem Inf Model 46(1):401–415
130. Kontoyianni M, McClellan LM, Sokol GS (2004) Evaluation of docking performance: comparative data on docking algorithms. J Med Chem 47(3):558–565
131. Kirchmair J et al (2008) Evaluation of the performance of 3D virtual screening protocols: RMSD comparisons, enrichment assessments, and decoy selection–what can we learn from earlier mistakes? J Comput Aided Mol Des 22(3–4):213–228
132. Bender A, Glen RC (2005) A discussion of measures of enrichment in virtual screening: comparing the information content of descriptors with increasing levels of sophistication. J Chem Inf Model 45(5):1369–1375
133. Truchon J-F, Bayly CI (2007) Evaluating virtual screening methods: good and bad metrics for the "early recognition" problem. J Chem Inf Model 47(2):488–508

134. Scsibrany H et al (2003) Clustering and similarity of chemical structures represented by binary substructure descriptors. Chemom Intell Lab Syst 67(2):95–108
135. Davis AM et al (2005) Components of successful lead generation. Curr Top Med Chem 5(4):421–439
136. Sams-Dodd F (2006) Drug discovery: selecting the optimal approach. Drug Discov Today 11(9–10):465–472
137. Fink T, Reymond JL (2007) Virtual exploration of the chemical universe up to 11 atoms of C, N, O, F: assembly of 26.4 million structures (110.9 million stereoisomers) and analysis for new ring systems, stereochemistry, physicochemical properties, compound classes, and drug discovery. J Chem Inf Model 47(2):342–353
138. Lahana R (1999) How many leads from HTS? Drug Discov Today 4(10):447–448
139. Goode DR et al (2008) Identification of promiscuous small molecule activators in high-throughput enzyme activation screens. J Med Chem 51(8):2346–2349
140. Foloppe N et al (2006) Identification of chemically diverse Chk1 inhibitors by receptor-based virtual screening. Bioorg Med Chem 14(14):4792–4802
141. Richardson CM et al (2007) Discovery of a potent CDK2 inhibitor with a novel binding mode, using virtual screening and initial, structure-guided lead scoping. Bioorg Med Chem Lett 17(14):3880–3885
142. Pellecchia M et al (2004) NMR-based techniques in the hit identification and optimisation processes. Expert Opin Ther Targets 8(6):597–611
143. Galperin MY, Koonin EV (2010) From complete genome sequence to 'complete' understanding? Trends Biotechnol 28(8):398–406
144. Tucker CL (2002) High-throughput cell-based assays in yeast. Drug Discov Today 7(18 Suppl):S125–S130
145. Lee YH et al (2005) Gene knockdown by large circular antisense for high-throughput functional genomics. Nat Biotechnol 23(5):591–599
146. Joshi T et al (2004) Genome-scale gene function prediction using multiple sources of high-throughput data in yeast Saccharomyces cerevisiae. OMICS 8(4):322–333
147. del Val C et al (2004) High-throughput protein analysis integrating bioinformatics and experimental assays. Nucleic Acids Res 32(2):742–748
148. Laurie AT, Jackson RM (2006) Methods for the prediction of protein-ligand binding sites for structure-based drug design and virtual ligand screening. Curr Protein Pept Sci 7(5):395–406
149. Blundell TL et al (2006) Structural biology and bioinformatics in drug design: opportunities and challenges for target identification and lead discovery. Philos Trans R Soc Lond B Biol Sci 361(1467):413–423
150. Vajda S, Guarnieri F (2006) Characterization of protein-ligand interaction sites using experimental and computational methods. Curr Opin Drug Discov Devel 9(3):354–362
151. Mercier KA, Germer K, Powers R (2006) Design and characterization of a functional library for NMR screening against novel protein targets. Comb Chem High Throughput Screen 9(7):515–534
152. Mercier KA, Shortridge MD, Powers R (2009) A multi-step NMR screen for the identification and evaluation of chemical leads for drug discovery. Comb Chem High Throughput Screen 12(3):285–295
153. Powers R et al (2006) Comparison of protein active site structures for functional annotation of proteins and drug design. Proteins 65(1):124–135
154. Park K, Kim D (2008) Binding similarity network of ligand. Proteins 71(2):960–971
155. Mercier KA et al (2009) Structure and function of Pseudomonas aeruginosa protein PA1324 (21-170). Protein Sci 18(3):606–618
156. Shortridge MD, Powers R (2009) Structural and functional similarity between the bacterial type III secretion system needle protein PrgI and the eukaryotic apoptosis Bcl-2 proteins. PLoS One 4(10):e7442
157. McCoy MA, Wyss DF (2000) Alignment of weakly interacting molecules to protein surfaces using simulations of chemical shift perturbations. J Biomol NMR 18(3):189–198

158. Schieborr U et al (2005) How much NMR data is required to determine a protein-ligand complex structure? Chembiochem 6(10):1891–1898
159. Wang B, Westerhoff LM, Merz KM Jr (2007) A critical assessment of the performance of protein-ligand scoring functions based on NMR chemical shift perturbations. J Med Chem 50(21):5128–5134
160. Gonzalez-Ruiz D, Gohlke H (2009) Steering protein-ligand docking with quantitative NMR chemical shift perturbations. J Chem Inf Model 49(10):2260–2271
161. Xu X-P, Case DA (2001) Automated prediction of 15N, 13CÎ±, 13CÎ and 13C′ chemical shifts in proteins using a density functional database. J Biomol NMR 21(4):321–333
162. Nilges M (1995) Calculation of protein structures with ambiguous distance restraints. Automated assignment of ambiguous NOE crosspeaks and disulphide connectivities. J Mol Biol 245(5):645–660
163. Guntert P, Wuthrich K (2001) Sampling of conformation space in torsion angle dynamics calculations. Comput Phys Commun 138(2):155–169
164. Gohlke H, Hendlich M, Klebe G (2000) Predicting binding modes, binding affinities and "hot spots" for protein-ligand complexes using a knowledge-based scoring function. Perspect Drug Discov Des 20:115–144
165. Osapay K, Case DA (1991) A new analysis of proton chemical shifts in proteins. J Am Chem Soc 113(25):9436–9444
166. Branson KM et al (2009) Discovery of inhibitors of lupin diadenosine 5′,5‴-P(1), P(4)-tetraphosphate hydrolase by virtual screening. Biochemistry 48(32):7614–7620
167. Jacobsson M et al (2008) Identification of Plasmodium falciparum spermidine synthase active site binders through structure-based virtual screening. J Med Chem 51(9):2777–2786
168. Lee Y et al (2009) Identification of compounds exhibiting inhibitory activity toward the Pseudomonas tolaasii toxin tolaasin I using in silico docking calculations, NMR binding assays, and in vitro hemolytic activity assays. Bioorg Med Chem Lett 19(15):4321–4324
169. Veldkamp CT et al (2010) Targeting SDF-1/CXCL12 with a ligand that prevents activation of CXCR4 through structure-based drug design. J Am Chem Soc 132(21):7242–7243
170. Irwin JJ, Shoichet BK (2005) ZINC–a free database of commercially available compounds for virtual screening. J Chem Inf Model 45(1):177–182
171. Lorber DM, Shoichet BK (2005) Hierarchical docking of databases of multiple ligand conformations. Curr Top Med Chem 5(8):739–749
172. Bembenek SD, Tounge BA, Reynolds CH (2009) Ligand efficiency and fragment-based drug discovery. Drug Discov Today 14(5–6):278–283
173. Reynolds CH, Tounge BA, Bembenek SD (2008) Ligand binding efficiency: trends, physical basis, and implications. J Med Chem 51(8):2432–2438
174. Barelier S et al (2010) Discovery of fragment molecules that bind the human peroxiredoxin 5 active site. PLoS One 5(3):e9744
175. Declercq JP et al (2001) Crystal structure of human peroxiredoxin 5, a novel type of mammalian peroxiredoxin at 1.5 A resolution. J Mol Biol 311(4):751–759
176. Hajduk PJ, Huth JR, Fesik SW (2005) Druggability indices for protein targets derived from NMR-based screening data. J Med Chem 48(7):2518–2525
177. Hajduk PJ, Huth JR, Tse C (2005) Predicting protein druggability. Drug Discov Today 10(23–24):1675–1682
178. Peng JW et al (2001) Nuclear magnetic resonance-based approaches for lead generation in drug discovery. Methods Enzymol 338:202–230
179. Morris GM et al (1998) Automated docking using a Lamarckian genetic algorithm and an empirical binding free energy function. J Comput Chem 19(14):1639–1662
180. Wang JF et al (1992) Solution studies of staphylococcal nuclease H124L. 2. 1H, 13C, and 15N chemical shift assignments for the unligated enzyme and analysis of chemical shift changes that accompany formation of the nuclease-thymidine 3′,5′-bisphosphate-calcium ternary complex. Biochemistry 31(3):921–936

181. Wang JF et al (1990) Two-dimensional NMR studies of staphylococcal nuclease. 2. Sequence-specific assignments of carbon-13 and nitrogen-15 signals from the nuclease H124L-thymidine 3′,5′-bisphosphate-Ca2+ ternary complex. Biochemistry 29(1):102–113
182. Wang JF, LeMaster DM, Markley JL (1990) Two-dimensional NMR studies of staphylococcal nuclease. 1. Sequence-specific assignments of hydrogen-1 signals and solution structure of the nuclease H124L-thymidine 3′,5′-bisphosphate-Ca2+ ternary complex. Biochemistry 29(1):88–101

Top Curr Chem (2012) 326: 35–46
DOI: 10.1007/128_2011_216

Published online: 2 August 2011

NMR as a Unique Tool in Assessment and Complex Determination of Weak Protein–Protein Interactions

Olga Vinogradova and Jun Qin

Abstract Protein–protein interactions are crucial for a wide variety of biological processes. These interactions range from high affinity (K_d < nM) to very low affinity (K_d > mM). While much is known about the nature of high affinity protein complexes, our knowledge about structural characteristics of weak protein–protein interactions (wPPIs) remains limited: in addition to the technical difficulties associated with their investigation, historically wPPIs used to be considered physiologically irrelevant. However, emerging evidence suggests that wPPIs, either in the form of intact protein complexes or as part of large molecular machineries, are fundamentally important for promoting rapid on/off switches of signal transduction, reversible cell–cell contacts, transient assembly/disassembly of signaling complexes, and enzyme–substrate recognition. Therefore an atomic-level elucidation of wPPIs is vital to understanding a cornucopia of diverse cellular events. Nuclear magnetic resonance (NMR) is famous for its unique abilities to study wPPIs and, by utilization of the new technical developments combined with sparse data based computational analysis, it now allows rapid identification and structural characterization of wPPIs. Here we present our perspective on the NMR methods employed.

Keywords Chemical shifts · NMR · NOE · PCS · PRE · RDC

O. Vinogradova (✉)
Department of Pharmaceutical Sciences, University of Connecticut, 69 North Eagleville Rd, Unit 3092, Storrs, CT 06269-3092, USA
e-mail: olga.vinogradova@uconn.edu

J. Qin
Department of Molecular Cardiology, Cleveland Clinic, NB20, 9500 Euclid Ave., Cleveland, OH 44195, USA
e-mail: qinj@ccf.org

Contents

1 Introduction 36
1.1 Chemical Shift Perturbation Mapping 37
1.2 Nuclear Overhauser Effect 38
1.3 Residual Dipolar Couplings 40
1.4 Paramagnetic NMR 41
2 Conclusions 43
References 44

1 Introduction

Living organisms are very complex, highly structured, tightly regulated systems with precisely orchestrated communications at every level of organizational hierarchy. These communications are largely mediated by protein–protein interactions (PPIs). The complete genome sequencing now reveals that there are thousands of potential PPIs that may function as building blocks for these communication networks [1,2]. PPIs can be classified based upon the strength of interaction, which is often rendered by the equilibrium dissociation constant (K_d) equal to k_{off}/k_{on}, where k_{off} is the rate constant of the complex dissociation reaction and k_{on} is the rate constant of the association reaction. The window of biologically relevant K_d values is extremely wide and can cover 12 orders of magnitude [3]. PPIs can be very loosely divided into three major subclasses [4]: (1) strong, with $K_d < 10^{-9}$ M, and permanent association, (2) strong and transient, where the change in the quaternary state can be triggered, for example, by ligand binding, and (3) weak, with $K_d > 10^{-4}$ M, and transient association with k_{off} rates of up to 10^4 s^{-1}, which results in lifetimes as short as 100 μs [5]. Decades of extensive studies have illuminated structural and functional features for the PPIs from the first two subclasses characterized by strong binding with $K_d < 10^{-6}$ M, which are summarized in numerous reviews [6–8]. The weak PPIs and their physiological importance (wPPIs, with $K_d > 10^{-4}$ M), on the other hand, are less well understood. This could be attributed in part to the technical difficulties encountered during attempts to characterize them directly in vitro or in vivo. The other reason relates to a common prejudice that wPPIs might not be found in living cells, especially considering low ($<10^{-7}$ M) protein concentrations estimated by the whole cell volumes. However, it is now being increasingly appreciated that wPPIs are crucial for promoting diverse biologically important processes such as reversible cell–cell contacts, transient assembly and/or disassembly of large signaling complexes, and dynamic regulation of enzymes [9]. Figure 1 provides three possible scenarios of wPPIs: (1) wPPI between two intact proteins, (2) wPPI as part of multi-domain interactions between two intact proteins, and (3) wPPI as part of a multi-protein complex. Conventional methods such as X-ray crystallography, surface Plasmon resonance (SPR), and isothermal titration calorimetry (ITC) often fail to study these wPPIs accurately. In contrast, nuclear magnetic resonance (NMR) has been proven as a particularly powerful tool to examine them at atomic level resolution

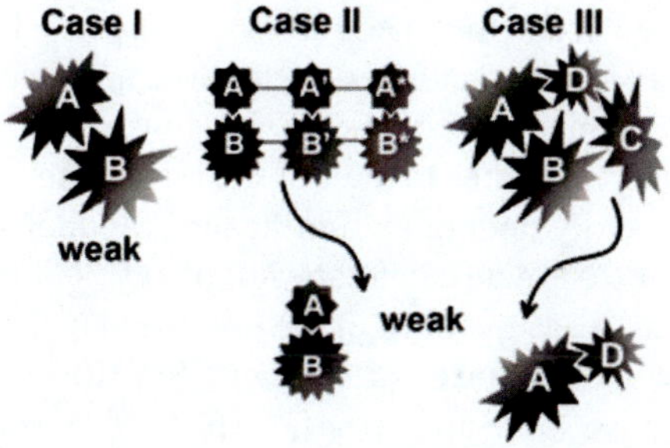

Fig. 1 Three representative cases of wPPIs. Case I: a weak protein–protein interaction found in a locally highly crowded manner. Case II: a weak domain–domain interaction, exemplified by A–B pair, as part of a tight multi-domain complex. Such weak binary domain–domain interaction may be undetectable by many conventional methods including deletion mapping, yeast-hybrid approach, immunoprecipitation, etc., but become apparent when the tertiary structure of the tight complex is challengingly determined. However, NMR may be able to pick this interaction at early stage of the characterization. Case III: a weak protein–protein interaction as a part of multi-protein complex. Similar to (*II*), a weak A–D pair may not be detectable in isolated manner by any conventional techniques except NMR

and at near physiological conditions [7,8,10,11]. In this chapter we present the various NMR approaches to assess and characterize these three types of wPPIs structurally.

1.1 Chemical Shift Perturbation Mapping

The resonance frequencies, also known as chemical shifts, of individual atoms in a particular protein strongly depend on the local environment and, because of that, are often considered as fingerprints of its structure. The chemical shift patterns of ^{15}N and directly attached amide ^{1}H are especially sensitive in this respect. Thus their perturbation, as the result of complex formation, provides a highly sensitive tool for mapping the binding interface. Binding surface on the target protein is identified by titrating the unlabeled target into the solution of the ^{15}N-labeled protein and monitoring the resultant spectral changes in its ^{1}H–^{15}N HSQC (heteronuclear single quantum correlation) spectrum or, for the larger proteins, its TROSY-based (transverse relaxation-optimized spectroscopy) version [12]. The utility and popularity of this experiment are based upon its straightforward nature and high sensitivity – the spectrum can be recorded in 5–40 min on a typical protein sample (~0.1–1 mM). The spectral changes, also denoted as chemical shift perturbations (CSP), are usually associated with a particular set of amino acids that either (1) directly participate in interaction or are situated very closely to the binding site or (2) reflect binding-induced conformational rearrangements (e.g., a disorder–order transition). The former happens more frequently for wPPIs, which have characteristic small CSP with little or no conformational change, at least within the backbone of well folded domains. Hence the binding interface can be qualitatively deduced from the spectrally perturbed residues. In addition, for wPPIs with fast exchange, characterized by high k_{off} rates, K_d can be estimated from concentration-dependent titrations [13]. However, the potential problem associated with this analysis relates to the low affinity of the complex: an interfacial residue might not necessarily undergo a significant CSP upon binding, meaning that interfaces derived from CSP data alone are not always complete [6]. One way to avoid this caveat is to increase the ratio of the unlabeled titrant, which can drive the equilibrium towards the bound form with bigger CSP. However, the ratio cannot be too high since it may cause some non-specific effects. The weakest PPI analyzed by CSP demonstrated a K_d of 10^{-2} M for the flavodoxin/flavodoxin reductase complex [14]. Both (1) and (2) might occur for strong PPIs, which often undergo significant conformational rearrangements upon binding, especially if a disorder–order transition occurs. In such cases, analysis of binding interfaces is more challenging and less straightforward, and one has to rely on additional techniques, such as incorporation of a cross-saturation [15] or inter-molecular Nuclear Overhauser Effects/Enhancements (NOEs) (described in detail in the next section). To conclude, although for wPPIs the CSP method provides fast and robust assessment of the residues forming an intermolecular interface, the mutual orientation of the two partners remains elusive. Thus, if the goal is to generate the structural model of the complex, additional experiments have to be performed and/or novel computational approaches have to be employed.

1.2 *Nuclear Overhauser Effect*

NOE, a relaxation mechanism based upon magnetic dipole–dipole interactions of the nuclei, allows measurement of interproton distances with the basic r^{-6} distance proportionality. This provides major distance restraints for structural calculations. Supplemented with additional data, such as original dihedral angle restraints obtained from J-couplings or more recent information about the orientation of the bond vectors connecting magnetically active nuclei with respect to the external magnetic field, this approach has been the foundation for NMR-based protein structure determination since its dawn in 1984 [11].

1.2.1 NOE in the Determination of the Structure of wPPIs

From the perspective of wPPIs structural characterization, two particular applications of NOE are proven most fruitful.

Transferred NOE Experiments

Transferred NOE Experiments (trNOESY) is a quick two-dimensional ^{1}H NMR experiment that allows the observation of intramolecular proton contacts (<5 Å) for the small peptide bound to its target protein [16]. In a nutshell, if the affinity of the interaction in question is low due to the fast dissociation rate, cross relaxation between protons of the peptide in the bound state, which is governed by the large correlation time of the complex, is transferred to the free state through chemical exchange. This phenomenon is manifested by the appearance of additional peaks at the original (corresponding to free state) frequencies in the NOESY spectrum of the peptide when it is mixed with a small molar portion of the large, typically over 50 kDa in molecular weight, target protein. Protein–peptide ratios for trNOESY may vary from 1:10 to 1:200 with mixing times ranging from 50 to 500 ms. Both parameters need to be optimized for each particular case. Substantially increased number of NOEs should be observed for the peptide in the presence of its target protein comparatively to the peptide free form. The method requires no isotope-labeling and is suitable for examination of protein–ligand interactions over a wide range of K_ds (micromolar–millimolar) [7]. The method can be applied to study initial lead compounds weakly bound to the target protein, which allows the structure determination of the bound compounds for further optimization leading to high affinity binding.

Half-Filtered NOESY (Intermolecular NOEs)

Half-filtered NOESY approach was developed to detect inter-molecular contacts in the form of NOEs only between protons pairs in which one of the protons is attached to ^{15}N or ^{13}C nuclei while the other is attached to magnetically inactive ^{14}N or ^{12}C nuclei. Thus it requires the preparation of ^{15}N–^{13}C-labeled protein mixed with its unlabeled binding partner, and/or vice versa. Two types of half-filtered NOESY experiment are commonly used [17]: (1) three dimensional ^{13}C-separated–^{15}N,^{13}C-filtered NOESY, which detects NOEs between protons attached to ^{13}C atoms of the doubly labeled protein and those attached to ^{12}C and ^{14}N on the unlabeled protein, and (2) three-dimensional ^{15}N-separated-^{15}N,^{13}C-filtered NOESY, which detects NOEs between ^{15}N-attached and ^{12}C-,^{14}N-attached protons. However, there are pulse sequences available with smart combinations of both, when separation in ^{13}C and ^{15}N dimensions can be achieved simultaneously, significantly reducing the experimental time. The sensitivity of this experiment crucially depends on the lifetime of a protein complex. For weak PPIs characterized by high k_{off} the actual portion of the complex within the sample might not be high enough for observing the intermolecular NOEs. However, in favorable cases, when the concentrations of both binding partners are high [18], the complexed state could be detectible with the help of high-sensitivity NMR instruments, such as those equipped with cryo-probes. Another relatively more sensitive experiment is the ^{15}N-edited NOESY on a ^{15}N/100% deuterated protein bound to the target, which will detect the NOEs between the amide proton of the ^{15}N-labeled protein and any nearby protons of the target [19]. This experiment can be complementary to those in (1) and (2) but provides a very unambiguous assignment of the amide protons, which are usually well resolved in the HSQC spectrum, to the protons of the unlabeled target. This experiment can detect very weak NOEs, possibly up to 7 Å in distance, due to the deuteration effect.

1.3 Residual Dipolar Couplings

Although NMR-based applications on dipolar coupling have been mainly associated with solid-state NMR or NMR of oriented samples, they have been recently applied to solution NMR for studying macromolecular structure and function in an aqueous solution [20,21]. The dipolar coupling describes a through-space interaction that arises between any two magnetically active nuclei. It depends upon the distance between two atoms, which is constant for the nuclei connected by covalent bonds such as ^{1}H–^{15}N or ^{1}H–^{13}C, and the orientation of the connecting vector with respect to the external magnetic field. In solution, dipolar interactions are averaged because of fast isotropic tumbling. However, if the macromolecules experience obstacles in some directions due to partial alignment with orienting media, for example, bacteriophage, bicelles or polyacrylamide gels, the dipolar couplings are not completely canceled out and whatever is left is designated as residual dipolar couplings (RDC). By measuring RDC the orientation of the molecular alignment tensor could be defined, providing the information about mutual orientation between the domains within single macromolecule or between binary units of the complex. Thus the quest began to find robust ways to orient the media weakly without significant increase in viscosity of the system or generation nucleation points for aggregation. The general idea is based upon a fact that certain media, possessing sufficiently large magnetic susceptibility anisotropy, can be aligned spontaneously by high magnetic field. In earlier 1990s, bicelles, disk-shaped particles made from the lipid/detergent mixtures, were introduced [22] for this purpose, followed by rod-shaped viruses [23], mechanically orientable systems [24,25], and G-tetrad DNA [26]. As compared to the more conventional NOEs approach, RDC carry complementary information: while NOEs provide only local distance restraints, RDC contain long range orientational information (e.g., see review by [27]), thus delivering powerful long-range geometric constraints for proper subunits orientation during the structure determination of the complex. In the case of wPPIs that may undergo fast exchange between the bound and free forms of the binding partners, measured RDC will represent the population weighted average values of those in bound and free form. Theoretically, knowing the molar ratio of bound and free forms (from K_d and molar concentrations), and after measuring RDC in the free form, one can calculate RDC in bound form [27]. From the RDC of weakly bound subunits, their alignment tensors can be calculated and matched for defining the structure of a weak complex. One example using this strategy to determine the structure of weak complex is α-methyl mannose bound to mannose-binding protein with a $K_d \sim 1$ mM [28]. In practice, however, it is not always straightforward for small ligands to determine the accurate fraction of bound form and, thus, full saturation could be beneficial to utilize the RDCs of the fully bound form.

1.4 Paramagnetic NMR

As NMR spectroscopists are constantly on the quest to improve the line-shape and to reduce the width of the peaks in their spectra, elimination of the paramagnetic species, often causing significant line-broadening effects, has been considered as paramount in sample preparations. However, the usually undesirable line-broadening effect can provide unique structural long range information [29] when the effect is specific and paramagnetic center is localized to a particular site of the macromolecules. Historically, this understanding has been mainly applied to proteins containing metal-binding sites (reviewed in [30]). The idea to utilize surface exposed cysteines for introduction of paramagnetic tags came later with a cornucopia of chemical agents and procedures developed for reliable conjugation at specific sites (reviewed in [31]). This approach not only provides information about intramolecular distances but can also help in defining alignment of binding subunits within complexes characterized by wPPIs [32], although actual quantification of the distance dependence in such systems is not always straightforward as we discuss below. Two distinguished NMR phenomena, based upon the specific nature of the magnetic moment of an attached paramagnetic center, present the basis for structural investigation. These are paramagnetic relaxation enhancement (PRE) characteristic for all paramagnetic moieties and the pseudocontact shifts (PCS) effect specific for the subclass of paramagnetic ions with an anisotropic electron g-tensor.

1.4.1 The PRE Effect

The large magnetic dipolar interaction of the unpaired electron from a paramagnetic atom with the neighboring NMR-active nucleus results in an increase of the relaxation rate of the above nucleus [33]. Similar to NOE, this effect has basic r^{-6} distance proportionality, but, because of the larger magnetic moment of the electron, PRE effect is observable at longer, up to 25–35 Å, distances depending upon the nature of the particular paramagnetic group [34,35]. Thus PRE measurements can provide much longer range distance restraints for structural calculations in comparison to the classical NOE approach. The caveat of PRE application for short distances determination comes from the same original source and relates to the fact that nuclei situated very close to a paramagnetic center are often broadened beyond detection. However, the data sets acquired by NOE and PRE approaches are at least complementary. The PRE measurements are based on the correlation of the increased transverse relaxation rate with the distance between the introduced paramagnetic moiety and the affected nucleus [36]. Simply speaking, we are measuring the distance-dependent reduction in peak intensities in a ^{15}N-HSQC spectrum of the target protein when a single paramagnetic tag is attached to it at the specific site, usually through a thioether bond formed with the side chain of the cysteine residue. Nitroxide stable radicals or metal chelators, such as EDTA-Mn^{2+}, which are characterized by an unpaired electron with an isotropic g-tensor, are especially useful since they do not give rise to pseudo-contact shifts, and Curie-spin relaxation is negligible [37]. The advantage of their employment for pure PRE measurements is the fast and straightforward nature of the method: the resonance assignments in the correlation spectra, known from through-bond scalar triple resonance experiments, are not perturbed by paramagnetic modification and the high sensitivity results in an experimental time of 5–40 min on a typical protein sample (the same advantages we discussed for CSP method, as both are based on the same types of data acquisition – HSQC experiments). The potential problems are associated with intrinsic flexibility/rotation of either the paramagnetic tag itself or its attachment to the protein, resulting in the time average distances sampled over all possible conformations. Thus certain caution is required for incorporation of the derived distances as the restraints for structure calculations, where a paramagnetic center, for example, can be treated as an ensemble average rather than a fixed point [38]. The other possibility for highly flexible systems is to use PRE data loosely, as a guide, rather than major geometric restraints, for example, in structure determination of the complex when the orientation of the peptide, which could be labeled by a paramagnetic tag, in a particular binding site needs to be addressed [39,40]. The potential ability to study transient low population intermediates in macromolecular interactions is conceivably one of the most exciting PRE implementations in structural biology. These illusive species are rarely accessible by other than NMR biophysical techniques. In an exchanging system the observed PRE measured on the resonance of the major species can be modulated by the minor species to the extent depending upon the rate of exchange [41], with the fast exchange allowing one to characterize structurally populations comprising as low as 1%. The example illustrating PRE potential to demonstrate the existence and visualize the distribution of an ensemble of transient non-specific intermediates in addition to specific complex formation has been presented by Clore and colleagues for a bacterial phosphotransferase system [42]. Thus, it has been proven that PRE data is highly sensitive asserting weak interactions characterized by large k_{off} rate and is salutary for structural analysis of weak PPIs.

1.4.2 The PCS Effect

PCS is a phenomenon that is only observable for paramagnetic systems with anisotropic unpaired electrons such as those found in Dy^{3+}, Tb^{3+}, and Fe^{3+} lanthanide ions, characterized by an anisotropic electron *g*-tensor. In general, if the *g*-tensor is anisotropic, than the magnetic susceptibility tensor (usually referred to as the χ-tensor) is anisotropic as well. The magnitude of the PCS depends on the orientations of the vectors connecting the lanthanide ion and affected nuclei with respect to direction of the external magnetic field. These orientations are not averaged because tumbling in aqueous solution appears to be non-isotropic due to the effect of large χ-tensor intrinsic for these paramagnetic species. This large magnetic susceptibility tensor provides enough energy to overcome random Brownian motion and to generate preferable orientation or alignment of the macromolecule containing the paramagnetic tag. This phenomenon causes changes in chemical shift of the affected nuclei which are sufficiently close to the paramagnetic center (reviewed in [43]). Importantly, the PCS displays basic r^{-3} distance proportionality in contrast to the r^{-6} dependence for the PRE or NOE. In theory, this will define a relatively longer experimentally attainable distance range, extending it up to, for example, ~40 Å for Dy^{3+}in metalloproteins. In practice, the principal axis of the χ-tensor is not rigidly fixed within the molecular frame when an extrinsic metal ion is attached to a macromolecule using a chelator with a flexible linker, causing significant reduction in the magnitude of the PCS because of χ-tensor principal axes fluctuations within the frame of the macromolecule. From the perspective of studying wPPIs, PCS restraints can be generated using a ^{15}N-labeled and/or ^{13}C-labeled protein bound to an unlabeled but paramagnetically tagged partner. ^{15}N and/or ^{13}C-HSQC experiments then need to be recorded for both the paramagnetic and diamagnetic states of a sample and chemical shift changes should be extracted from the spectra [31]. However, to use PCS data, one first has to define the tensor describing the anisotropic magnetic moment of the paramagnetic center [44]. When the structures of the individual proteins are known, PCS data can be combined with rigid-body docking to produce a model for a protein complex. This approach has been proven successful in determination of a 30-kDa complex between the θ and ε subunits of *Escherichia coli* polymerase III [45], where the active-site bound Mg^{2+}/Mn^{2+} pairs were exchanged with paramagnetic Dy^{3+} or Er^{3+} and corresponding ^{15}N-HSQC spectra of the diamagnetic apo-complex and paramagnetic-ion-bound complexes were compared. An analogous approach taking advantage of the intrinsic iron-binding capability of cytochrome f has been used earlier to define the structure of its transient complex with plastocyanin: conveniently, iron in its oxidized Fe^{3+} form is paramagnetic while in the Fe^{2+} form it displays a diamagnetic nature [46].

2 Conclusions

While tight protein interactions can be addressed experimentally by many techniques, including X-ray crystallography, the vast majority of these approaches fail or become unreliable when the interactions are weak. Solution NMR spectroscopy is unique among the structural techniques, permitting the characterizing of weak interactions and providing structures of weak protein-target complexes. If such interactions involve small molecules, NMR can be employed for optimization and development of drug-leads. In the current post-genomic era, the NMR methods we have highlighted in combination with functional and computational approaches hold significant promise for characterizing the plethora of weak protein complexes that regulate cellular events, thereby providing an unbiased and comprehensive view of how proteins function in living cells.

References

1. Gavin AC, Bosche M et al (2002) Functional organization of the yeast proteome by systematic analysis of protein complexes. Nature 415(6868):141–147
2. Ho Y, Gruhler A et al (2002) Systematic identification of protein complexes in Saccharomyces cerevisiae by mass spectrometry. Nature 415(6868):180–183
3. Nooren IM, Thornton JM (2003) Diversity of protein-protein interactions. EMBO J 22(14): 3486–3492
4. Perkins JR, Diboun I et al (2010) Transient protein-protein interactions: structural, functional, and network properties. Structure 18(10):1233–1243
5. Prudencio M, Ubbink M (2004) Transient complexes of redox proteins: structural and dynamic details from NMR studies. J Mol Recognit 17(6):524–539
6. O'Connell MR, Gamsjaeger R et al (2009) The structural analysis of protein-protein interactions by NMR spectroscopy. Proteomics 9(23):5224–5232
7. Qin J, Vinogradova O et al (2001) Protein-protein interactions probed by nuclear magnetic resonance spectroscopy. Meth Enzymol 339:377–389
8. Zuiderweg ER (2002) Mapping protein-protein interactions in solution by NMR spectroscopy. Biochemistry 41(1):1–7
9. Vaynberg J, Qin J (2006) Weak protein-protein interactions as probed by NMR spectroscopy. Trends Biotechnol 24(1):22–27
10. Clore GM, Gronenborn AM (1998) Determining the structures of large proteins and protein complexes by NMR. Trends Biotechnol 16(1):22–34
11. Wuthrich K (1986) NMR of proteins and nucleic acids. Wiley, New York
12. Pervushin K, Riek R et al (1997) Attenuated T2 relaxation by mutual cancellation of dipole-dipole coupling and chemical shift anisotropy indicates an avenue to NMR structures of very large biological macromolecules in solution. Proc Natl Acad Sci USA 94(23): 12366–12371
13. Velyvis A, Vaynberg J et al (2003) Structural and functional insights into PINCH LIM4 domain-mediated integrin signaling. Nat Struct Biol 10(7):558–564
14. Hall DA, Vander Kooi CW et al (2001) Mapping the interactions between flavodoxin and its physiological partners flavodoxin reductase and cobalamin-dependent methionine synthase. Proc Natl Acad Sci USA 98(17):9521–9526
15. Takahashi H, Nakanishi T et al (2000) A novel NMR method for determining the interfaces of large protein-protein complexes. Nat Struct Biol 7(3):220–223
16. Clore GM, Gronenborn AM (1982) The two-dimensional transferred nuclear Overhauser effect. J Magn Reson 48:402–417
17. Zwahlen C, Legault P et al (1997) Methods for measurement of intermolecular NOEs by multinuclear NMR spectroscopy: application to a bacterio-phage l N-peptide/*box*B RNA complex. J Am Chem Soc 119:711–721
18. Vaynberg J, Fukuda T et al (2005) Structure of an ultraweak protein-protein complex and its crucial role in regulation of cell morphology and motility. Mol Cell 17(4):513–523
19. Walters KJ, Matsuo H et al. (1997) A simple method to distinguish intermonomer nuclear Overhauser effects in homodimeric proteins with C2 symmetry. J Am Chem Soc 119:5958–5959
20. Bax A, Kontaxis G et al (2001) Dipolar couplings in macromolecular structure determination. Meth Enzymol 339:127–174
21. Prestegard JH, Bougault CM et al (2004) Residual dipolar couplings in structure determination of biomolecules. Chem Rev 104(8):3519–3540
22. Sanders CR, Hare BJ et al. (1994) Magnetically-oriented phospholipid micelles as a tool for the study of membrane-associated molecules. Prog Nucl Magn Reson Spectrosc 26:421–444
23. Clore GM, Starich MR et al. (1998) Measurement of residual dipolar couplings of macromolecules aligned in the nematic phase of a colloidal suspension of rod-shaped viruses. J Am Chem Soc 120:10571–10572

24. Chou JJ, Kaufman JD et al (2002) Micelle-induced curvature in a water-insoluble HIV-1 Env peptide revealed by NMR dipolar coupling measurement in stretched polyacrylamide gel. J Am Chem Soc 124(11):2450–2451
25. Ishii Y, Markus MA et al (2001) Controlling residual dipolar couplings in high-resolution NMR of proteins by strain induced alignment in a gel. J Biomol NMR 21(2):141–151
26. Lorieau J, Yao L et al (2008) Liquid crystalline phase of G-tetrad DNA for NMR study of detergent-solubilized proteins. J Am Chem Soc 130(24):7536–7537
27. Lipsitz RS, Tjandra N (2004) Residual dipolar couplings in NMR structure analysis. Annu Rev Biophys Biomol Struct 33:387–413
28. Bolon PJ, Al-Hashimi HM et al (1999) Residual dipolar coupling derived orientational constraints on ligand geometry in a 53 kDa protein-ligand complex. J Mol Biol 293(1):107–115
29. Kosen PA (1989) Spin labeling of proteins. Meth Enzymol 177:86–121
30. Bertini I, Luchinat C et al (2005) NMR spectroscopy of paramagnetic metalloproteins. Chembiochem 6(9):1536–1549
31. Otting G (2008) Prospects for lanthanides in structural biology by NMR. J Biomol NMR 42(1):1–9
32. Mahoney NM, Rastogi VK et al. (2000) Binding orientation of proline-rich peptides in solution: polarity of the profilin–ligand interaction. J Am Chem Soc 122:7851–7852
33. Bloembergen N, Morgan LO (1961) Proton relaxation times in paramagnetic solutions. Effects of electron spin relaxation. J Chem Phys 34:842–850
34. Battiste JL, Wagner G (2000) Utilization of site-directed spin labeling and high-resolution heteronuclear nuclear magnetic resonance for global fold determination of large proteins with limited nuclear Overhauser effect data. Biochemistry 39(18):5355–5365
35. Iwahara J, Anderson DE et al (2003) EDTA-derivatized deoxythymidine as a tool for rapid determination of protein binding polarity to DNA by intermolecular paramagnetic relaxation enhancement. J Am Chem Soc 125(22):6634–6635
36. Iwahara J, Tang C et al (2007) Practical aspects of (1)H transverse paramagnetic relaxation enhancement measurements on macromolecules. J Magn Reson 184(2):185–195
37. Clore GM, Tang C et al (2007) Elucidating transient macromolecular interactions using paramagnetic relaxation enhancement. Curr Opin Struct Biol 17(5):603–616
38. Iwahara J, Schwieters CD et al (2004) Ensemble approach for NMR structure refinement against (1)H paramagnetic relaxation enhancement data arising from a flexible paramagnetic group attached to a macromolecule. J Am Chem Soc 126(18):5879–5896
39. Deshmukh L, Gorbatyuk V et al (2010) Integrin beta3 phosphorylation dictates its complex with Shc PTB domain. J Biol Chem 285:34875–34884
40. Wang X, Fukuda K et al (2008) The structure of alpha-parvin CH2-paxillin LD1 complex reveals a novel modular recognition for focal adhesion assembly. J Biol Chem 283 (30):21113–21119
41. Iwahara J, Clore GM (2006) Detecting transient intermediates in macromolecular binding by paramagnetic NMR. Nature 440(7088):1227–1230
42. Tang C, Iwahara J et al (2006) Visualization of transient encounter complexes in protein-protein association. Nature 444(7117):383–386
43. Clore GM, Iwahara J (2009) Theory, practice, and applications of paramagnetic relaxation enhancement for the characterization of transient low-population states of biological macromolecules and their complexes. Chem Rev 109(9):4108–4139
44. Schmitz C, John M et al (2006) Efficient chi-tensor determination and NH assignment of paramagnetic proteins. J Biomol NMR 35(2):79–87
45. Keniry MA, Park AY et al (2006) Structure of the theta subunit of *Escherichia coli* DNA polymerase III in complex with the epsilon subunit. J Bacteriol 188(12):4464–4473
46. Ubbink M, Ejdeback M et al (1998) The structure of the complex of plastocyanin and cytochrome f, determined by paramagnetic NMR and restrained rigid-body molecular dynamics. Structure 6(3):323–335

Top Curr Chem (2012) 326: 47–68
DOI: 10.1007/128_2011_215

Published online: 28 September 2011

The Use of Residual Dipolar Coupling in Studying Proteins by NMR

Kang Chen and Nico Tjandra

Abstract The development of residual dipolar coupling (RDC) in protein NMR spectroscopy, over a decade ago, has become a useful and almost routine tool for accurate protein solution structure determination. RDCs provide orientation information of magnetic dipole–dipole interaction vectors within a common reference frame. Its measurement requires a nonisotropic orientation, through a direct or indirect magnetic field alignment, of the protein in solution. There has been recent progress in the developments of alignment methods to allow the measurement of RDC and of methods to analyze the resulting data. In this chapter we briefly go through the mathematical expressions for the RDC and common descriptions of the alignment tensor, which may be represented using either Saupe order or the principal order matrix. Then we review the latest developments in alignment media. In particular we looked at the lipid-compatible media that allow the measurement of RDCs for membrane proteins. Other methods including conservative surface residue mutation have been invented to obtain up to five orthogonal alignment tensors that provide a potential for de novo structure and dynamics study using RDCs exclusively. We then discuss approximations assumed in RDC interpretations and different views on dynamics uncovered from the RDC method. In addition to routine usage of RDCs in refining a single structure, novel applications such as ensemble refinement against RDCs have been implemented to represent protein structure and dynamics in solution. The RDC application also extends to the study of protein–substrate interaction as well as to solving quaternary structure of oligomer in equilibrium with a monomer, opening an avenue for RDCs in high-order protein structure determination.

Keywords RDC · Alignment medium · Ensemble · Dynamics · Oligomer

---

K. Chen and N. Tjandra (✉)
Laboratory of Molecular Biophysics, National Heart, Lung, and Blood Institute, National Institutes of Health, Building 50, Room 3503, Bethesda, MD 20892, USA
e-mail: tjandran@nhlbi.nih.gov

Contents

1 Introduction 48
2 Theoretical Expressions 49
3 Alignment Methods 52
3.1 Charged Polyacrylamide Gel 53
3.2 DNA Based Media 53
3.3 Collagen Gel 54
3.4 Composite Media 54
3.5 Conservative Mutation 55
3.6 RDC/RCSA Accuracy Improvement 55
4 Interpretation and Implementation 57
4.1 Approximations 57
4.2 Common Applications of RDC 58
4.3 Ensemble Minimization 59
4.4 Structure of a Ligand in a Bound State 61
4.5 Structure of Oligomeric State 62
5 Conclusion 63
References 64

1 Introduction

Solution NMR spectroscopy is a powerful technique to study protein structure and dynamics at the atomic level. The method relies on a variety of experimental restraints to determine protein structure. These include the nuclear Overhauser effects (NOEs) that provide interproton distances, the *J*-coupling constants that depend on dihedral angles [1–3], the paramagnetic relaxation enhancement (PRE) that is distance dependent with respect to the paramagnetic center [4], and the residual dipolar couplings (RDC) that report on internuclei vector orientations. NOE typically measures interproton distances of less than 5 Å and *J*-coupling probes spin nuclear interactions within a few bonds away and they are therefore local in nature. In contrast, PRE can measure distances up to 20–30 Å from the paramagnetic center. In this respect RDC is unique. It can provide relative orientations among internuclei vectors irrespective of their distance separations. This unique property of RDC has opened up new possibilities in using NMR to study phenomena that were previously unattainable.

In the presence of a magnetic field, RDCs arise when the proteins in solution weakly align relative to the field, thus creating an anisotropic condition. The direction of the alignment of the protein molecules in the magnetic field is commonly referred to as the alignment tensor frame. Under such anisotropic condition, with the presence of an external field, a magnetic dipole–dipole interaction does not average to zero and yields a measurable dipolar coupling. The magnitude of the dipolar coupling depends on the angle between the internuclei vector and the external magnetic field as well as the internuclei distance. If the dipolar interaction is between two covalently bonded nuclei, then the internuclear distance is fixed and only the orientation dependence remains. A typical measurement may report hundreds of

RDCs within a protein that correspond to bond directions within the alignment tensor frame, providing orientation restraints for protein structure determination.

There are at least two approaches to create weak alignment conditions for measuring RDCs. One is to take advantage of the large magnetic susceptibility of a protein where its interaction with the magnetic field could produce a weak alignment [5, 6]. The other is to mix the protein sample with a medium that can be mechanically manipulated to create an anisotropic matrix or one containing supramolecules with substantially large susceptibility anisotropy that can be aligned under an external magnetic field. The interaction between the proteins and the media in turn could induce weak alignment of the proteins [7]. The latter approach creates a degree of alignment that is roughly one order of magnitude stronger than the former one, significantly larger than the experimental error, and thus is more practical for general applications.

A common usage of RDCs is to include them with other NMR restraints in refining a protein structure. For studies of large or membrane associated proteins, where high level of deuteration is required to achieve narrower line-widths, the number of observed NOEs will be reduced greatly. Therefore RDC restraints are necessary [8]. In this chapter we will focus on a short description on how RDC was developed, its practical mathematical expressions, and novel methods used in creating different alignment conditions that would allow more proteins to be studied using RDCs. We will describe RDC data interpretations and some common approximations. Finally we will discuss the most recent RDC applications in ensemble structure refinement, protein–ligand, and protein high-order structure determinations.

2 Theoretical Expressions

Dipolar coupling measures the interaction between two magnetic nuclei in an external field $\boldsymbol{B}_0$. If the vector connecting nuclei A and B is parallel to the field $\boldsymbol{B}_0$, the coupling is at its strongest with a magnitude $D_{\max}$, which is given by Eq. (1), where μ_0 is the vacuum permeability, h is Planck's constant, γ_A and γ_B are the gyromagnetic ratios of nuclei A and B, respectively, and r_{AB} is the distance between nuclei A and B. In some expressions vacuum permeability μ_0 was assumed and therefore omitted in the $D_{\max}$ expression, and there will be a factor of 4π difference in the denominator of Eq. (1). $D_{\max}$ is bond type dependent and usually on the order of 10^3 Hz, e.g., $D_{\max}$ for protein backbone amide ^{1}H–^{15}N spin nuclei pair is 21.66 kHz with an assumed bond length of 1.04 Å. Because of diffusive motion the direction of the internuclei vector fluctuates relative to the $\boldsymbol{B}_0$ direction and therefore the dipolar coupling has to be evaluated with respect to every possible orientation. This orientation dependency follows the second order Legendre polynomials. Shown in Eq. (2) is the expression for the dipolar coupling D of an internuclei vector with a fixed distance (which is the case for bonded nuclei), where Θ is the instantaneous angle between the dipole–dipole or bond vector and $\boldsymbol{B}_0$ (Fig. 1), and the angular bracket indicates time or population average. The average

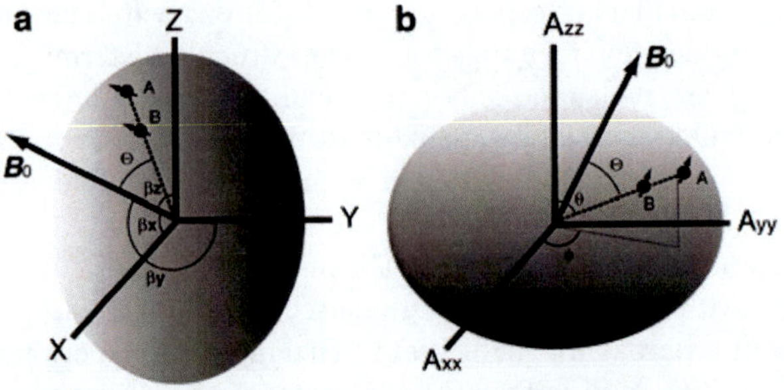

Fig. 1 Illustrations of relationships between RDC internuclei vector *AB* and an arbitrary molecular frame (**a**) and the alignment tensor frame (**b**). A protein molecule, carrying spin nuclei *A* and *B*, is represented using an ellipsoid. $\boldsymbol{B}_0$ is the external field. Θ is the instantaneous angle between the internuclei vector *AB* and $\boldsymbol{B}_0$. $\beta_{x,y,z}$ specify the projection angles of $\boldsymbol{B}_0$ onto each axis of a molecular frame. Polar angle θ and azimuth angle ϕ are spherical coordinates of the vector *AB* in the alignment tensor frame

for *D* will be zero if Eq. (2) is integrated over the variable Θ that covers the surface of a sphere, corresponding to all possible orientation for an isotropic diffusing molecule. However, due to a weak alignment causing a nonisotropic sampling of orientations, the average of *D* is not zero. In practice, adjusting the concentration of the alignment media can allow between an equivalent of 0.1–1% fractions of protein molecules being aligned and it results in a value for *D* to be within ±20 Hz for ^{1}H–^{15}N vectors, referred to as RDC.

$$D_{\max} = \frac{-\mu_0 h \gamma_A \gamma_B}{8\pi^3 r_{AB}^3}, \tag{1}$$

$$D = D_{\max}\left\langle \frac{3\cos^2\Theta - 1}{2} \right\rangle, \tag{2}$$

$$D = D_{\max}\begin{pmatrix} x & y & z \end{pmatrix}\begin{pmatrix} S_{xx} & S_{xy} & S_{xz} \\ S_{yx} & S_{yy} & S_{yz} \\ S_{zx} & S_{zy} & S_{zz} \end{pmatrix}\begin{pmatrix} x \\ y \\ z \end{pmatrix}, \tag{3}$$

$$S_{ij} = \left\langle \frac{3\cos\beta_i \cos\beta_j - \delta_{ij}}{2} \right\rangle,\ ij = \{x, y, z\}. \tag{4}$$

The equation for RDC, Eq. (2), could be rewritten in the form of the expectation value for a vector in a Saupe order matrix, i.e., Eqs. (3, 4) [9]. The *x*, *y*, *z* in Eq. (3) are directional cosines of a bond vector in an arbitrary molecular frame (Fig. 1a), the most convenient would be an existing PDB coordinate frame; Saupe element S_{ij} is the averaged projection of axes of the molecular frame onto the direction of $\boldsymbol{B}_0$ with $\beta_{x,y,z}$ (Fig. 1a) specifying the projection angle for each axis; δ_{ij} is Kronecker delta. The Saupe matrix is symmetric and traceless and contains five unknown variables, i.e., S_{xx}, S_{yy}, S_{xy}, S_{xz}, and S_{yz}. The product on the right side of Eq. (3) is a scalar, the value of which equals the average of the second order Legendre polynomials in Eq. (2). The aforementioned 10^{-3} to 10^{-4} (0.1–1%) scaling in the dipolar coupling is contained within all Saupe order elements. Since the five unknowns in the Saupe matrix are common to all bond vectors or RDC measurements in one aligned protein sample, in theory with more than five RDC bond vectors pointing in different directions one could solve the Saupe matrix. Prestegard and coworkers have written a protocol for solving the five Saupe unknowns using the singular-value-decomposition (SVD) method to obtain alignment parameters [10].

$$D = \frac{3}{2} D_{\max} \begin{pmatrix} x & y & z \end{pmatrix} T^{*}(\alpha,\beta,\gamma) \begin{pmatrix} A_{xx} & 0 & 0 \\ 0 & A_{yy} & 0 \\ 0 & 0 & A_{zz} \end{pmatrix} T(\alpha,\beta,\gamma) \begin{pmatrix} x \\ y \\ z \end{pmatrix}, \tag{5}$$

$$\begin{pmatrix} x' \\ y' \\ z' \end{pmatrix} = T(\alpha,\beta,\gamma) \begin{pmatrix} x \\ y \\ z \end{pmatrix}, \tag{6}$$

$$D = D_a \left(3\cos^2\theta - 1 + \frac{3}{2} R \sin^2\theta \cos 2\phi \right). \tag{7}$$

One can diagonalize the Saupe matrix to obtain the alignment parameters. Diagonalization of the Saupe matrix results in the principal order matrix and Euler rotation matrices $\boldsymbol{T}(\alpha,\beta,\gamma)$ and $\boldsymbol{T}^*(\alpha,\beta,\gamma)$, where α, β, and γ are Euler angles and * indicates conjugate transpose (Eq. (5)). The Euler rotation of Cartesian coordinates in the molecular frame xyz of Eq. (3) results in a set of new coordinates $x'y'z'$ Eq. (6) for the bond vector within the principle order frame (Fig. 1b). The principal order matrix equals the alignment tensor matrix with its Eigenvalues, A_{xx}, A_{yy}, A_{zz} in Eq. (5), representing the alignment order in the corresponding tensor direction. There are different conventions in describing the alignment order parameters. For instance, one can keep the S representation for Eigenvalues in the alignment tensor matrix, i.e., S_{zz}, etc. Alternatively, the A_{zz}, notation which is equivalent to $2S_{zz}/3$ in the diagonalized Saupe matrix can be used [11].

Following the convention in Eq. (5), and in analogy to S_{xx} and S_{yy} of the diagonalized Saupe matrix, there are the following relationships: $A_{xx} = A_{zz}(-1/2 + 3R/4)$ and $A_{yy} = A_{zz}(-1/2-3R/4)$, where R is the rhombicity which can be in the range of 0–2/3, with the convention that $|A_{zz}| \geq |A_{yy}| \geq |A_{xx}|$. Because A_{zz}, on the order of 10^{-3} to 10^{-4}, is not a convenient number, A_{zz} can be replaced with a more convenient D_a ($= 3D_{\max}A_{zz}/4$) representation, which allows for an easier comparison among different alignment conditions. Here is an alternative to using the five Saupe unknowns. We use D_a to specify the alignment order of a sample, rhombicity R to describe asymmetry of the alignment tensor, and three Euler angles to define the tensor directions instead. With Euler angles one could conveniently visualize the tensor within a molecular frame. Further simplifications can be made by using spherical coordinates, i.e., polar angle θ and azimuth angle ϕ, to replace

$x'y'z'$ (Fig. 1b). After substituting D_a, rhombicity R, and the spherical coordinates in the alignment frame into Eqs. (5, 6), one will end up with a more familiar expression for RDC Eq. (7). It is essential to keep in mind that Eq. (2) and Eq. (7) are the similar formula, and the only difference is that the latter carries no ensemble averages.

3 Alignment Methods

Dipolar coupling manifests itself as an additional coupling to the scalar or J-coupling. RDC measurements normally require two NMR samples prepared in parallel, one with and the other without the presence of an aligning medium, corresponding to an isotropic and an anisotropic samples, respectively. An identical NMR pulse sequence is applied to both samples to measure the J and $J + D$ values. The difference yields the dipolar coupling D. Therefore any pulse sequences developed to measure J-coupling are applicable for measuring dipolar coupling. There are two general methods to determine $J + D$ value under anisotropic conditions, measuring direct splitting on NMR spectra and fitting a J-modulated intensity [12]. Methods for measuring $D_{\text{H–N}}$, $D_{\text{H}\alpha\text{–C}\alpha}$, $D_{\text{C}\alpha\text{–C}'}$, $D_{\text{N–C}'}$, $D_{\text{H–C}'}$, and $D_{\text{H–H}}$ were extensively reviewed [11, 13, 14]. Here we focus on ways of adjusting alignment conditions and improvement in measurement accuracy.

A number of liquid-crystalline media have been employed to generate weak alignment for solution protein samples. The utilization and mechanisms of the commonly used alignment media have been reviewed previously [11, 15] and they include bicelles made of dimyristoylphosphatidylcholine (DMPC) and dihexanoylphosphatidylcholine (DHPC) [16, 17], filamentous phages *Pf*1 [18] or *fd* [19], stretched [20, 21] or compressed [22] polyacrylamide gel, and poly(ethylene glycol)/hexanol mixture [23]. In addition, one can also add charged molecules into the alignment media to generate different alignment tensors for protein molecules [24, 25]. Though the pool of alignment media seems large, the demand for new media still exists for several reasons. One reason is that when there are more choices of alignment media the chance of finding a compatible one for challenging proteins or membrane proteins increases. The second reason is that any additional orthogonal alignment tensors from new alignment media add new information and can potentially better define bond vector orientation and dynamics. This has a promising potential for de novo protein structure determination. Recent additions in alignment media include charged gel [26], novel DNA based media [27, 28], and collagen gel [29]. The physical interactions between proteins and the medium are almost exclusively to be either steric and/or electrostatic, which could potentially limit the ability to obtain a complete set of five orthogonal alignment tensors. However, recent developments such as conservative mutation on protein surface [30] and use of composite media [31] may overcome these limitations. To improve measurement accuracy by providing more consistent reference (isotropic) data, approaches such as extracting RDCs from two samples at different concentrations of alignment media [32] and application of a new two-stage NMR tube [33] have been proposed.

3.1 Charged Polyacrylamide Gel

Mechanically stretched [21] or compressed [22] polyacrylamide gel medium is a promising medium for measuring RDCs on membrane proteins that are reconstituted in micelles or bicelles. This is due to its inertness and inability to react with detergents. However, to establish weak alignment that is practical, it is necessary to polymerize at least 7% (w/v) acrylamide in a sample [34]. At such concentration the narrow pore size of the gel matrix limits protein diffusion, resulting in peak line-width broadening. Meier et al. [35] initially showed a sufficient alignment order was achieved by copolymerizing only 2% (w/v) acrylamide and acrylic acid, leading to an anionic polymer. Cierpicki and Bushweller [26] used acrylamide (<5% w/v) with different charged polymer units to generate alignment order. In addition to anions, positive charges were introduced by addition of (3-acrylamidopropyl)-trimethylammonium chloride or *N*-(2-acryloamidoethyl)-triethylammonium iodide. With such a charged gel, satisfactory sample stability and NMR spectra quality were obtained using integral membrane protein OmpA dissolved in dodecylphosphocholine (DPC) micelles. The RDCs obtained from the charged gel were directly used for membrane protein structure determination.

3.2 DNA Based Media

Nucleic acids carry a relatively stronger magnetic susceptibility than proteins and seldom react with detergent. In the continuous effort to develop detergent compatible alignment media, Douglas et al. [27] initially exploited DNA nanotubes as alignment media. Two bundles of six DNA strands in 7,000-base length were linked through base-pairing sticky ends to form a micron-long DNA filament. The transmembrane helices of a T-cell receptor reconstituted in DPC/SDS bicelles were aligned in a cosolvent of DNA nanotubes at a concentration of 28 mg/mL. The measured $D_{\mathrm{H-N}}$ and $D_{\mathrm{H\alpha-C\alpha}}$ were shown to be consistent with the protein structure determined without RDC.

To ease the DNA nanotube medium preparation and reduce the cost, Lorieau et al. [28] used potassium salt of dinucleotide (d(GpG)) that would tetramerize through guanosine hydrogen bonds at a concentration of 10 mg/mL. The G-tetrad DNA, similar to bacteriophage *Pf*1, is strongly negatively charged. However, different from phage *Pf*1, its liquid crystalline phase is chiral nematic and the director can run perpendicular to the external field. Analysis of RDCs collected on a mutant of protein GB3 aligned in G-tetrad DNA showed the obtained alignment tensor carried the same directions as those in phage *Pf*1, but the sign of D_a

switched [28]. In this case, it is the electrostatic interaction between alignment medium and protein that determine the uniqueness of the alignment tensor. Nevertheless, G-tetrad DNA allowed RDC measurements on the fusion peptide of the influenza viral hemagglutinin solubilized in DPC micelles [36].

3.3 Collagen Gel

Collagen proteins, abundant in mammals, are made of trimer of polypeptide chains. Each chain is rich in proline or hydroxyl-proline, glycine, and others, which together form an extended left-handed polyproline-II structure [37]. Three copies of such chains form a right-handed helix, carrying a weak magnetic moment. For in vitro use the rat tail tendon type I collagen monomers were prepared [38] and stored in acidic buffer at 4 °C. At pH 6–8 and temperatures over 30 °C the collagen helices will polymerize and cross-link through amino acid side chains, e.g., lysine. When such reactions occur in the presence of magnetic field, weak alignment can be achieved [29]. At a concentration of 13 mg/mL, a sample with collagens polymerized in the presence of magnetic field yielded deuteron splitting of 20 Hz, usually sufficient for dipolar coupling [29].

3.4 Composite Media

As mentioned before, the nature of alignment forces, either steric or electrostatic, could limit the number of orthogonal alignment tensors we can observe experimentally. However, Ruan and Tolman [31] showed the interference of the two alignment forces could produce one additional orthogonal tensor. They polymerized 5% (w/v) polyacrylamide gel together with 3–4 mg/mL of bacteriophage *Pf*1 in the presence of the magnetic field. A special tube with a cross section size of 7 $\times$ 5 mm was filled with both media and positioned at a maximal angle of 55° to field $\boldsymbol{B}_0$ (Fig. 2). The gel was dried, later soaked with protein sample, and stretched into a 5-mm NMR tube. Phages were assumed to be field aligned and trapped at the direction because their motions were inhibited by polymerized acrylamide. Essentially, the direction of phages formed a tilting angle with the direction of gel stretch. Proteins were subject to interfered alignment forces, resulting in both strong steric and electrostatic interactions. Surprisingly, the measured alignment tensors at different tilting angles are not a simple linear combination of those obtained from individual alignment using phage or acrylamide gel alone. As reported [31], at least three orthogonal tensors were solved for a ubiquitin sample subjected to composite alignment media. The same group also reported that the unambiguous bond direction might be determined with only three orthogonal alignment tensors [39].

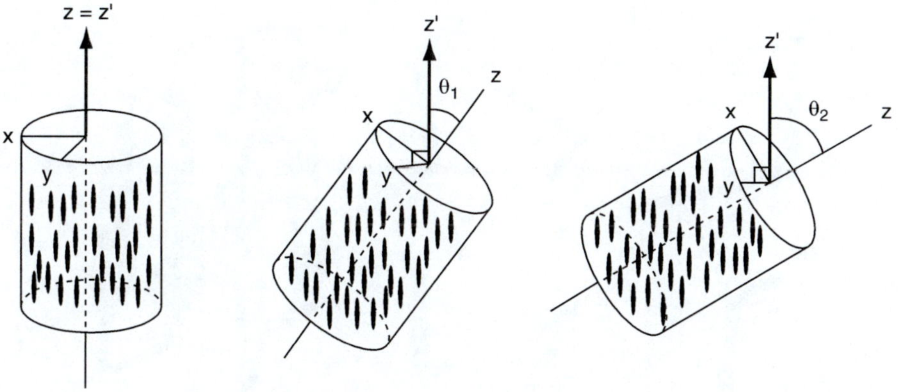

Fig. 2 Variation of alignment using stretched polyacrylamide gels (SAG) and bacteriophage *Pf*1, which has been embedded and aligned (along *z'*) at different angles relative to the long axis (*z*) of the sample. The gels were cast in an approximately ellipsoidal (squashed cylinder) geometry, with dimensions of 5, 7, and 10 mm along the *x*, *y*, and *z* axes, respectively. The gels were then dried, rehydrated, and stretched to fit within a 4.2 mm i.d. NMR tube. (Reprinted with permission from [31])

3.5 Conservative Mutation

The surface charge of *Pf*1 phage is heavily negative and proteins will be aligned according to its surface electrostatic potentials [40]. Yao and Bax [30] carefully modified the surface charge distribution of the 6-kDa protein GB3 by either conservatively mutating one or two residues at each time, e.g., K to E, or keeping the histidine-tag at terminus of the native GB3. The backbone structure was later found to be unperturbed in those modified proteins. A total of six mutant proteins were found to align quite differently with respect to the $\boldsymbol{B}_0$ field and five singular values were obtained (Fig. 3). With this well-defined system, the amplitude and direction of RDC bond dynamics up to the millisecond (ms) were unambiguously obtained and compared to nanosecond to picosecond (ns–ps) dynamics from spin relaxation measurements (Fig. 4) [41]. Interestingly, Yao et al. found both RDC and relaxation measurements showed the same amount of flexibility for residues in regular secondary structures. Bond vectors within loop regions, however, were shown by RDC to have larger amplitude of motions compared to what were suggested from the relaxation data [41].

3.6 RDC/RCSA Accuracy Improvement

In addition to RDC, protein alignment creates slight chemical shift changes for the aligned sample relative to the isotropic sample. The chemical shift difference, named residual chemical shift anisotropy (RCSA), comes from the projection of the chemical shift tensor, which is not averaged to zero, onto the alignment tensor. RCSA is also a long range structural restraint, providing orientation dependences

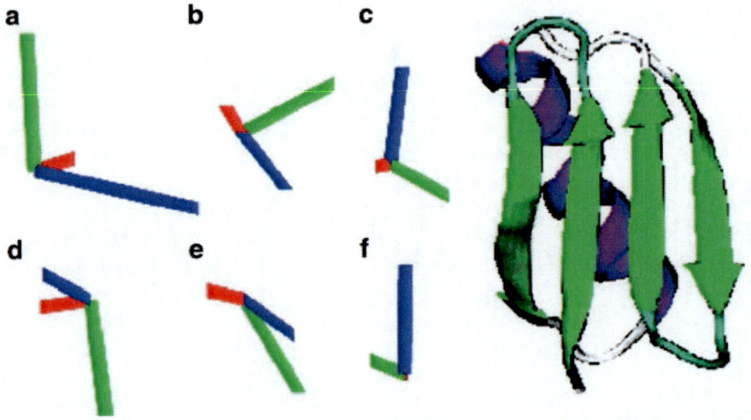

Fig. 3 Alignment tensor orientations relative to the ribbon backbone structure of GB3 for six mutants, all in liquid-crystalline *Pf*1 medium. The six tensors are for *A* – K19AD47K; *B* – K19ED40N; *C* – K19EK4A-C-His-tag; *D* – K19EK4A-N-His-tag; *E* – K19AT11K; *F* – K19EK4A. Diagonalized tensor elements, Dxx (*red*), Dyy (*green*), and Dzz (*blue*) have magnitudes proportional to the length of the corresponding lines. (Reprinted with permission from [30])

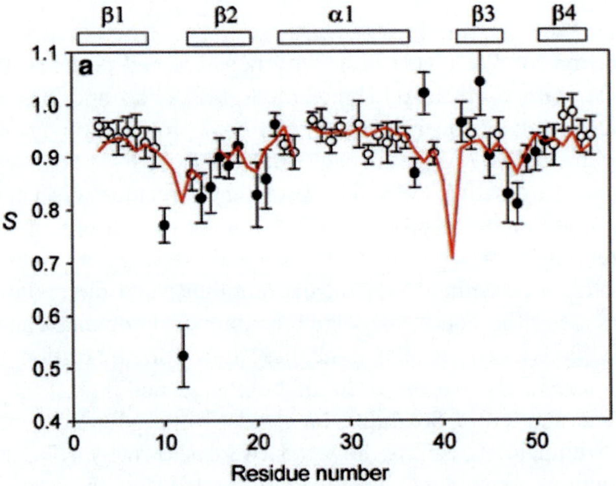

Fig. 4 Experimental order parameters, *S*, of NH bond vectors in GB3 derived from iterative Direct Interpretation of Dipolar Couplings (DIDC) using all six sets of RDCs. The *red line* marks the order parameters derived from ^{15}N relaxation. *Filled symbols* represent residues for which the fully anisotropic model was required to get a satisfactory fit to the data, while, for *open symbols*, the isotropic internal motion model was able to fit the RDCs to within the experimental noise. (Reprinted with permission from [41])

generally complementary to RDCs [42–44]. The solvent used in the isotropic protein sample usually differs from that in the anisotropic sample. This can introduce errors in measuring both dipolar couplings and chemical shift differences. Such error may be below the measurement accuracy for RDCs, on the order of 0.1–1 Hz. However,

RCSA, which is on the order of parts per billion (ppb), may be greatly affected. The solvent effect can be removed by measuring several aligned samples at increasing alignment orders and extrapolating the RCSA values to zero concentration [32]. The methods were applied to backbone ^{15}N [32] and ^{13}C′ [45] RCSA measurements.

Another way to overcome this is to measure isotropic sample under the same unaligned medium. Liu and Prestegard [33] developed a two-stage NMR tube with two different internal diameters (I.D.) at upper and lower parts of an open-ended NMR tube. For isotropic condition the protein sample was soaked into polyacrylamide gel in 5 mm I.D. part, then the same piece of gel was pulled and stretched into the 3 mm I.D. part by vacuum created using a syringe at the other end of the two-stage tube, resulting anisotropic condition. The method allows for higher accuracy measurements for both RDC and RCSA.

Aside from using both aligned and unaligned samples for RDC and RCSA, one may also keep a single sample that contains alignment media for both isotropic and anisotropic measurements. Upon the application of magic angle spinning, similar to what is used in solid state NMR, the alignment effect for bicelles [46] or phage [47] was removed. In this way the measurement will not contain any solvent effects.

4 Interpretation and Implementation

Measured RDC values are representative averages of the whole ensemble of dipolar interactions within protein molecules in solution. Such an ensemble should include all protein conformers interconverting at time scales faster than the inverse of RDC values ($1/D$). For instance, the observed dipolar coupling is affected by the internuclei or bond vectors that stretch and vibrate on a femtosecond to nanosecond (fs–ns) time scale, the protein domain reorientation on a nanosecond to microsecond (ns–μs) time scale, and conformational change that ranges from nanoseconds, e.g., unstructured terminus, to milliseconds. It is nearly impossible to describe protein structure and dynamics using RDC values without any assumptions. Some approximations have to be made in interpreting RDC measurements.

4.1 *Approximations*

Most RDCs except for $D_{\mathrm{H\text{–}H}}$ [48] are measured on fixed internuclei distances such as bond vectors or geometries within the peptide plane so that in Eqs (1) and (2) a constant internuclei distance r_{AB} in $D_{\max}$ is assumed. This in itself is an approximation because the effective bond lengths vary due to dynamic processes. For instance, Yao et al. [49] determined the average protein backbone H–N bond length to be 1.01–1.02 Å. However, for deriving the true alignment order an effective value of 1.04 Å was proven proper and used extensively to account for H–N bond libration dynamics [50]. A slight increase in effective bond length for RDC analysis also applies to Hα–Cα bonds. On the other hand, bond lengths for heavy nuclei, e.g., Cα–C′, are consistent with values reported from crystallography studies [50].

By expressing RDC using Eq. (3), where the Saupe matrix is multiplied by bond directional cosines within a local molecular frame, we assume protein conformational change does not affect the alignment tensor. Thus Eq. (3) is an approximation to Eq. (2), in which the average bracket over the second order Legendre polynomial is applied. This approximation is valid when RDCs from the structured part of the proteins are analyzed. For small proteins GB3 [41, 51] and ubiquitin [52] the model free (MF) order parameter from ^{15}N spin relaxation analysis and the order parameter derived from RDC analysis were strongly correlated. These indicate that the effect of slowmotion dynamics on microsecond to millisecond time scales on RDC may be negligible and a single structure representation is sufficient for the structured part of proteins within the current experimental uncertainty [53].

When the dynamics involve large scale amplitudes of motion, e.g., the MF order parameter $S^2 < 0.6$, multiple structures may exist and each one of them is subjected to its own alignment tensor that may vary significantly due to different steric, and electrostatic interactions of each conformer to the alignment media. Specifically for unfolded proteins it becomes challenging to separate intrinsic bond dynamics from Saupe order parameters in the laboratory frame because local motion and the overall perturbed diffusive motions are coupled [54]. Monte-Carlo simulation of an ensemble of conformers and the following comparison to measurements remain the only option to interpret RDCs in a flexible system [55, 56]. The application of RDCs to study protein in an unfolded state is an active research field [57–59]. For instance, efforts have been made to represent the conformational space of urea denatured ubiquitin with as few as 200 conformers, which seem to reproduce measured RDCs [59]. However, additional specific bond type RDC scaling factors had to be applied for H–N, Hα–Cα, etc., which indicate different amounts of motion along bond vectors within a peptide plane [57, 59].

4.2 Common Applications of RDC

The most common use of RDCs is for structure validation and refinement. Given structure coordinates determined either from X-ray crystallography or solution NMR, one can readily fit RDC data to the corresponding bond vector directions within the molecular frame. Any programs that utilize Eqs (3) or (5) in a chi-square minimization routine or formal software packages such as PALES [60, 61] and REDCAT [62] which implement the SVD algorithm can be employed to carry out the numerical fitting. The fitting will result in the optimized alignment tensor and values that best matched the measured RDCs. To quantify the agreement between a structure and measured dipolar couplings, Cornilescu et al. [63] proposed quality factor Q. This factor can better determine the quality of the fit than Pearson's correlation R. Shown in (8) is the expression for the Q factor where rms refers to root mean square. It provides an estimate of average disagreement in percentage between measured and calculated dipolar couplings. A Q factor of 40%, roughly corresponding to Pearson's R of 0.9, are commonly found on structures with 2–3 Å resolution; and a Q value of 20% or less indicates the structure is at high resolution (1.5 Å) and accuracy. Clore and Garrett [64] suggested an alternative form $[2D_a^2(4+3R^2)/5]^{1/2}$ to replace the denominator of (8) in the case of a limited RDC sampling over all of the possible orientations. A thorough discussion on Q factor can be found elsewhere [65].

$$Q = \frac{\mathrm{rms}(D_{\mathrm{meas}} - D_{\mathrm{calc}})}{\mathrm{rms}(D_{\mathrm{meas}})}, \tag{8}$$

$$E = k(D_{\mathrm{meas}} - D_{\mathrm{calc}})^2. \tag{9}$$

For solution structure determination, RDC is normally not included as a potential term (9) during the initial structure calculation in a simulated annealing protocol in a program such as Xplor-NIH [66, 67]. Due to directional degeneracy associated with RDC restraints, as described in the theoretical expression section, multiple local potential minima will hinder the successful search for the right conformation. Typically a rough tertiary fold of protein is obtained first with the use of NOE restraints and such a fold can be used as a starting structure for the next simulated annealing procedure with the combined NOE and RDC restraints. The procedure to include RDC restraints in a program such as Xplor-NIH uses four pseudo-atoms (*OXYZ*) [68] to represent the alignment tensor directions [69]. The RDC force constant, k in (9), can be increased gradually as the temperature is being lowered in the simulated annealing. In the end it is ideal to adjust the force constant k so that the deviation from the measured RDCs matches the experimental error.

4.3 Ensemble Minimization

As discussed before, RDC reflects the ensemble averaged dipolar coupling and incorporates a wide range of time scales up to milliseconds. During RDC restrained structure calculations we made approximations by assuming a single conformer. The question remains whether it is reasonable to keep this assumption in all cases. Clore and Schwieters [52] initially employed a two-member ensemble minimization to test whether a better agreement between measured and calculated RDC data could be achieved, and whether it was statistically significant. The ensemble algorithm proposed by Clore and Schwieters [52] kept a user-defined number of conformers during the course of a simulated annealing, and any evaluated physical quantity, e.g., D_{calc}, is linearly averaged among individual conformers. After refining the ubiquitin structure with $D_{\mathrm{H\text{-}N}}$ data sets collected in 11 different alignment media, and a couple of other heteronuclear RDCs, they found a single conformer in most cases is sufficient to yield a good quality factor Q less than 20%. The structure was validated against the $D_{\mathrm{H\alpha\text{-}C\alpha}}$

data set that was not being applied as a restraint. A 15% reduction in Q value was obtained with an ensemble size of two, which was statistically significant. In addition, several residues were found to be undergoing anisotropic motion and can be better represented by the two-member ensemble (Fig. 5a) [52]. Later a simulated annealing with an ensemble size of four to eight on protein GB3 had been found to be optimal [71]. The consensus is that other than residues having large amplitude or anisotropic motions, most structured residues can be represented with a single structure for monomeric small proteins because the accuracy of NMR structural coordinates is well within the measurement uncertainty of the RDCs.

Other studies on ensemble minimization reveal a relatively larger conformational sampling for ubiquitin [72]. In another study, an ensemble of around 100 ubiquitin structures was generated using NMR restrained simulated annealing and molecular dynamics. Interestingly, such an ensemble covers structural heterogeneity observed in 40 or so ubiquitin–ligand complex crystal structures (Fig. 5b) [70]. RDC restraints used in such ensemble minimizations were measured in over 30 different alignment media [73]. The biological implication is that ligand induced ubiquitin conformers preexist in its structural ensemble. Further, the ubiquitin N–H order parameters derived from RDC analysis [73–76] were overall lower by 0.1–0.2 than that of MF order parameter derived from relaxation data [77], indicating an appreciable amount of microsecond to millisecond motion that was not observed in spin relaxation measurement. It is nevertheless an interesting and different conclusion from other findings on ubiquitin [52] and GB3 [41].

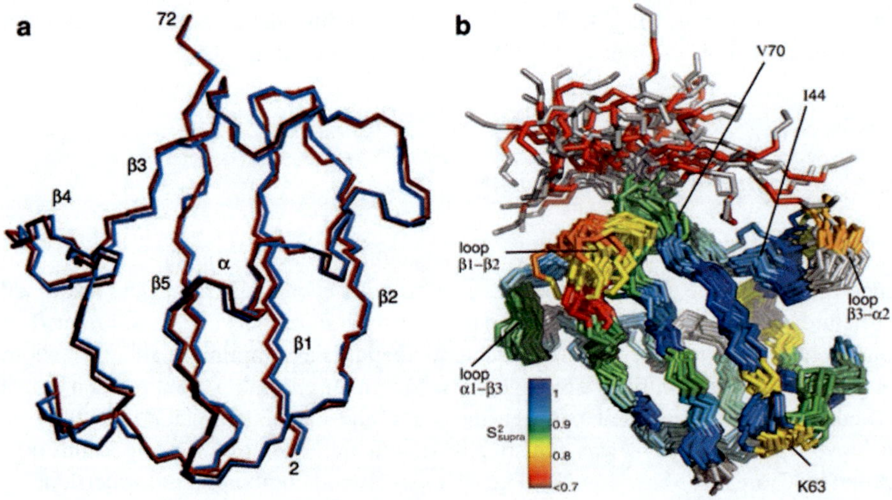

Fig. 5 Structure ensemble of Ubiquitin. (**a**) Two members (shown in *red* and *blue*) of a typical ensemble from the two-member size calculation. (Reprinted with permission from [52].) (**b**) Backbone trace of 40 randomly chosen structures from the ensemble. Residues are colored by the amount of additional (slower than-τc) mobility as compared with the Lipari-Szabo order parameters. (Reprinted with permission from [70])

Besides backbone dynamics for monomeric proteins, ensemble minimization may be better suited for studying relatively slower and larger scale protein domain dynamics and structure. Liu et al. [78] implemented ensemble minimization to characterize the dynamic structure of a membrane-anchored ADP ribosylation factor (Arf). The yeast Arf is composed of flexibly joined C-terminal domain and the short N-terminal helix that interacts with the membrane. The Arf protein anchored on DMPC/DHPC bicelles were aligned in a negatively charged polyacrylamide gel, and the resulting alignment order parameters D_a and R for C- and N-terminal domains are 8.6 Hz and 0.29, and 4.7 Hz and 0.61, respectively. The differences in alignment order indicate significant interdomain motions. A three-member ensemble minimization significantly increased the linear correlation coefficient between measured and back-calculated RDC (Fig. 6) and PRE data on the N-terminal helix. Such dynamics could be crucial for Arf to function as a regulator of effectors GTP/GDP.

4.4 *Structure of a Ligand in a Bound State*

The conformational ensemble described previously represents a continuous distribution of conformers and no significant protein fold or high-order structural change within the ensemble members was taken into account. Other systems may also be composed of discrete conformers in equilibrium; for instance, large structural differences could exist between free and bound states of ligand molecules in the presence of their receptors. Generally the NMR signals from the ligands in this type of equilibrium are dominated by the ligands in the free states. RDCs can play a role in solving ligand structure in the bound form not generally accessible by other techniques.

Photo-activated retina rhodopsin, a G-protein coupled receptor (GPCR), is a major component in purple membranes (PM). Due to the high structural content of helices in GPCR, GPCR rich PM disks align in the presence of a magnetic field

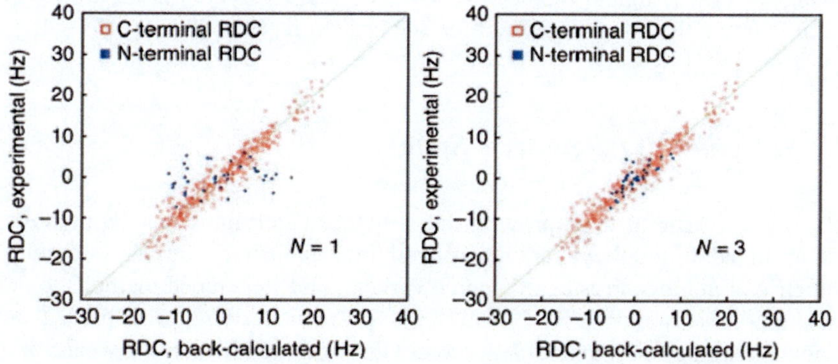

Fig. 6 Ensemble structural fitting to RDCs. Agreement between experimental and back-calculated RDCs for a one-state ensemble (*left*) and a three-state ensemble (*right*). Data include NH, NC′, and phenyl CH (the latter two are normalized to NH). RDCs are collected in positive and negative gels. (Reprinted with permission from [78])

with the membrane normal being parallel to the field [79, 80]. The binding of the GDP-bound form of the heterotrimeric G protein transducin (Gt) to the light activated MII intermediate of the rhodopsin activates the visual signal transduction pathway. A peptide consisting of the C terminal ten residues of the α-subunit of transducin (Gtα) is capable of competing with Gt for binding and was studied using NMR spectroscopy [81]. In the presence of light, Gtα transiently binds activated GPCR that is embedded in membrane disks of bovine retina. Since these disks were aligned in the magnetic field, RDCs for Gtα could be observed. The measured RDCs followed a time-decay course after light activation, while GPCR returned to its inactive state. The decay time constant is on the order of 1 h that allowed rapid acquisition of 2D ^{1}H–^{15}N and ^{1}H–^{13}C spectra on the Gtα peptide sample. The RDC values at zero time were obtained from extrapolation of multiple RDCs collected along a time decay curve. The structure and orientation of the peptide in the bound state to GPCR were derived from RDCs and transferred NOEs [81]. The validity of such use of RDCs relies on two conditions. The first condition is the fast exchange between the free and bound states of Gtα that allows the observation of only one set of resonances of the free state. The other condition is that the free Gtα peptides are not aligned in GPCR enriched PM media so that alignment contribution all comes from the bound state. Indeed the alignment order of the system was weak with a D_a value of only 1.6 Hz.

In the above case, the ligand receptor GPCR is naturally buried in bicelles, which by itself is the alignment medium. This ensures strong alignment order for bound ligand molecules. However, for a general study protein receptors are not necessarily buried in any disks that could be aligned; thus the alignment order for ligand molecules in equilibrium with its free form would be too weak to be observed. Seidel et al. [82] demonstrated a method of anchoring polyhistidine-tagged protein receptors onto the bicelles that were doped with histidine-tag binding lipid molecules. This increased the alignment order of the protein–ligand complex and prevented measurements of the protein RDCs. The RDCs of the bound ligand in fast exchange with free ligand molecules, however, were observed and they could be used to map the ligand configuration on the protein receptors.

4.5 Structure of Oligomeric State

RDCs are very useful in improving the structure determination of a monomeric protein. In addition, they can be utilized to establish quaternary structure of symmetric oligomers. Prestegard and coworkers have developed methods to determine dimer structure using RDC data [83, 84]. The underlying principle is that the rotational symmetric C_2 axis of any protein dimer should be parallel to one axis of the alignment tensor of the dimer. By solving alignment tensor axes from different alignment media, one could identify the common tensor axis that should be parallel to the symmetric axis [85, 86]. *Bacillus subtilis* proteins YkuJ dimerize with strong affinity and its K_d is on the order of 10^{-9} M [83]. When working at the typical NMR protein concentration (~0.1–1.0 mM) the measured RDC would primarily reflect the dimer structure. The crystal structure of YkuJ is a tetramer and there are two possible configurations for a dimer structure. Wang et al. [83] aligned the protein using two alignment media, 5% poly(ethylene glycol)/hexanol and 10 mg/mL phage *Pf*1. Both yielded different alignment tensor directions. However, the x axes from the two tensors were close to each other and were believed to be the symmetric axis of the dimer. The further rotation and docking procedure with energy minimization yielded dimer structure very close to one possible dimer structure observed in the crystal structure of the tetramer. Prestegard and colleagues further extended this approach to study structures of weakly associated dimer. The *Staphylococcus epidermidis* proteins SeR13 weakly dimerize at a K_d on the order of 10^{-3} M [84]. The measured RDCs were the weighted average of monomer and dimer states and the resulting alignment tensor from direct fitting would be the average of both monomer and dimer alignment tensors. Lee et al. [84] circumvented this by deriving the exact K_d value from concentration dependent chemical shift values, then extrapolating the RDCs as a function of protein concentration together with the K_d to yield pure RDC data set for the dimer. The rest of the work to determine the dimer structure was similar to the previous approach [83], but with the additional restraints from chemical shift perturbation and paramagnetic surface perturbation data that helped identifying newly buried surface residues upon dimerization.

5 Conclusion

Some additional reviews [11, 65, 87] can be found on RDC topics with different emphasis such as methods [69, 88, 89], theories [13, 14, 90], and dynamics [15]. In this chapter we have briefly summarized the theoretical expressions for the widely used RDCs and common descriptions for magnetic alignment under either Saupe order matrix or alignment tensor representations. RDCs have become notably more relevant in solution structure and dynamics studies for larger system due to the fact that the number of NOEs diminishes. In some cases RDCs may be the sole major experimental NMR restraints for determining domain positions in large systems of over 100 kDa [8]. With its essentiality in mind, more alignment media are being explored. The more choices available, the higher the chance biologically important proteins and membrane proteins reconstituted in micelles or bicelles can be aligned without any interference with the media. More RDC measurements on large and multidomain proteins can reveal their quaternary structure and dynamics manifested through differences in alignment order. However, caution should be exercised in RDC data interpretations because a single RDC data set may not be able to separate differences in domain positions and dynamics simultaneously. Cross-validating the results with different RDCs under another orthogonal alignment tensor, rotational diffusion tensors from spin relaxation, and small angle X-ray scattering that are sensitive to the protein overall shape should help overcome this problem.

Acknowledgment We thank Nils-Alexander Lakomek for helpful discussion. This work was supported by the Intramural Research Program of the NIH, National Heart, Lung, and Blood Institute.

References

1. Wüthrich K (1989) Determination of 3-dimensional protein structures in solution by nuclear-magnetic-resonance – an overview. Methods Enzymol 177:125–131
2. Bax A (1989) Two-dimensional NMR and protein-structure. Annu Rev Biochem 58:223–256
3. Clore GM, Gronenborn AM (1994) Multidimensional heteronuclear nuclear-magnetic-resonance of proteins. Nucl Magn Reson Pt C 239:349–363
4. Clore GM, Tang C, Iwahara J (2007) Elucidating transient macromolecular interactions using paramagnetic relaxation enhancement. Curr Opin Struct Biol 17(5):603–616
5. Tolman JR et al (1995) Nuclear magnetic dipole interactions in field-oriented proteins – information for structure determination in solution. Proc Natl Acad Sci USA 92(20):9279–9283
6. Tjandra N, Grzesiek S, Bax A (1996) Magnetic field dependence of nitrogen-proton J splittings in ^{15}N-enriched human ubiquitin resulting from relaxation interference and residual dipolar coupling. J Am Chem Soc 118(26):6264–6272
7. Tjandra N, Bax A (1997) Direct measurement of distances and angles in biomolecules by NMR in a dilute liquid crystalline medium. Science 278(5340):1111–1114
8. Schwieters CD et al (2010) Solution structure of the 128 kDa enzyme I dimer from Escherichia coli and its 146 kDa complex with HPr using residual dipolar couplings and small- and wide-angle X-ray scattering. J Am Chem Soc 132(37):13026–13045
9. Saupe A (1968) Recent results in field of liquid crystals. Angew Chem Int Ed 7(2):97
10. Losonczi JA et al (1999) Order matrix analysis of residual dipolar couplings using singular value decomposition. J Magn Reson 138(2):334–342
11. Bax A, Kontaxis G, Tjandra N (2001) Dipolar couplings in macromolecular structure determination. Nucl Magn Reson Biol Macromol B 339:127–174
12. Bax A et al (1994) Measurement of homonuclear and heteronuclear J-couplings from quantitative J-correlation. Nucl Magn Reson C 239:79–105
13. Prestegard JH, Al-Hashimi HM, Tolman JR (2000) NMR structures of biomolecules using field oriented media and residual dipolar couplings. Quart Rev Biophys 33(4):371–424
14. Prestegard JH et al (2005) Determination of protein backbone structures from residual dipolar couplings. Nucl Magn Reson Biol Macromol C 394:175
15. Tolman JR, Ruan K (2006) NMR residual dipolar couplings as probes of biomolecular dynamics. Chem Rev 106(5):1720–1736
16. Bax A, Tjandra N (1997) High-resolution heteronuclear NMR of human ubiquitin in an aqueous liquid crystalline medium. J Biomol NMR 10(3):289–292
17. Ottiger M, Bax A (1998) Characterization of magnetically oriented phospholipid micelles for measurement of dipolar couplings in macromolecules. J Biomol NMR 12(3):361–372
18. Hansen MR, Mueller L, Pardi A (1998) Tunable alignment of macromolecules by filamentous phage yields dipolar coupling interactions. Nat Struct Biol 5(12):1065–1074
19. Clore GM, Starich MR, Gronenborn AM (1998) Measurement of residual dipolar couplings of macromolecules aligned in the nematic phase of a colloidal suspension of rod-shaped viruses. J Am Chem Soc 120(40):10571–10572
20. Sass HJ et al (2000) Solution NMR of proteins within polyacrylamide gels: diffusional properties and residual alignment by mechanical stress or embedding of oriented purple membranes. J Biomol NMR 18(4):303–309
21. Chou JJ et al (2001) A simple apparatus for generating stretched polyacrylamide gels, yielding uniform alignment of proteins and detergent micelles. J Biomol NMR 21(4):377–382

22. Tycko R, Blanco FJ, Ishii Y (2000) Alignment of biopolymers in strained gels: a new way to create detectable dipole-dipole couplings in high-resolution biomolecular NMR. J Am Chem Soc 122(38):9340–9341
23. Ruckert M, Otting G (2000) Alignment of biological macromolecules in novel nonionic liquid crystalline media for NMR experiments. J Am Chem Soc 122(32):7793–7797
24. Losonczi JA, Prestegard JH (1998) Improved dilute bicelle solutions for high-resolution NMR of biological macromolecules. J Biomol NMR 12(3):447–451
25. Wang H et al (1998) A liquid crystalline medium for measuring residual dipolar couplings over a wide range of temperatures. J Biomol NMR 12(3):443–446
26. Cierpicki T, Bushweller JH (2004) Charged gels as orienting media for measurement of residual dipolar couplings in soluble and integral membrane proteins. J Am Chem Soc 126(49): 16259–16266
27. Douglas SM, Chou JJ, Shih WM (2007) DNA-nanotube-induced alignment of membrane proteins for NMR structure determination. Proc Natl Acad Sci USA 104(16):6644–6648
28. Lorieau J, Yao LS, Bax A (2008) Liquid crystalline phase of G-tetrad DNA for NMR study of detergent-solubilized proteins. J Am Chem Soc 130(24):7536
29. Ma JH, Goldberg GI, Tjandra N (2008) Weak alignment of biomacromolecules in collagen gels: an alternative way to yield residual dipolar couplings for NMR measurements. J Am Chem Soc 130(48):16148
30. Yao LS, Bax A (2007) Modulating protein alignment in a liquid-crystalline medium through conservative mutagenesis. J Am Chem Soc 129(37):11326
31. Ruan K, Tolman JR (2005) Composite alignment media for the measurement of independent sets of NMR residual dipolar couplings. J Am Chem Soc 127(43):15032–15033
32. Burton RA, Tjandra N (2006) Determination of the residue-specific ^{15}N CSA tensor principal components using multiple alignment media. J Biomol NMR 35(4):249–259
33. Liu YZ, Prestegard JH (2010) A device for the measurement of residual chemical shift anisotropy and residual dipolar coupling in soluble and membrane-associated proteins. J Biomol NMR 47(4):249–258
34. Ishii Y, Markus MA, Tycko R (2001) Controlling residual dipolar couplings in high-resolution NMR of proteins by strain induced alignment in a gel. J Biomol NMR 21(2):141–151
35. Meier S, Haussinger D, Grzesiek S (2002) Charged acrylamide copolymer gels as media for weak alignment. J Biomol NMR 24(4):351–356
36. Lorieau JL, Louis JM, Bax A (2010) The complete influenza hemagglutinin fusion domain adopts a tight helical hairpin arrangement at the lipid:water interface. Proc Natl Acad Sci USA 107(25):11341–11346
37. Bella J et al (1994) Crystal-structure and molecular-structure of a collagen-like peptide at 1.9-Angstrom resolution. Science 266(5182):75–81
38. Saffarian S et al (2004) Interstitial collagenase is a Brownian ratchet driven by proteolysis of collagen. Science 306(5693):108–111
39. Ruan K, Briggman KB, Tolman JR (2008) De novo determination of internuclear vector orientations from residual dipolar couplings measured in three independent alignment media. J Biomol NMR 41(2):61–76
40. Zweckstetter M, Hummer G, Bax A (2004) Prediction of charge-induced molecular alignment of biomolecules dissolved in dilute liquid-crystalline phases. Biophys J 86(6):3444–3460
41. Yao L et al (2008) Simultaneous NMR study of protein structure and dynamics using conservative mutagenesis. J Phys Chem B 112(19):6045–6056
42. Sanders CR et al (1994) Magnetically-oriented phospholipid micelles as a tool for the study of membrane-associated molecules. Prog Nucl Magn Reson Spectrosc 26:421–444
43. Lipsitz RS, Tjandra N (2003) ^{15}N chemical shift anisotropy in protein structure refinement and comparison with NH residual dipolar couplings. J Magn Reson 164(1):171–176
44. Tjandra N, Suzuki M, Chang SL (2007) Refinement of protein structure against non-redundant carbonyl ^{13}C NMR relaxation. J Biomol NMR 38(3):243–253

45. Burton RA, Tjandra N (2007) Residue-specific $^{13}C'$ CSA tensor principal components for ubiquitin: correlation between tensor components and hydrogen bonding. J Am Chem Soc 129(5):1321–1326
46. Kurita J et al (2003) Measurement of ^{15}N chemical shift anisotropy in a protein dissolved in a dilute liquid crystalline medium with the application of magic angle sample spinning. J Magn Reson 163(1):163–173
47. Yao LS et al (2010) Site-specific backbone amide ^{15}N chemical shift anisotropy tensors in a small protein from liquid crystal and cross-correlated relaxation measurements. J Am Chem Soc 132(12):4295–4309
48. Boisbouvier J, Delaglio F, Bax A (2003) Direct observation of dipolar couplings between distant protons in weekly aligned nucleic acids. Proc Natl Acad Sci USA 100(20):11333–11338
49. Yao LS et al (2008) NMR determination of amide N-H equilibrium bond length from concerted dipolar coupling measurements. J Am Chem Soc 130(49):16518
50. Ottiger M, Bax A (1998) Determination of relative N-H^N N-C′, C^{α}-C′, and C^{α}-H^{α} effective bond lengths in a protein by NMR in a dilute liquid crystalline phase. J Am Chem Soc 120 (47):12334–12341
51. Clore GM, Schwieters CD (2004) Amplitudes of protein backbone dynamics and correlated motions in a small alpha/beta protein: correspondence of dipolar coupling and heteronuclear relaxation measurements. Biochemistry 43(33):10678–10691
52. Clore GM, Schwieters CD (2004) How much backbone motion in ubiquitin is required to account for dipolar coupling data measured in multiple alignment media as assessed by independent cross-validation? J Am Chem Soc 126(9):2923–2938
53. Bax A, Tjandra N (1997) Are proteins even floppier than we thought? Nat Struct Biol 4(4):254–256
54. Fredriksson K et al (2004) On the interpretation of residual dipolar couplings as reporters of molecular dynamics. J Am Chem Soc 126(39):12646–12650
55. Louhivuori M et al (2003) On the origin of residual dipolar couplings from denatured proteins. J Am Chem Soc 125(50):15647–15650
56. Jha AK et al (2005) Statistical coil model of the unfolded state: resolving the reconciliation problem. Proc Natl Acad Sci USA 102(37):13099–13104
57. Meier S, Grzesiek S, Blackledge M (2007) Mapping the conformational landscape of urea-denatured ubiquitin using residual dipolar couplings. J Am Chem Soc 129(31):9799–9807
58. Esteban-Martin S, Fenwick RB, Salvatella X (2010) Refinement of ensembles describing unstructured proteins using NMR residual dipolar couplings. J Am Chem Soc 132(13): 4626–4632
59. Nodet G et al (2009) Quantitative description of backbone conformational sampling of unfolded proteins at amino acid resolution from NMR residual dipolar couplings. J Am Chem Soc 131(49):17908–17918
60. Zweckstetter M, Bax A (2000) Prediction of sterically induced alignment in a dilute liquid crystalline phase: aid to protein structure determination by NMR. J Am Chem Soc 122(15):3791–3792
61. Zweckstetter M (2008) NMR: prediction of molecular alignment from structure using the PALES software. Nat Protoc 3(4):679–690
62. Valafar H, Prestegard JH (2004) REDCAT: a residual dipolar coupling analysis tool. J Magn Reson 167(2):228–241
63. Cornilescu G et al (1998) Validation of protein structure from anisotropic carbonyl chemical shifts in a dilute liquid crystalline phase. J Am Chem Soc 120(27):6836–6837
64. Clore GM, Garrett DS (1999) R-factor, free R, and complete cross-validation for dipolar coupling refinement of NMR structures. J Am Chem Soc 121(39):9008–9012
65. Bax A (2003) Weak alignment offers new NMR opportunities to study protein structure and dynamics. Protein Sci 12(1):1–16
66. Schwieters CD, Kuszewski JJ, Clore GM (2006) Using Xplor-NIH for NMR molecular structure determination. Prog Nucl Magn Reson Spectrosc 48(1):47–62

67. Schwieters CD et al (2003) The Xplor-NIH NMR molecular structure determination package. J Magn Reson 160(1):65–73
68. Tjandra N et al (1997) Defining long range order in NMR structure determination from the dependence of heteronuclear relaxation times on rotational diffusion anisotropy. Nat Struct Biol 4(6):443–449
69. de Alba E, Tjandra N (2004) Residual dipolar couplings in protein structure determination. Methods Mol Biol 278:89–106
70. Lange OF et al (2008) Recognition dynamics up to microseconds revealed from an RDC-derived ubiquitin ensemble in solution. Science 320(5882):1471–1475
71. Clore GM, Schwieters CD (2006) Concordance of residual dipolar couplings, backbone order parameters and crystallographic B-factors for a small alpha/beta protein: a unified picture of high probability, fast atomic motions in proteins. J Mol Biol 355(5):879–886
72. Lindorff-Larsen K et al (2005) Simultaneous determination of protein structure and dynamics. Nature 433(7022):128–132
73. Lakomek NA et al (2008) Self-consistent residual dipolar coupling based model-free analysis for the robust determination of nanosecond to microsecond protein dynamics. J Biomol NMR 41(3):139–155
74. Lakomek NA et al (2006) A thorough dynamic interpretation of residual dipolar couplings in ubiquitin. J Biomol NMR 34(2):101–115
75. Peti W et al (2002) Model-free analysis of protein backbone motion from residual dipolar couplings. J Am Chem Soc 124(20):5822–5833
76. Meiler J et al (2001) Model-free approach to the dynamic interpretation of residual dipolar couplings in globular proteins. J Am Chem Soc 123(25):6098–6107
77. Chang SL, Tjandra N (2005) Temperature dependence of protein backbone motion from carbonyl ^{13}C and amide ^{15}N NMR relaxation. J Magn Reson 174(1):43–53
78. Liu YZ, Kahn RA, Prestegard JH (2010) Dynamic structure of membrane-anchored Arf·GTP. Nat Struct Mol Biol 17(7):876
79. Lewis BA et al (1985) Magnetic birefringence studies of dilute purple membrane suspensions. Biophys J 47(2):143–150
80. Koenig BW et al (1999) NMR measurement of dipolar couplings in proteins aligned by transient binding to purple membrane fragments. J Am Chem Soc 121(6):1385–1386
81. Koenig BW et al (2002) Structure and orientation of a G protein fragment in the receptor bound state from residual dipolar couplings. J Mol Biol 322(2):441–461
82. Seidel RD, Zhuang TD, Prestegard JH (2007) Bound-state residual dipolar couplings for rapidly exchanging ligands of His-tagged proteins. J Am Chem Soc 129(15):4834–4839
83. Wang X et al (2008) RDC-assisted modeling of symmetric protein homo-oligomers. Protein Sci 17(5):899–907
84. Lee HW et al (2010) Three-dimensional structure of the weakly associated protein homodimer SeR13 using RDCs and paramagnetic surface mapping. Protein Sci 19(9):1673–1685
85. Al-Hashimi HM, Bolon PJ, Prestegard JH (2000) Molecular symmetry as an aid to geometry determination in ligand protein complexes. J Magn Reson 142(1):153–158
86. Bewley CA, Clore GM (2000) Determination of the relative orientation of the two halves of the domain-swapped dimer of cyanovirin-N in solution using dipolar couplings and rigid body minimization. J Am Chem Socety 122(25):6009–6016
87. Bax A, Grishaev A (2005) Weak alignment NMR: a hawk-eyed view of biomolecular structure. Curr Opin Struct Biol 15(5):563–570
88. Tjandra N (1999) Establishing a degree of order: obtaining high-resolution NMR structures from molecular alignment. Struct Fold Des 7(9):R205–R211
89. Lipsitz RS, Tjandra N (2004) Residual dipolar couplings in NMR structure analysis. Annu Rev Biophys Biomol Struct 33:387–413
90. Prestegard JH, Bougault CM, Kishore AI (2004) Residual dipolar couplings in structure determination of biomolecules. Chem Rev 104(8):3519–3540

Top Curr Chem (2012) 326: 69–98
DOI: 10.1007/128_2011_214

Published online: 2 August 2011

NMR Studies of Metalloproteins

Hongyan Li and Hongzhe Sun

Abstract Metalloproteins represent a large share of the proteomes, with the intrinsic metal ions providing catalytic, regulatory, and structural roles critical to protein functions. Structural characterization of metalloproteins and identification of metal coordination features including numbers and types of ligands and metal-ligand geometry, and mapping the structural and dynamic changes upon metal binding are significant for understanding biological functions of metalloproteins. NMR spectroscopy has long been used as an invaluable tool for structure and dynamic studies of macromolecules. Here we focus on the application of NMR spectroscopy in characterization of metalloproteins, including structural studies and identification of metal coordination spheres by hetero-/homo-nuclear metal NMR spectroscopy. Paramagnetic NMR as well as ^{13}C directly detected protonless NMR spectroscopy will also be addressed for application to paramagnetic metalloproteins. Moreover, these techniques offer great potential for studies of other non-metal binding macromolecules.

Keywords C direct detection · Metal coordination · Metalloprotein · NMR spectroscopy · Paramagnetic metalloprotein

Contents

1 Introduction 70
2 The Contribution of NMR to Structural Metalloproteins 71
  2.1 Conventional Method for Structure Determination of Metalloproteins 71
  2.2 Utilization of Chemical Shifts to Generate Structures 76
3 Identification of Metal Coordination 76
  3.1 Homonuclear and Heteronuclear Metal NMR Spectroscopy 76
  3.2 Chemical Shift Perturbation 82

H. Li (✉) and H. Sun
Department of Chemistry, The University of Hong Kong, Pokfulam Road, Hong Kong, P. R. China
e-mail: hylichem@hku.hk

4 NMR in Studies of Paramagnetic Metalloproteins 84
4.1 Paramagnetism-Based Structural Restraints 84
4.2 ^{13}C-Detected Protonless NMR 87
5 Perspectives 89
References 90

1 Introduction

Metal ions play important roles in life science and the molecular mechanism of metal-dependent life processes and all matters regarding metal within a cell or tissue/organ are topics of the emerging fields of metallomics and metalloproteomics [1–3]. Metal ions must usually associate with proteins (and other biomolecules), i.e., metalloproteins, to prevent the toxic effects of metal excess. Metalloproteins are one of the most diverse classes of proteins with the intrinsic metal ions providing catalytic, regulatory, and structural roles critical to protein function, and are found in plants, animals, and many microorganisms. It has been estimated that metalloproteins account for approximately one-quarter to one-third of all the proteins in the human body [4]. A recent study revealed that many of metalloproteomes still remain uncharacterized [5]. A systematic bioinformatics survey of 1,371 metalloenzymes with known structures showed that about 40% of enzyme-catalyzed reactions involve metal ions, e.g., magnesium, zinc, and iron [6]. Enormous effort has been devoted toward understanding the structure and function of metalloproteins and such knowledge has been used to design a new functional metalloprotein [7] and to rationalize and to search for new metalloproteins by a bioinformatics approach [6, 8]. However, it is impossible currently from gene sequences to predict the numbers and types of metals an organism assimilates from its environment or uses in its metalloproteins because the geometry and composition of metal binding site are diverse and poorly recognized [9, 10]. Therefore, understanding of the function of metalloproteins comes from individual characterization of the structures of the proteins and chemical states of the metal centers by various spectroscopic techniques including NMR spectroscopy, circular dichroism (CD), electronic absorption spectroscopy (UV), small angle X-ray absorption, as well as extended X-ray absorption fine structure (EXAFS).

Over the last three decades, NMR spectroscopy has been developed into a very important and versatile analytical technique both in the chemical and biological sciences. It has been used within the framework of Structural Genomic (SG) projects worldwide for determination of structures of proteins at the atomic level under physiologically relevant conditions [11–13]. Moreover, NMR spectroscopy is applicable to study of the interactions of proteins with other molecules including proteins, nucleic acids, and even small molecules which are mainly based on the sensitivity of the chemical shifts towards changes in chemical environments [14, 15]. Application of this technique to structural studies is limited to small proteins (30–35 kDa) even with the aid of isotopic labeling (^{13}C, ^{15}N and ^{2}H) although backbone assignments of a 723-residue enzyme with a molecular weight of 81.4 kDa were achieved [16]. Recently, a few breakthroughs have been made in this field. This includes the first NMR structure of a seven-helix transmembrane protein determined in membrane-mimetic environments [17] and the first structure determined in living cells by in-cell NMR [18]. Apart from its application in structural studies, NMR spectroscopy is also able to monitor the internal motion of biomolecules ranging over times from subnanoseconds to beyond seconds. Characterization of dynamics of biomolecules, such as folding transition, will be a great help for our understanding of the biological function of biomolecules.

Application of NMR spectroscopy in studies of metalloproteins is in principle the same as other proteins if the proteins contain diamagnetic metal ions. In the case of paramagnetic metalloproteins, things become more challenging since paramagnetic metals affect longitudinal and transverse relaxation rates of observed nuclei. However, recent advances in the hardware and methodology have enable structures of such metalloproteins to be determined, these advances including ^{13}C-detected experiments, solid-state NMR, and the discovery of paramagnetic relaxation enhancement (PRE) [19–21]. In this review, we will address mainly the application of NMR spectroscopy in studies of metalloproteins including the contribution of NMR to structural characterization of metalloproteins, and special attention will be devoted to the utilization of NMR in characterization/identification of the metal binding sites and their coordination environments as well as to probing conformational changes of metalloproteins upon metal binding and release. The techniques used for paramagnetic metalloproteins will also be included briefly since a number of reviews in this field can be found [19–22]. A systematic review of the application of metal NMR spectroscopy will not be made and interested readers are directed elsewhere [23].

2 The Contribution of NMR to Structural Metalloproteins

2.1 Conventional Method for Structure Determination of Metalloproteins

NMR spectroscopy can be applied to characterize structurally diamagnetic metal containing metalloproteins similar to other proteins. Amongst the structures deposited in the Brookhaven Protein Data Bank (PDB), ca. 15% of the structures were resolved by NMR. Here, we will give a brief summary since the detailed methodology can be found in numerous references [11, 24–26].

In structural studies of proteins/metalloproteins, concentrations of about 1 mM are typically required and proteins must be soluble and stable over a period of time (weeks). For small proteins with several tens of amino acids, e.g. metallothionine [27, 28], it is sufficient to use ^{15}N-labeled samples to determine structures of the proteins. However, if proteins can be overexpressed in a bacterial system (e.g., *Escherichia coli*), it is desirable to overexpress the protein with uniform enrichment of ^{15}N and ^{13}C even for small proteins. This will make full use of multidimensional heteronuclear NMR experiments to increase the spectral resolution. Unlike other proteins, expression of metalloproteins in a bacterial system usually requires specific metal ions to be supplemented in the medium to induce overexpression of the targeted proteins [14, 29]. Alternatively, metal ions have to be incorporated into the proteins after purification, particularly if the metal ions play structural roles, as otherwise the proteins may not be stable for structural characterization. For example, HypA from *Helicobacter pylori* precipitates easily in the absence of zinc which serves a structural role [15]. However, caution has to be taken during metal incorporation since excess metal ions may also cause protein aggregation.

A major bottleneck in solving protein structures by NMR is the highly peak-picking and assignment of chemical shifts and NOEs. The strategy of the assignment process and structure calculation can be found in an excellent review [30]. In general, for a $^{15}N/^{13}C$-labeled protein, a series of double/triple resonance experiments are recorded for resonance assignments. Backbone assignments are derived from HNCA, HNCOCA, HNCACB, CBCACONH, HNCO, and HNCACO, whereas side-chain protons and carbon atoms are assigned from HCCH-TOCSY, HCCH-COSY, HBHACONH, C(CO)NH, and H(CCO)NH [24, 25]. The chemical shifts of backbone and side-chain are then used to assign NOEs (^{15}N-/^{13}C-HSQC-NOESY) to derive inter-protein distance restraints. Usually the structure determination process goes though several iterations of compiling a NOESY peak list, assignment of NOE cross-peaks to sequence-specific interactions, structure generations and assessment, refinement of NOESY peak lists, and reassignment of the cross-peaks, which can be carried out automatically [31, 32]. In addition to distance restraints, dihedral angle restraints are usually obtained from several experiments, e.g., HNHA [33] or HNHB [34], or predicated from TALOS, a program that empirically predicts backbone angles (ϕ,Ψ) based on the chemical shifts of H^{α}, C^{α}, C^{β}, C', and N [35], as well as the H-bond restraints derived from H-D exchange experiments. For elongated macromolecules, residual dipolar couplings (RDC) as additional restraints are necessary for structure determination. Information about RDC can be found in this book. In order to get relatively good quality for the structures, the numbers of NMR restraints used for structural determination are usually of the order of 10–20 independent interatomic distances per amino acid plus some dihedral restraints and as well as atom–atom vector directions. The quality of calculated structures has to be evaluated using programs PROCHECK, WHATIF, etc., and detailed description can be found in a recent review [13].

In addition to the general strategies described above, metal-based NMR parameters are also of great help in the evaluation of structures of metalloproteins, especially for those metalloproteins whose folding is highly metal-dependent. To incorporate metal cluster constraints into structural calculation, residues that coordinate to metal ions (e.g., Zn^{2+}) must be identified first either by mutagenesis studies or by physical characterization such as UV absorption spectroscopy, EXAFS, and NMR spectroscopy [36–38]. Providing that metal coordination residues and geometries are unveiled, metal cluster restraints can be obtained based on relevant crystal structures of either macromolecules or small molecules. Usually, metal cluster restraints are not used in initial structure generation to avoid bias, especially during the automated assignment procedure in CYANA [39]; instead, metal ligand restraints are incorporated in the last step of structural calculation as well as in the refinement stage [40, 41].

The application of NMR spectroscopy in protein structure determination actually started with a small metalloprotein, metallothionine (MT) [42]. Metallothionines are a class of low molecular weight (typically 6–7 kDa) cysteine-rich proteins. The proteins lack a well-defined secondary structure and their folding is dictated mostly by a clustered network of cysteine residues and metal ions usually represented by Zn^{2+}, Cu^{+}, and Cd^{2+} [43, 44]. Since the first solution of the structure of rabbit liver Cd_7MT2 [42], numerous three dimensional structures of metallothiones from different isoforms (MT1/MT2/MT3) or different species such as blue crab and mammalian (rabbit, rat, and human) have been resolved by NMR spectroscopy [45] with only one structure (rat liver Cd_5Zn_2MT2) determined by X-ray crystallography [46]. The protein consists of two dynamic metal-thiolate clusters folded into two domains (α, β) and the structural mobility of the protein makes it difficult to be crystallized. The metal cluster restraints, e.g., Cd–S bond lengths, as well as Cd–S–Cd, S–Cd–S, and $CysC^{\beta}$–S–Cd bond angles from the X-ray crystal structures of model cadmium complexes and rat liver Cd_5Zn_2MT2 were often incorporated with other distance and angle restraints in structure calculation. Recently, a new member of metallothionine MT3 with the conserved CPCP motif in the N-termini has been involved in the growth inhibitory activity and is down-regulated in the brain of Alzheimer's patients [47]. The solution structures of both human [27] and mouse MT3 [48] resolved by NMR spectroscopy for the C-terminal α-domain, Fig. 1a, revealed a similar Cd_4Cys_{11} cluster as well as very similar tertiary folds to MT1/2. However, a loop in the acidic hexapeptide insertion is found and is slightly longer in human MT3 than in mouse MT3. The first solution structure of Cd_7MT-nc of the Antarctic fish *Notothenia coriiceps* was also determined [28]. The position of the ninth cysteine of Cd_7MT-nc is different from mammalian MT which results in a structural change of the domain, in particular in the orientation of the loop (Lys50–Thr53), Fig. 1b, and in turn a different charge distribution with respect to mammalian MT [28]. Interestingly, an intriguing class of histidine-containing metallothionines has also been identified in fungi and bacterial [49]. The histidine residue has been thought to be able to modulate zinc affinity and reactivity. Solution structure of one of this class of MTs, Zn_4SmtA from cyanobacterium *Synechococcus* PCC 7942 was determined [41], Fig. 1c, revealing a $Zn_4Cys_9His_2$ cluster with a topology similar to that of the Zn_4Cys_{11} cluster of the α-domain of mammalian MT. However, the two ZnCysHis sites and one $ZnCys_4$ site readily exchange Zn^{2+} for exogenous Cd^{2+}. Moreover, SmtA contains a short α-helix and two small antiparallel β-sheets surround the inert zinc site, which resemble zinc finger portions of GATA and LIM proteins. Such a structure of SmtA probably produces its function of specific protein and/or DNA recognition [41].

NMR spectroscopy makes an enormous contribution to structural biology of metalloproteins, particularly in zinc-binding proteins. Zinc, the second most abundant metal found in eukaryotic organisms, plays important catalytic and structural roles in a variety of biological processes. Binding of zinc is able to stabilize the folded

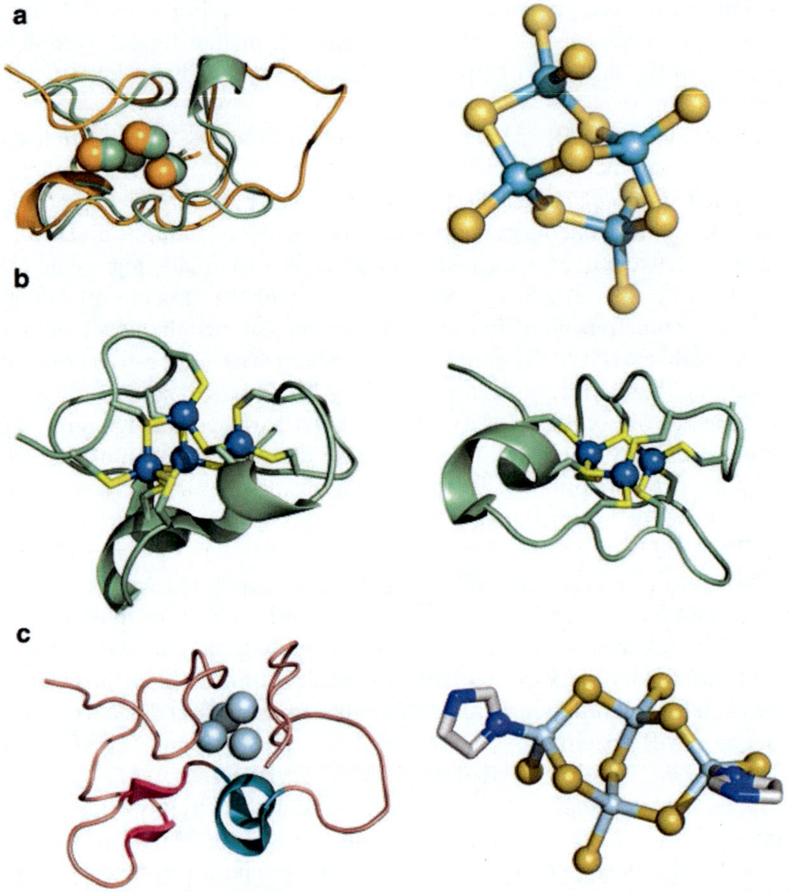

Fig. 1 (**a**) Overlaid view of structures of human MT3 (PDB: 2f5h) in *orange* and mouse MT3 (PDB: 1JI9) in *pale green* with the cadmium ions shown as spheres (*left*), and the four metal-thiolate cluster of human MT3 in the α-domain with the metal ions and sulfur atoms from cysteines shown in *cyan* and *yellow* respectively (*right*). (**b**) NMR structures of the *Notothenia coriiceps* Cd_7MT-nc with the α-domain (PDB: 1m0g) (*left*) and the β-domain (PDB: 1m0j) (*right*). The metal-thiolate clusters are also shown with the cadmium ions shown as spheres in *sky blue* and sulfurs as sticks in *yellow*. (**c**) Solution structure of bacterial SmtA (PDB: 1jjd) with the zinc ions shown in pale cyan as spheres (*left*) and the metal cluster $Zn_4Cys_9His_2$ (*right*) with the zinc ions shown in *pale cyan*, sulfurs in *yellow*, and the nitrogens of histidines in *blue*

conformations of domains, which renders a proper function of the protein [50]. Zinc finger proteins are the most abundant class of zinc-binding proteins that contain conserved cysteines and histidines coordinated to zinc. Intensive structural and functional studies have established the invariance of the ββα framework of the Cys_2His_2 zinc finger module and provided a sound basis for understanding the nature of DNA recognition [40, 51–55]. Diverse structures of zinc fingers also account for their diverse functions such as DNA recognition, RNA packaging,

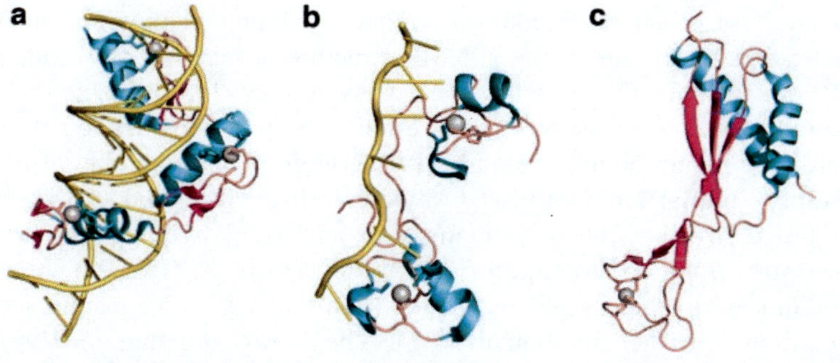

Fig. 2 (**a**) Ribbon diagram of the Gfizf35-DNA complex (PDB: 2kmk) with the zinc ions shown as *gray spheres* and the side-chains of two coordinated histidines and two cysteines shown in sticks and DNA shown in *yellow sticks*. (**b**) Solution structure of the RNA complex of TIS11d (PDB: 1rgo) with the zinc ions in *gray spheres* coordinated to three cysteines and one histidine and RNA shown as *yellow sticks*. (**c**) Solution structure of HypA from *Helicobacter pylori* (PDB: 2kdx) with the zinc ions in *gray sphere* coordinated to four cysteine sulfurs

transcriptional activation, regulation of apoptosis, protein folding and assembly, as well as lipid binding [52, 56, 57]. Comprehensive reviews in this area can be found [56–58]. Here, we will only highlight some of the recent studies. Solution structure of Gfi-1 zinc finger 3–5 complex with a 16-mer consensus DNA (Fig. 2a) demonstrated that zinc fingers 3–5 bind into the major groove of the target DNA, reminiscent of canonical Cys_2His_2 zinc-finger domains, which provide valuable insight into the structure determinants for DNA binding specificity as well as molecular rationales for a naturally occurring mutation that causes acute myeloid leukemia [59]. Poly(ADP-ribose) (PAR) is an important post-translational modification in higher eukaryotes. Solution structures of two PBZ modules (PAR-binding zinc finger) of PNK-like factor (APLF) and the PDB domain of *Drosophila melanogaster* CG1218-PA reveal a novel type of Cys_2His_2 zinc finger and provide a structural basis for PBZ-PAR recognition. Intriguingly, Cys_2His_2 zinc coordination of the PBZ modules is structurally and functionally dissimilar from canonical double stranded DNA-binding TFIIIA-type zinc fingers; rather they resemble single-stranded RNA-binding Cys_3His_1 tandem zinc fingers (TZFs). Both of them lack secondary structures but have rigid backbone conformations as a result of zinc binding [60, 61]. Zinc finger proteins are also able to bind to RNA. The NMR structure of tandem zinc finger (TZF) domain of the protein TIS11d bound to the RNA sequence 5′-UUAUUUAUU-3′ (Fig. 2b) reveals a pair of novel Cys_3His_1 fingers which independently recognizes the four nucleotide sequence UAUU and the sequence specificity in RNA recognition is achieved by a network of intermolecular hydrogen bonds [62]. This structure provides insights into RNA-binding function of this family of Cys_3His_1 zinc finger proteins [62]. The Cys_3His_1 zinc finger motif is also found in the structure of SAP30 polypeptide of the Sin3 corepressor complex which adopts a novel fold comprising two β-strands and two α-helices with the zinc organized center. Such a structure may also function as a double-stranded DNA-binding motif [63]. The zinc finger CW (zf-CW) domain with a motif of about 60 residues is frequently found in proteins involved in epigenetic regulation. Interestingly, NMR structure of human zf-CW domain and PWWP domain containing proteins reveal a new fold in which a zinc is coordinated tetrahedrally by four conserved Cys residues [64]. Such a structure partially resembles the plant homeo domain (PHD) finger bound to the histone tail, implicating a similar function of zf-CW domain [64]. This kind of Cys_4 motif is widely found in other metalloproteins such as [NiFe] hydrogenases accessory protein HypA. Solution structure of HypA from *H. pylori* (Fig. 2c) showed that zinc coordinated to four cysteines donated from loops and no apparent secondary structure found in the zinc-domain [15]. The X-ray structure of HypA from *Thermococcus kodakaraensis KOD1* further confirmed such a zinc coordination sphere [65].

2.2 Utilization of Chemical Shifts to Generate Structures

Protein NMR chemical shifts are highly sensitive to local structure and reflect a wide array of structure factors including backbone and side-chain conformation, secondary structure, hydrogen bonds, and the orientation/position of aromatic rings. Chemical shift data can be used in conjunction with protein sequence information and reasonable force field to generate 3D structure models using the method of CHEMSHIRE or CS-ROSETTA [66–68]. The Chemical-Shift-ROSETTA (CS-ROSETTA) is a robust protocol available for de novo protein structure generation. The method uses experimental chemical shifts of $^{13}C^{\alpha}$, $^{13}C^{\beta}$, $^{13}C'$, ^{15}N, $^{1}H^{\alpha}$, and $^{1}H^{N}$ as an input to select polypeptide fragments in existing protein structures (e.g., PDB data bank) in conjunction with the standard ROSETTA Monte Carlo fragment assembly and energy minimization protocol [67, 68]. The CS-ROSETTA has been further combined with CYANA using unassigned NOESY data to direct Rosetta trajectories toward the native structure and produces a more accurate models than CS-ROSEAAR alone [69]. Moreover, chemical shifts have been further extended in determination of protein–protein complex structures via the CamDock method [70]. The method that utilizes chemical shifts to generate structures may potentially provide a new direction for high-throughput NMR structure determination of proteins including metalloproteins, although such a method has not yet been applied in metalloproteins so far.

3 Identification of Metal Coordination

3.1 Homonuclear and Heteronuclear Metal NMR Spectroscopy

Metalloprotein functionality depends on subtle interaction between properties of the metal ion, dictated by its coordination chemistry. Our present knowledge in terms of structure-function of metalloproteins, in particular the role of metal ions involved, varies considerably from protein to protein. Therefore, identification of metal coordination parameters including numbers and types of ligands and metal-ligand geometry and mapping the structural and dynamic changes upon metal binding are of significance towards understanding biological functions of metalloproteins. Today, NMR spectroscopy is one of the leading techniques for this purpose. Applicability of homonuclear metal NMR and heteronuclear ^{1}H-metal HMQC to monitor directly protein–metal interactions rely greatly on the properties of the nuclei. Some of the metal (e.g., ^{113}Cd) NMR has been used extensively to identify coordination spheres of the metal ions. Moreover, the coupling constants between the NMR active metals and nuclei of the protein provide insight into the identity and geometry of the metal ligands [71]. Many metal (e.g., ^{43}Ca and ^{67}Zn) NMRs are less powerful and hardly used owing to the fact that these nuclei have the spin quantum number I greater than 1/2 which leads to lower sensitivity, poor resolution, and broadening due to large quadrupolar moments although ultrahigh fields improve it. Several reviews have systematically summarized the application of heteronuclear NMR spectroscopy in biological and medicinal chemistry as well as in the study of metalloproteins [23, 71–73]. Here, we will highlight some of the recent progresses as a snapshot of using metal NMR to identify metal coordination.

Cadmium is one of the most widely used metal nuclei for probing metal–protein interactions, despite its toxic properties. It has two NMR active nuclei, ^{113}Cd and ^{111}Cd (spins of 1/2), with the former being slightly more sensitive and therefore usually used as a preferred nucleus. At natural abundance, the sensitivity of ^{113}Cd is very low (ca. 7.6-fold of ^{13}C), and therefore isotopic enrichment (ca. 96%) of ^{113}Cd is usually needed to ensure reasonable quality of spectra to be acquired in a relatively short period of time (a few hours for ca. 0.5 mM samples). ^{113}Cd or ^{1}H-^{113}Cd NMR spectroscopy has been utilized in the study of a variety of metalloproteins where the native Zn^{2+}, Ca^{2+}, Mg^{2+}, Mn^{2+}, Fe^{2+}, and Cu^{2+} can be substituted by ^{113}Cd, given that the adaptable ligand coordination number and geometry of Cd^{2+} is similar to Zn^{2+} and the ionic radius of Cd^{2+} (0.97 Å) is similar to that of Ca^{2+} (0.9 Å) [27, 74–77]. Moreover, the substitution of the native zinc from metalloenzymes and DNA-binding proteins by cadmium caused almost no changes in their structures and functions [78, 79].

^{113}Cd chemical shifts are very sensitive to the nature, number, and geometric arrangement of the coordinated ligands [71], as shown in Fig. 3. Such wide chemical shift dispersion not only provides information about the types and numbers of ligand at a particular metal site, but also discriminates multiple metal sites with identical ligand coordination environments. ^{113}Cd NMR and ^{1}H-^{113}Cd HMQC have been employed exclusively in identification of metal-thiolate clusters in a family of small proteins, e.g., metallothionines [27, 48, 71, 80]. Both homonuclear 1D ^{113}Cd decoupling studies (Fig. 4a) and 2D ^{113}Cd-^{113}Cd COSY (Fig. 4b) of ^{113}Cd$_7$-MTs established the existence of two metal-thiolate clusters in this protein, while ^{1}H-^{113}Cd HMQC (Fig. 4c) was used to identify sequence-specific cysteine–cadmium coordination bonds. The chemical shift patterns for the two clusters Cd_3Cys_9 and Cd_4Cys_{12} of human MT3 as shown in Fig. 4a showed seven resonances at analogous positions compared with MT1/2 with chemical shift ranging from 600 to 690 ppm [27].

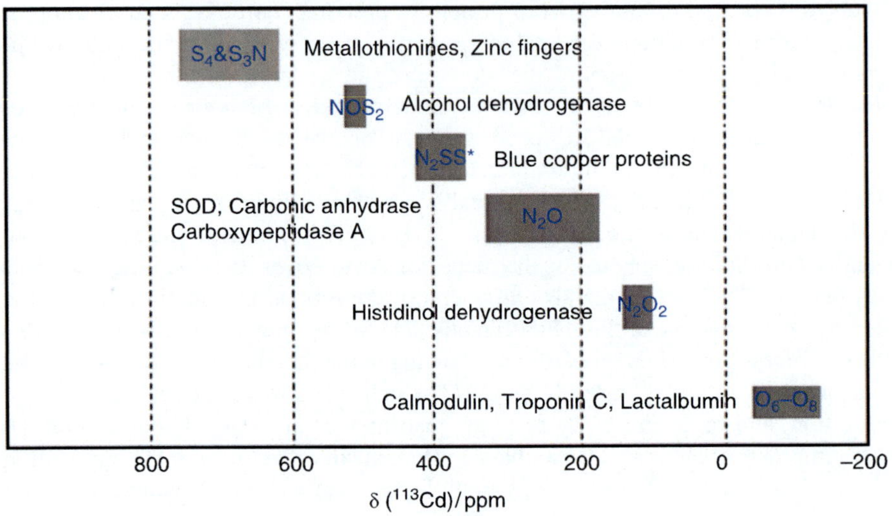

Fig. 3 Chemical shifts of ^{113}Cd for structurally characterized ^{113}Cd-substituted metalloproteins relative to external 0.1 M [Cd(ClO_4)]. The chemical shift positions are represented by *gray bars* and coordinating atoms are highlighted in *blue* with typical proteins listed besides. Here S represents sulfur from cysteine, S* represents sulfur from methionine, O represents oxygen from carboxylate or water, and N represents nitrogen from histidine [71]

However, the resonances from the α-domain (I, V, VI, and VII) are slightly shifted, particularly for resonance VII, probably attributable to the hexa-peptide insertion in this domain. The homonuclear 2D ^{113}Cd-^{113}Cd, Fig. 4b, clearly shows the correlation of cadmium signals which confirms the existence of two Cd-thiolate clusters [81]. The 2D ^{1}H-^{113}Cd HMQC as shown in Fig. 4c is normally used to obtain detailed metal-thiolate connectivity within each of these clusters [27]. Recently, metallothionines from different species, such as sea mussel *Mytilus galloprovincialis* (Cd_7MT10) [81] and blue crab *Callinectes sapidus* (MTc) [82], have been studied by combined use of ^{113}Cd NMR, ^{113}Cd-^{113}Cd COSY, and ^{1}H-^{113}Cd HMQC. Both proteins have 21 cysteine residues with the position of cysteines distinct from mammalian MTs. The unique structure and dynamic features of the metal-thiolate cluster in these proteins are obviously seen from their distinct NMR parameters of Cd_7MTC [81, 82]. ^{111}Cd NMR has also been applied in studies of His-containing metallothionine, e.g., SmtA. The metal cluster of CdS_4 and CdN_2S_2 were identified [83], and ^{1}H-^{111}Cd HMQC of Cd_7SmtA unequivocally demonstrated couplings of two Cd^{2+} to both $H^{\varepsilon 1}$ or $H^{\delta 2}$ protons of two histidine residues [41]. Apart from metallothionines, cadmium NMR has also been employed to identify metal coordination environments in various metalloproteins, such as zinc finger [40] and [NiFe] hydrogenase accessory protein HypA [15], where, in the latter case, zinc ions (substituted by ^{113}Cd) are coordinated to four cysteine side-chains tetrahedrally. Moreover, it was also used to investigate major zinc binding sites on human albumin [74, 84]. The chemical shifts of ^{111}Cd of human albumin

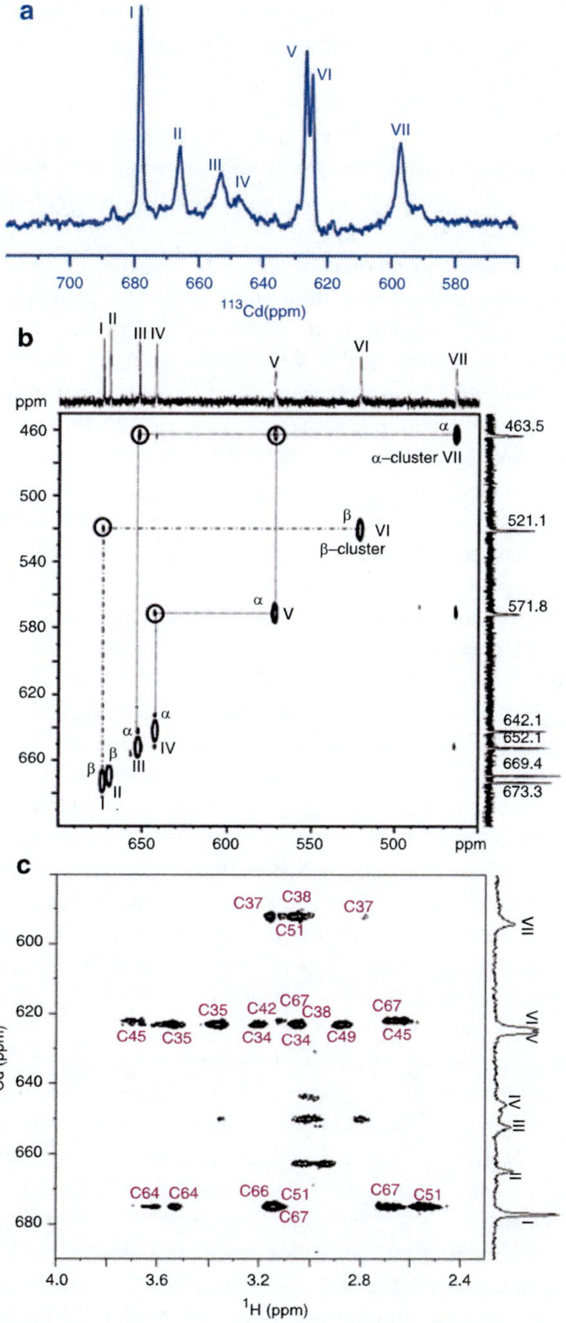

Fig. 4 (**a**) The 500-MHz NMR spectrum of human $^{113}Cd_7$-MT3 in 15 mM phosphate buffer, pH 7.3 and 310 K. Resonances I, V, VI, and VII are assigned to the α-domain and II, III, and IV to the β-domain [27]. (**b**) 2D ^{113}Cd-^{113}Cd COSY NMR of Cd_7-MT10 in 17 mM Tris-d_{11} buffer, pH 7.0, 298 K with the one-dimensional proton-decoupled ^{113}Cd NMR spectrum at the *top*. The cadmium connectivity in the α- and β-domains is also shown [81]. (**c**) The 500 MHz two-dimensional ^1H-^{113}Cd HMQC spectrum of human $^{113}Cd_7$-MT3 in 15 mM phosphate buffer with ^3J (^1H, ^{113}Cd) of 30 Hz. The assignments were also labeled with one-letter amino acid code [27]

(ca. 130 and 30 ppm) in combination with other techniques (EXAFS and mutagenesis studies) clearly demonstrated that the major zinc is a five-coordinate site with residues of histidine and aspartate.

Similarly, heteronuclear magnetic resonance spectroscopy with nuclei of spins 1/2 (e.g., $^{109}Ag/^{107}Ag$, ^{207}Pb, and ^{199}Hg) has also been extensively employed in studies of metal coordination environments or the active site structures of metalloproteins. $^{109}Ag/^{107}Ag$ NMR has so far found limited application in biological systems due to low sensitivity and extremely long spin-lattice relaxation times. The only known metalloprotein studied by 2D ^{1}H-^{109}Ag HMQC spectroscopy to date is the silver-substituted yeast metallothionine [85]. The toxic lead (^{207}Pb), although not directly biological relevant, is an excellent substitute for Ca^{2+} in calcium-binding, allowing them to retain similar structures and function. ^{207}Pb has a moderate resonance frequency, a vast chemical shift range (over thousands), and potentially large spin–spin coupling to neighboring nuclei which limits its application in studies of metalloproteins. Although ^{207}Pb has been used to study active sites of model compounds or peptides of metalloenzymes [86, 87], there appears to be only one application using ^{207}Pb as a probe to study Pb^{2+} binding to the Ca^{2+} site of calcium-binding proteins including calmodulin (CaM) [88]. Binding of ^{207}Pb to both carp and pike parvalbumins gave rise to two ^{207}Pb signals from 750 to 1,260 ppm downfield relative to aqueous $[Pb(NO_3)_2]$. Similarly, four ^{207}Pb signals, which fall in the same chemical shift window, could be observed for CaM. Both ^{207}Pb and ^{1}H have demonstrated that Pb^{2+} binds to all four sites simultaneously, in contrast to the behavior of the protein in the presence of Ca^{2+} [88]. The large chemical shift dispersion and remarkable sensitivity to the chemical environment of ^{207}Pb signals promote extensive studies using model complexes or peptides on mining the relationship of chemical shifts of ^{207}Pb with the numbers and types of coordinating ligands [86, 87, 89]. A recent study again demonstrated that chemical shifts of ^{207}Pb can be used to discriminate between PbS_3 (from Cys of thiolate-rich peptides) with other ligands such as PbS_3O and other O, S, and N donor ligands [89]. Therefore, based on ^{207}Pb chemical shift maps, Fig. 5a, both homonuclear and heteronuclear ^{207}Pb spectroscopy should provide a useful tool for investigation of Pb^{2+} coordination in more complex biological systems.

In spite of being widely known as a protein-modifying agent, the toxic metal Hg^{2+} ion can be used as a probe to substitute native metal ions for several metalloenzymes in a manner that preserves catalytic activity. Mercury has two NMR-active isotopes, ^{199}Hg and ^{201}Hg, with the latter having the spin of 1/2 and a natural abundance of 16.84%. ^{199}Hg exhibits several favorable NMR properties for structural and functional study, such as large coupling constants, a wide range of chemical shift dispersions (ca. 5,000 ppm), and a relative sensitivity of 5.4 times that of ^{13}C and 8 times that of ^{113}Cd for an equal number of nuclei. Therefore, ^{199}Hg NMR techniques (^{199}Hg and ^{1}H-^{199}Hg HMQC) have been used as useful tools to probe the metal coordination environment in biological relevant complexes [91, 92] as well as regulatory proteins, copper enzymes, and zinc transcription factor complexes as large as 50 kDa [90, 93–96]. The chemical shift of ^{199}Hg is very sensitive to the primary coordination spheres including numbers, types of ligands, as well as

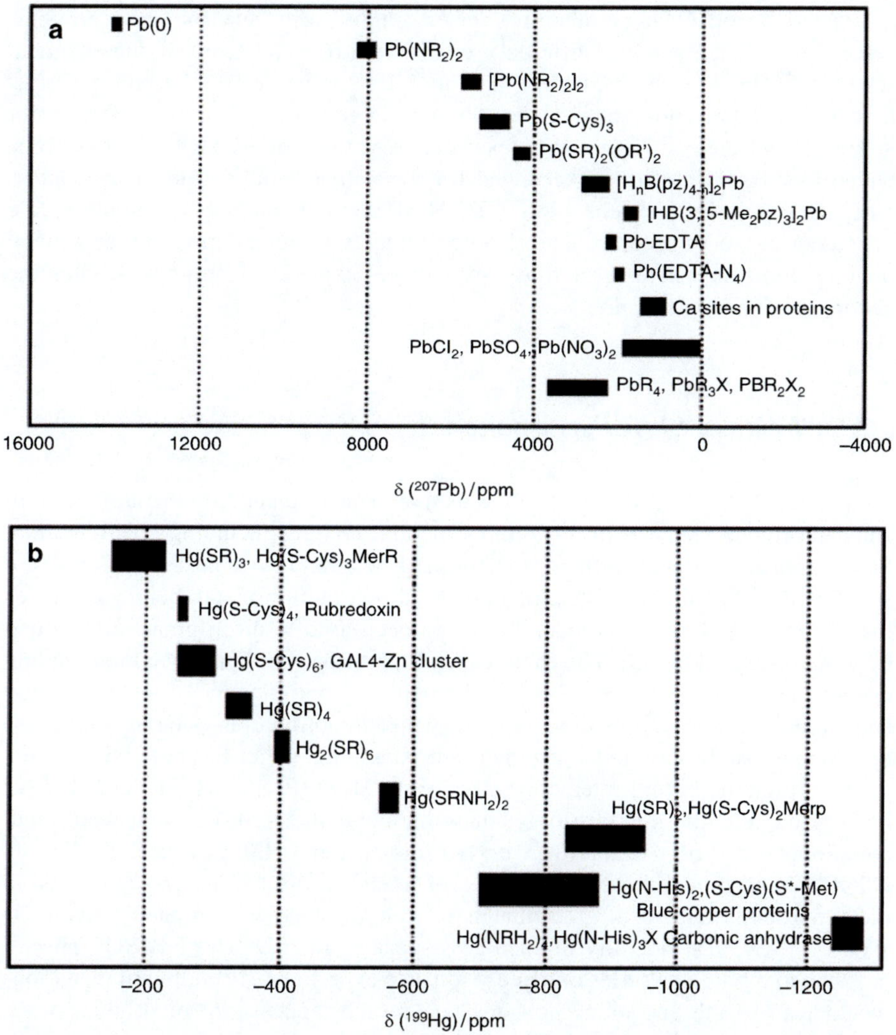

Fig. 5 (**a**) ^{207}Pb chemical shifts of various lead model complexes and ^{207}Pb-substituted proteins relative to external 1 M [$Pb(NO_3)_2$] in 99.9% D_2O pH* 3.3. The pz represents the pyrazolyl ring and *S*-Cys represents a thiolate from cysteine (adapted from [87, 89]). (**b**) ^{199}Hg chemical shifts of aliphatic amine/thiol model complexes and ^{199}Hg-substituted proteins relative to $Hg(CH_3)_2$ at 298 K. The NRH_2 represents a primary amine and SR represents a thiolate, and the coordination environments of the Hg proteins include His imidazole nitrogen, Cys thiolate S-Cys, and Met thioether S*-Met [90]

coordination geometry. For examples, ^{199}Hg bound blue copper proteins gave rise to signals at ca. −880 ppm for azurin, −749 ppm for plastocyanin, and −706 ppm for rusticyanin [93, 95]. These blue proteins have very similar coordination environments, e.g., two histidines and one cysteine and the fourth weakly associated ligand (Met or others) which is slightly different for these proteins. Nevertheless, such a subtle difference can be faithfully reflected by the chemical shifts of ^{199}Hg. The chemical shift map of ^{199}Hg can be derived based on various model complexes and proteins; Fig. 5b. The large chemical shift dispersion for ^{199}Hg allows clear differentiation between a variety of $M(SR)_n$ environments. Given that Hg^{2+} is readily exchanged for the native metal ion in many copper, zinc, and iron metalloproteins [96], ^{199}Hg NMR methods can play an important role in structural, spectroscopic, and chemical studies of metalloproteins and metal binding domains where the tertiary structure of the folded proteins dictates the geometry of the metal ion.

3.2 Chemical Shift Perturbation

Protein NMR chemical shift is highly sensitive to the exact environment of the atom and can provide valuable insights into structural features including metal ligation. For examples, chemical shifts of $^{13}C^{\beta}$ for the zinc bound cysteines (ca. 34 ppm) are significantly downfield shifted relative to those of non-metal bound cysteines (ca. 27 ppm) [97]. Such an index has often been used to discriminate zinc bound cysteine residues [15, 98]. The chemical shift perturbations upon metal ion binding can be used to estimate the affinity, stoichiometry, and kinetics of metal binding, and, moreover, they can be used to identify metal coordination environments. This approach is usually denoted as chemical shift mapping, which has been widely used to study protein–protein interaction [99, 100]. ^{1}H NMR and ^{1}H-^{1}H TOCSY have been used to identify the types of residues binding to metal ions for unlabeled small metalloproteins [101–103]. However, two dimensional HSQC, especially ^{1}H-^{15}N HSQC, is often employed in studies of metal coordination environments in metalloproteins due to the fact that it is well resolved in comparison with 2D ^{1}H-^{13}C HSQC. The identity of each cross-peak in the 2D ^{1}H-^{15}N HSQC spectra is assigned based on a series of triple resonance experiments. Binding of metal ions would lead to the appearance of new peaks or disappearance of original peaks depending on the exchange rates of the apo- and metal-bound forms on the NMR time scales [99, 104]. The chemical shifts perturbation (CSP) can be followed in titration experiments, where the concentration of diamagnetic metal ion is increased gradually. When heteronuclear data are available, the binding site is usually predicated by the combined chemical shift perturbation $\Delta\delta_{\mathrm{comb}}$, which has been shown to be a more reliable approach to evaluate titration data quantitatively [105]. Although several approaches are available to obtain this value as summarized previously [105], in practice, the weighting for ^{1}H and ^{15}N is considered to be the same, and the $\Delta\delta_{\mathrm{comb}}$ is usually quantified by the following equation:

$$\Delta\delta_{\mathrm{comb}} = \sqrt{\frac{\Delta\delta_{\mathrm{HN}}^2 + \frac{1}{25}\Delta\delta_N^2}{2}}.$$

Chemical shift mapping provides information about the location of the metal binding sites and has been used to identify the metal coordination for numerous metalloproteins [15, 106–108]. For example, addition of Cu^{+} to human Cox17 induced significant chemical shift variations over residues Lys20 and Ala24, and the appearance of the NH signals of Cys22–Cys23, which is thought to serve as the Cu^{+} binding motif [107]. However, this approach usually has to be used in combination with other physical techniques or biological approaches, e.g., mutagenesis to specify the metal binding sites, since the chemical shift perturbations mainly stem from either direct binding or conformational changes caused by the metal ions. Binding of Ni^{2+} to *H. pylori* HypA led to disappearance of signals of Glu3 and Asp40 in the 2D ^{1}H-^{15}N HSQC spectrum. When combining with mutagenesis study, side-chain 2D ^{1}H-^{15}N HMQC, UV absorption, as well as CD, it was proposed that Ni^{2+} coordinates with His2 (side-chain N^{δ}), His2 (backbone N), and the backbone nitrogens of Glu3, and Asp40 with a square-planar geometry [15]. Such a binding also induced structural changes which were thought to be important for its downstream receptor's recognition [15].

Histidine often serves as a metal binding ligand in metalloproteins and can provide both backbone and side nitrogens to coordinate with metal ions such as $Zn^{2+}$and Ni^{2+}. It has been shown that different tautomeric forms of histidine imadazole rings have different, distinguishable signal patterns in a long-range 2D ^{1}H-^{15}N HMQC spectrum [109], metal coordination often causes recognizable changes in the NMR spectrum of histidine side-chains, and imadazole nitrogen atoms involved in direct metal coordination have specific chemical shift [15, 40, 98, 110]. This technique has been extensively used to identify Zn^{2+} binding. The chemical shifts observed for the unprotonated imadazole nitrogen atoms of a zinc finger domain Hdm2(429–491) appeared at ca. 215 ppm. The cross-peak pattern in ^{1}H-^{15}N HMQC spectrum (with $^{2}J_{HN}$) of zinc bound Hdm2 (429–491) showed that His452 and His457 assumed different tautomeric forms with the former being $N^{\varepsilon 2}$-protonated and the latter $N^{\delta 1}$-protonated with chemical shifts around 170 ppm, which demonstrated that Zn^{2+} coordinated to both His452 and His457 via the $N^{\varepsilon 2}$ and $N^{\delta 1}$ respectively [98]. More interestingly, a comparison of the 2D ^{1}H-^{15}N HMQC spectra of Zn^{2+}-bound proteins with $^{113}Cd^{2+}$-bound proteins, as shown in Fig. 6, allows one to observe the ^{15}N-^{113}Cd coupling, which assists identification of overlapping of histidine side-chains that bind to metal ions [40]. The HMQC spectrum of the zinc-bound domain, shown as black in Fig. 6a, clearly shows two of the ^{15}N resonances His42 and His40 shifted downfield as a result of zinc coordination. Based on the pattern of the cross-peaks, His42 is in the ε tautomeric state indicative of zinc coordination to the $N^{\delta 1}$ of His42; Fig. 6b. Such a method cannot be used to assign the nitrogen atom of His40 due to overlapping of the $H^{\delta 2}$ and $H^{\varepsilon 1}$ resonances of His40. The HMQC spectrum of ^{113}Cd bound protein, shown as red in Fig. 6a, clearly shows coupling between ^{113}Cd and the $N^{\delta 1}$ of His42 observed on $H^{\varepsilon 1}$, suggesting a covalent bond between them. Importantly, the pattern of the connectivities and the coupling observed to both $H^{\delta 2}$ and $H^{\varepsilon 1}$ from the ^{113}Cd allows the unambiguous assignment of the δ tautomer for His40, with the metal coordination via the $N^{\varepsilon 2}$ of the side-chain; Fig. 6b [40]. The side-chain 2D ^{1}H-^{15}N HMQC spectrum of histidine has also been used to identify Ni^{2+} binding

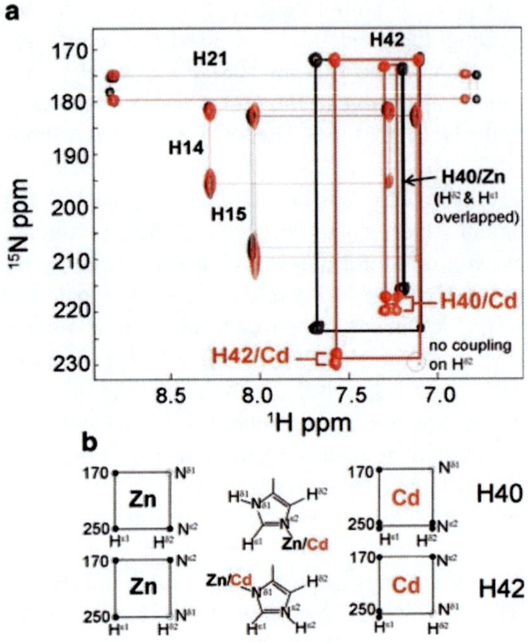

Fig. 6 (**a**) Comparison of the 500-MHz ^{1}H-^{15}N HMQC spectrum of Zn-bound ZZ domain in *black* with the ^{113}Cd-bound form in *red*. The coupling of ^{113}Cd to $N^{\varepsilon 2}$ is observed on $H^{\delta 2}$ and $H^{\varepsilon 2}$ for His40; whereas ^{113}Cd coupling to $N^{\delta 1}$ is observed on $H^{\varepsilon 1}$ for His 42. (**b**) Diagrams showing the connectivities of two zinc-ligand histidine residues H40 and H42 based on the above HMQC spectrum. The coupling of ^{113}Cd to the attached ^{15}N is useful for identification of metal bound histidine residues and the location of the metal coordination on the histidine rings (adapted from [40])

atoms in *H. pylori* HypA and significant downfield shift of His2 upon Ni^{2+} binding indicates that Ni^{2+} binds to His2 through the $N^{\delta 1}$ atom [15].

4 NMR in Studies of Paramagnetic Metalloproteins

Metalloproteins represent a large share of a proteome. A large number of them contain paramagnetic metal ions, which possess unpaired electrons. The presence of a paramagnetic center causes pronounced effects in NMR spectra and reduces dramatically the intensity of NOEs and the efficiency in the transfer of scalar couplings both in homonuclear and heteronuclear experiments, which hampers spectrum assignment and structural determination through standard approaches. However, with the advances in novel experiment design and development of software protocol in recent years, the presence of paramagnetic centers has been used as a precious source of structural information [19, 20]. Extensive reviews regarding this topic can be referred to [19–22, 111–113]. Here we will give a very brief description of paramagnetism-based restraints as well as application of ^{13}C-detected experiments in the structural examination of paramagnetic metalloproteins.

4.1 Paramagnetism-Based Structural Restraints

In paramagnetic systems, where paramagnetic metal ions are either intrinsic or extrinsic, there are three NMR experimental observables that yield long-range structural information, i.e., paramagnetic resonance enhancement (PRE), pseudocontact shifts (PCS), and RDC induced by anisotropic paramagnetic centers. In addition, cross-correlated relaxation (CCR) effects between anisotropic paramagnetic centers and anisotropic parameters of the nuclear spins can also be exploited to generate long-range restraints [19, 111, 112]. Paramagnetic centers with isotropic electron spin distribution (Mn^{2+} and Gd^{3+}) produce large PREs due to slow electron relaxation. In contrast, paramagnetic centers with anisotropic electron spin distribution for most paramagnetic metal ions, including most of the lanthanides, create all four long-range paramagnetic effects, which contain rich structural information [114]. Here, we will focus on PRE and PCS and their applications. Information about RDC can be found in this book.

The PRE arises from magnetic dipolar interactions between a nucleus and the unpaired electrons of the paramagnetic center, resulting in an increase in nuclear relaxation rates. In contrast to NOE, where the effects are limited to short range interaction (<6 Å), the PRE effects are relatively large and can be detected up to 35 Å owing to the large magnetic moment of an unpaired electron. There are two mechanisms, i.e., the Solomon mechanism and the Curie spin mechanism, that give the PREs, with the former being predominant for slowly tumbling molecules with long lifetimes of electronic spin state (such as Mn^{2+} and Gd^{3+}). However, the Curie relaxation becomes important when the electronic relaxation is much faster than the rotational tumbling of the molecules, which is the case for the majority of paramagnetic metal ions. Theoretical and experimental aspect of PRE as well as its application in studies of structures of proteins and protein–protein complexes can be found in recent reviews [114–116].

At high magnetic fields (over 500 MHz for ^{1}H frequency), the PRE rate, Γ_2, which arises from the dipole–dipole interaction between a nucleus and unpaired electrons with an isotopic g-tensor, is conventionally calculated by the Solomon–Bloembergen (SB) equation:

$$\Gamma_2 = \frac{1}{15}\left(\frac{\mu_0}{4\pi}\right)^2 \gamma_I^2 g^2 \mu_B^2 S(S+1) r^{-6} \left\{ 4\tau_c + \frac{3\tau_c}{1+(\omega_H \tau_c)^2} \right\}$$

where r is the distance between the paramagnetic center and the observed nucleus, μ_0 the permeability of vacuum, γ_I the nuclear gyromagnetic ratio, g the electron g-factor, μ_B the electron Bohr magneton, S the electron spin quantum number, and τ_c the PRE total correlation time. In practice, Γ_2 is measured as a difference in transverse relaxation rates between the paramagnetic ($R_{2,para}$) and diamagnetic ($R_{2,dia}$) states. A two-time point measurement is recommended as a simple approach for obtaining Γ_2 rates and their corresponding errors without fitting procedures. In this approach, the Γ_2 rates are determined from two time points ($T = 0$ and ΔT) for transverse relaxation as shown by the following equation [116]:

$$\Gamma_2 = R_{2,para} - R_{2,dia} = \frac{1}{T_b - T_a} \ln \frac{I_{dia}(T_b) I_{para}(T_a)}{I_{dia}(T_a) I_{para}(T_b)}$$

where I_{dia} and I_{para} are the peak intensities for the diamagnetic and paramagnetic states, respectively. The choice of time points is important to minimize the error. For example, if the range of Γ_2 rates is 0–75 s^{-1}, a second time point T_b should be at ca. $1.15/(R_{2,dia} + 50)$ s, representing a reasonable choice [115, 116].

For isotropic metal ions such as Mn^{2+} and Gd^{3+}, the Curie-spin relaxation that could potentially exhibit significant cross-correlation with other relaxation mechanisms is negligible for medium-size macromolecules [116], and the Solomon relaxation is predominant. PRE analysis in such a system is thus simple. The PRE has been used extensively in metalloproteins that possess a rigid intrinsic paramagnetic center [117–119]. Such a strategy has also been extended not only in the NMR structure determination of non metalloproteins [120–122], in which paramagnetic metal ions (Mn^{2+} or Gd^{3+}) or nitroxide radicals were conjugated through appropriate chemical modification [123], but also in the characterization of protein–protein/nucleic acid complexes [124–126] and membrane-proteins [127], in particular in transient macromolecular interactions [115, 128–131].

PCSs are precious sources of structure information and are observed only in paramagnetic systems with anisotropic unpaired electrons, e.g., Dy^{3+}, Tb^{3+}, and Fe^{3+}. The magnitude of the PCS, δ^{pcs}, is calculated using the following equation [132]:

$$\delta^{pcs} = \frac{1}{12\pi} r^{-3} \left\{ \Delta\chi_{ax}(3\cos^2\theta - 1) + \frac{3}{2}\Delta\chi_{rh}\sin^2\theta \cos 2\varphi \right\}$$

$$\Delta\chi_{ax} = \Delta\chi_{zz} - \frac{1}{2}(\chi_{xx} + \chi_{yy}) \text{ and } \Delta\chi_{rh} = \chi_{xx} - \chi_{yy}$$

where r is the distance between the metal ion and the nuclear spin, θ and φ are the angles describing the position of the nuclear spin with respect to the principle axes of the magnetic susceptibility tensor χ, and $\Delta\chi_{ax}$ and $\Delta\chi_{rh}$ are the axial and rhombic components, respectively, of the magnetic susceptibility tensor.

The PCS are manifested by large changes in chemical shifts of the nuclear spins that are exposed to the paramagnetic metal ions and arise from through-space dipolar interactions with rapidly relaxing unpaired electrons. The PCS displays an r^{-3} distance dependence, in contrast to the r^{-6} dependence for the PRE, which results in a relatively long distance range for the PCS to be detected (ca. 40 Å for Dy^{3+}) [133]. In general, the δ^{pcs} values can be measured after the complete assignment is obtained for the ^1H-^{15}N HSQC spectra of the both the diamagnetic and the paramagnetic samples, and are calculated as the difference between the chemical-shift values observed for the nuclei in a paramagnetic system and in a diamagnetic analog. The δ^{pcs}-derived restraints alone cannot be used to solve the structures. Instead, the PCSs have to be incorporated with NOEs and dihedral-angle restraints to determine structures of proteins or to refine protein structures. The first example of using PCS in a structure refinement was reported on a low-spin Fe^{3+} heme protein [134]. Such a strategy has been extended not only in studies of paramagnetic proteins [135, 136], but also in non-metal binding proteins [137, 138], which were labeled by paramagnetic metal ions such as lanthanides [114, 139] or genetically encoded Co^{2+}-binding amino acid [140]. Moreover, PCSs can also be used as restraints in molecular dynamics [141, 142].

4.2 ^{13}C-Detected Protonless NMR

Direct detection of heteronuclei, in particular ^{13}C, offers a valuable alternative to ^{1}H detection in the study of biological macromolecules [143–145] as well as paramagnetic proteins [136, 145–147]. The recent development of high magnetic fields as well as the availability of cryogenically cooled probe heads has improved ^{13}C sensitivity significantly, which has greatly stimulated research using ^{13}C-detected experiments on enriched samples to study biological macromolecules [148]. The ^{13}C detection takes advantage of the slower relaxing ^{13}C spins and overcomes the drawbacks produced by the fast ^{1}H transverse relaxation, which leads ^{1}H signals to broaden beyond detection limits for large proteins. Such an approach is particularly useful in paramagnetic systems since ^{13}C direct detection is less affected by the paramagnetic center than ^{1}H owing to the lower ^{13}C gyromagnetic ratio, which decreases the paramagnetic dipolar contributions to its relaxation by a factor of around 16 $(\gamma_C/\gamma_H)^2$ [136, 146, 149].

In ^{13}C direct detection experiments, several approaches were used to achieve "virtual" decoupling (to remove homonuclear one-bond carbon–carbon couplings) such as IPAP schemes (in-phase anti-phase) [150–152], in which two FIDs for each increment are recorded and stored separately, one for in-phase and another for anti-phase, the two components being combined to remove the splitting. An alternative is S^{3}E schemes [152, 153] (spin-state selective excitation), in which two different experiments are performed with one being absorptive and another dispersive. One or more of these building blocks (IPAP and S^{3}E) can be implemented in any experiments based on ^{13}C direct detection. A set of ^{13}C based experiments, which can be used for the assignment of backbone and side-chains of ^{13}C/^{15}N labeled proteins, is now available and summarized in a recent review [144]. The sequence-specific assignment was achieved by CACO and CANCO, which provide the correlation of each CO to the two neighboring C^{α} nuclei; CACO, CBCACO, and ^{13}C-^{13}C TOCSY can provide spin-system assignment [143, 154, 155]. The CON-IPAP experiment is used to correlate backbone nitrogen with CO through the one-bond C′-N coupling [149]. The ^{13}C-^{13}C NOESY experiments based on dipole–dipole interaction with longitudinal magnetization transfer represent a valuable alternative to COSY experiments based on scalar couplings to detect C–C one-bond correlation for large macromolecules and paramagnetic metalloproteins [149, 156, 157]. The ^{13}C-^{13}C NOESY with direct ^{13}C detection is exploited as a valuable tool to extend the assignment to side-chains in large molecules such as $C'–C^{\beta}$ if mixing times is long enough (e.g., 800 ms) [156]. The correlations between nuclei not directly bound and not mediated by spin diffusion in ^{13}C-^{13}C NOESY spectra are identified, which would represent a breakthrough of structure determination of large macromolecules in solution by providing distance constraints.

However, the sensitivity for ^{13}C direct detection is required to be improved before long-range correlations to obtain ^{13}C–^{13}C distance constraints can be used in structural characterization of large macromolecules.

^{13}C direct detection has been successfully applied to paramagnetic proteins, where the contribution to line broadening coming from the paramagnetic center is so large that ^{1}H signals around the metal ion are beyond detectable limits [136, 144, 147, 149, 158]. Such a technique can also be used in generation of paramagnetism-based restraints including PCS, PRE, and RDC [148, 159–161]. It has been demonstrated that ^{13}C directly detected spectra provide an alternative method for the measurement of RDC with precision as good as that from ^{1}H detection, but with additional advantage for measuring those broad resonances in ^{1}H detection [148]. Direct detection of ^{13}C intrinsically offers a way to detect resonances close to the metal ion where ^{1}H resonances are too broad to be detected. Indeed, with the aid of a ^{13}C direct detection approach, ^{13}C resonances as close as 6 Å from the metal ion are detected for CopC, a Cu^{2+} binding protein involved in copper homeostasis, whereas no ^{1}H resonance can be detected within a sphere of 11 Å from the metal due to fast relaxation caused by paramagnetic Cu^{2+} [136]. Incorporation of heteronuclear paramagnetism-based restraints, e.g., PCSs and longitudinal relaxation rate enhancement, allows CopC structures to be resolved with the RMSD of Cu^{2+} determined only by the paramagnetism-based constraints of 1.1 Å [136]. The ^{13}C direct detection technique has also been used for residue-specific assignments of resonances, in particular those near paramagnetic centers (e.g., Ni^{2+} and Fe^{3+}) such as in a 20-kDa Ni-containing enzyme, acireductone dioxyhenase (ARD) [162], and oxidized human [2Fe-2S] ferredoxin [146], as well as a 19-kDa Fe^{3+} hemophore HasA [147]. In many paramagnetic systems, the longitudinal relaxation rates are influenced to a smaller extent than the transverse relaxation rates. The ^{13}C–^{13}C NOESY experiments are therefore a useful approach to overcome the quench of scalar coupling based transfer, in particular for large macromolecules. The use of ^{13}C direct detected experiments, e.g., ^{13}C-^{13}C COSY, ^{13}C-^{13}C NOESY, and ^{13}C-^{13}C COCAMQ, allows ^{13}C signals as close as 4 Å to Cu^{2+} to be detected in oxidized monomeric copper/zinc superoxide dismutase (SOD) [149]. The advantage of ^{13}C-^{13}C NOESY experiments for higher molecular weights was seen by comparison of the protein SOD [149, 156]. All of the expected Cα–CO connectivities were detected with higher intensity in the dimeric protein than in the monomeric state. In addition, most of the two bond CO–Cβ cross-peaks were observable for the dimeric SOD when the long mixing times were used [156]. Interestingly, the intrinsic asymmetry of a ^{13}C-^{13}C COSY experiment allows the coordinating residues of paramagnetic metal ions to be identified easily, providing a unique method to distinguish between monodentate and bidentate coordinating side-chain carbonyls [163]. Significantly, the ^{13}C-based strategy in combination with solid-state NMR led to partial sequence-specific (35%) and side-chain assignments for the iron storage protein, ferritin, a very large protein with a molecular mass of 480 kDa and 24 subunits [164, 165]. The solution ^{13}C-^{13}C NOESY spectra for side-chain observation has provided the identification of an iron channel that guides the direction transport of the multimeric

Fe^{3+} products from the active site toward the nanocage. The interior of the four-helix bundle is identified as the functional channel based on the observed paramagnetic effects on residues lining the internal face of the four-helix bundle. The NMR data provide a basis for the pathway of iron from the ferrous/dioxygen oxidoreductase site to the central cavity of ferritin [164]. Such studies open new avenues for the application of ^{13}C direct detection experiments to systems with molecular assemblies larger than 100 kDa.

5 Perspectives

Since the first protein solution structure was determined by high resolution NMR spectroscopy about 25 years ago [166], NMR has been established as the only experimental method that provides both structural and dynamical information at atomic resolution close to physiologically relevant conditions. Protein structure determination in living cells has also been achieved recently by in-cell NMR [18]. However, the limitation of this technique in structural studies lies in low sensitivities and poor resolution when the size of macromolecules increases. Some proteins, in particular metalloproteins, might not be stable for a period of time (days or weeks) or have limited solubility. Moreover, new challenges in life science have also promoted development of new NMR methods which will improve sensitivities and reduce acquisition times to fulfil the requirement of characterization of these proteins and their complexes.

Enormous effort has been made to improve NMR instrumentation in terms of experimental sensitivity, which results in availability of high-field magnets, cryogenically cooled probes. In the meantime, tremendous advances in methodology have contributed to an increased interest in the study of molecular systems of increasing size and complexity. The introduction of a nonlinear sampling scheme (instead of conventional uniform sampling) allows a very fast acquisition of multidimensional NMR [167–169]. Many schemes have been developed to reduce the spectral dimensionality and thus to speed up the experiments, which enables quick assignment of large proteins [170, 171]. The examples include the G-matrix Fourier transform ("GFT") NMR approach where sub-spectra from joint sampling of indirect dimensions are linearly recombined and analyzed [172]. In the projection reconstruction ("PR") method, the corresponding full-dimensional spectrum is reconstructed [171, 173, 174]. Moreover, various ultrafast NMR techniques including SOFAST/BEST NMR [175, 176] and Hadmard NMR [177] are also available for studying biomolecules even in real time. All these new schemes deliver appreciable improvement in the speed of data acquisition and show promise for speeding up multidimensional NMR of normal size proteins [170, 178, 179] and very large proteins [180, 181] as well as sequence assignment for intrinsically unstructured proteins [178].

^{13}C NMR spectroscopy is emerging as a powerful tool to complement ^{1}H NMR spectroscopy in the investigation of biomolecules, in particular for large molecules and paramagnetic metalloproteins and also for the study of short-lived molecules

[182]. However, data acquisition time is rather long even for samples with high concentrations (ca. mM). Implementation of fast NMR methods such as non-uniform sampling in the indirect dimension significantly reduced experimental times [183]. Such a strategy will open new avenues to applications of ^{13}C NMR to take advantage of the favorable heteronuclear chemical shift dispersion in biological systems, especially for systems of increasing size.

Chemical shifts of selective nuclei ($^{113/111}$Cd, ^{209}Pb, and ^{195}Pt) are sensitive towards types of ligands (N, O, and S), numbers, and geometries, and will continue to play a role in characterization of metal–protein local coordination. The ultra-field NMR facilities direct observation of biologically important metal ions with half integer, quadrupolar nuclei (e.g., ^{67}Zn and ^{25}Mg) [184, 185].

In spite of the availability of all these new techniques, their applications to metalloproteins are currently sparse. There is an urgent need to promote these advanced techniques in the scientific community through introducing integrated software packages for experimental set-up, data processing, and analysis. This will enable protein chemists and bioinorganic chemists, who are not NMR experts, to employ the new techniques in their research. The combination of fast NMR techniques, ^{13}C directly detected NMR with paramagnetic NMR, will offer great possibilities in tackling new challenges in life science and will open new avenues for NMR spectroscopy to be utilized not only in the characterization of single biomolecules, e.g., structural and dynamical studies of proteins/metalloproteins and paramagnetic proteins, short-lived macromolecules, and intrinsically unstructured proteins, but also in the investigation of more complex systems to give an integrated view of interacting molecular networks. In particular, ultra-fast NMR opens up new perspectives for NMR structural investigations of unstable protein/metalloprotein samples and real-time site-resolved studies of protein kinetics or monitoring folding/unfolding processes of proteins/metalloproteins caused by ligand or metal binding/release.

Acknowledgements This work is supported by the Research Grants Council of Hong Kong (HKU7043/06P, HKU2/06C, HKU7042/07P, HKU1/07C, HKU7038/08P, HKU7049/09P and N-HKU752/09), Croucher Foundation.

References

1. Ge R, Sun X, Gu Q et al (2007) A proteomic approach for the identification of bismuth-binding proteins in Helicobacter pylori. J Biol Inorg Chem 12:831–842
2. Sun XS, Tsang CN, Sun HZ (2009) Identification and characterization of metallodrug binding proteins by (metallo)proteomics. Metallomics 1:25–31
3. Sun HZ, Chai ZF (2010) Metallomics: an integrated science for metals in biology and medicine. Annu Rep Prog Chem A Inorg Chem 106:20–38
4. Waldron KJ, Robinson NJ (2009) How do bacterial cells ensure that metalloproteins get the correct metal? Nat Rev Microbiol 7:25–35
5. Cvetkovic A, Menon AL, Thorgersen MP et al (2010) Microbial metalloproteomes are largely uncharacterized. Nature 466:779–782
6. Andreini C, Bertini I, Cavallaro G et al (2008) Metal ions in biological catalysis: from enzyme databases to general principles. J Biol Inorg Chem 13:1205–1218

7. Lu Y, Yeung N, Sieracki N et al (2009) Design of functional metalloproteins. Nature 460:855–862
8. Andreini C, Bertini I, Rosato A (2009) Metalloproteomes: a bioinformatic approach. Acc Chem Res 42:1471–1479
9. Shu N, Zhou T, Hovmoller S (2008) Prediction of zinc-binding sites in proteins from sequence. Bioinformatics 24:775–782
10. Kasampalidis IN, Pitas I, Lyroudia K (2007) Conservation of metal-coordinating residues. Proteins 68:123–130
11. Wuthrich K (1989) Protein structure determination in solution by nuclear magnetic resonance spectroscopy. Science 243:45–50
12. Montelione GT, Arrowsmith C, Girvin ME et al (2009) Unique opportunities for NMR methods in structural genomics. J Struct Funct Genomics 10:101–106
13. Banci L, Bertini I, Luchinat C et al (2010) NMR in structural proteomics and beyond. Prog Nucl Magn Reson Spectrosc 56:247–266
14. Wang H, Li H, Cai B et al (2008) The effect of nitric oxide on metal release from metallothionein-3: gradual unfolding of the protein. J Biol Inorg Chem 13:411–419
15. Xia W, Li H, Sze KH et al (2009) Structure of a nickel chaperone, HypA, from *Helicobacter pylori* reveals two distinct metal binding sites. J Am Chem Soc 131:10031–10040
16. Tugarinov V, Muhandiram R, Ayed A et al (2002) Four-dimensional NMR spectroscopy of a 723-residue protein: chemical shift assignments and secondary structure of malate synthase G. J Am Chem Soc 124:10025–10035
17. Gautier A, Mott HR, Bostock MJ et al (2010) Structure determination of the seven-helix transmembrane receptor sensory rhodopsin II by solution NMR spectroscopy. Nat Struct Mol Biol 17:768–774
18. Sakakibara D, Sasaki A, Ikeya T et al (2009) Protein structure determination in living cells by in-cell NMR spectroscopy. Nature 458:102–105
19. Bertini I, Luchinat C, Parigi G et al (2005) NMR spectroscopy of paramagnetic metalloproteins. Chembiochem 6:1536–1549
20. Bertini I, Luchinat C, Parigi G et al (2008) Perspectives in paramagnetic NMR of metalloproteins. Dalton Trans 3782–3790
21. Bertini I, Luchinat C, Piccioli M (2001) Paramagnetic probes in metalloproteins. Methods Enzymol 339:314–340
22. Bertini I, Luchinat C (1999) New applications of paramagnetic NMR in chemical biology. Curr Opin Chem Biol 3:145–151
23. Ronconi L, Sadler PJ (2008) Applications of heteronuclear NMR spectroscopy in biological and medicinal inorganic chemistry. Coord Chem Rev 252:2239–2277
24. Rule GS, Hitchens TK (2006) Fundamentals of protein NMR spectroscopy. Springer, Dordrecht
25. Clore GM, Gronenborn AM (1999) In: Krishna NR, Berliner LJ (eds) Biological magnetic resonance. Kluwer Academic/Plenum Publishers, New York
26. Wuthrich K (2003) NMR studies of structure and function of biological macromolecules (Nobel lecture). Angew Chem Int Ed Engl 42:3340–3363
27. Wang H, Zhang Q, Cai B et al (2006) Solution structure and dynamics of human metallothionein-3 (MT-3). FEBS Lett 580:795–800
28. Capasso C, Carginale V, Crescenzi O et al (2003) Solution structure of MT_nc, a novel metallothionein from the Antarctic fish Notothenia coriiceps. Structure 11:435–443
29. Ge R, Watt RM, Sun X et al (2006) Expression and characterization of a histidine-rich protein, Hpn: potential for Ni^{2+} storage in *Helicobacter pylori*. Biochem J 393:285–293
30. Williamson MP, Craven CJ (2009) Automated protein structure calculation from NMR data. J Biomol NMR 43:131–143
31. Guntert P (2009) Automated structure determination from NMR spectra. Eur Biophys J 38:129–143
32. Lopez-Mendez B, Guntert P (2006) Automated protein structure determination from NMR spectra. J Am Chem Soc 128:13112–13122

33. Kuboniwa H, Grzesiek S, Delaglio F et al (1994) Measurement of HN-H alpha J couplings in calcium-free calmodulin using new 2D and 3D water-flip-back methods. J Biomol NMR 4:871–878
34. Dux P, Whitehead B, Boelens R et al (1997) Measurement of ^{15}N-^{1}H coupling constants in uniformly ^{15}N-labeled proteins: application to the photoactive yellow protein. J Biomol NMR 10:301–306
35. Cornilescu G, Delaglio F, Bax A (1999) Protein backbone angle restraints from searching a database for chemical shift and sequence homology. J Biomol NMR 13:289–302
36. Cun S, Sun H (2010) A zinc-binding site by negative selection induces metallodrug susceptibility in an essential chaperonin. Proc Natl Acad Sci USA 107:4943–4948
37. Scrofani SD, Wright PE, Dyson HJ (1998) The identification of metal-binding ligand residues in metalloproteins using nuclear magnetic resonance spectroscopy. Protein Sci 7:2476–2479
38. Bordiga S, Bonino F, Lillerud KP et al (2010) X-ray absorption spectroscopies: useful tools to understand metallorganic frameworks structure and reactivity. Chem Soc Rev 39:4885–4927
39. Herrmann T, Guntert P, Wuthrich K (2002) Protein NMR structure determination with automated NOE assignment using the new software CANDID and the torsion angle dynamics algorithm DYANA. J Mol Biol 319:209–227
40. Legge GB, Martinez-Yamout MA, Hambly DM et al (2004) ZZ domain of CBP: an unusual zinc finger fold in a protein interaction module. J Mol Biol 343:1081–1093
41. Blindauer CA, Harrison MD, Parkinson JA et al (2001) A metallothionein containing a zinc finger within a four-metal cluster protects a bacterium from zinc toxicity. Proc Natl Acad Sci USA 98:9593–9598
42. Arseniev A, Schultze P, Worgotter E et al (1988) Three-dimensional structure of rabbit liver [Cd_7]metallothionein-2a in aqueous solution determined by nuclear magnetic resonance. J Mol Biol 201:637–657
43. Kagi JH (1991) Overview of metallothionein. Methods Enzymol 205:613–626
44. Thirumoorthy N, Manisenthil Kumar KT, Shyam Sundar A et al (2007) Metallothionein: an overview. World J Gastroenterol 13:993–996
45. Oz G, Pountney DL, Armitage IM (1999) In: Klaassen CD (ed) Metallothionein IV. Birkhauser Verlag, Basel
46. Furey WF, Robbins AH, Clancy LL et al (1986) Crystal structure of Cd, Zn metallothionein. Science 231:704–710
47. Palmiter RD, Findley SD, Whitmore TE et al (1992) MT-III, a brain-specific member of the metallothionein gene family. Proc Natl Acad Sci USA 89:6333–6337
48. Oz G, Zangger K, Armitage IM (2001) Three-dimensional structure and dynamics of a brain specific growth inhibitory factor: metallothionein-3. Biochemistry 40:11433–11441
49. Blindauer CA (2008) Metallothioneins with unusual residues: histidines as modulators of zinc affinity and reactivity. J Inorg Biochem 102:507–521
50. Berg JM, Shi Y (1996) The galvanization of biology: a growing appreciation for the roles of zinc. Science 271:1081–1085
51. Lee MS, Gippert GP, Soman KV et al (1989) Three-dimensional solution structure of a single zinc finger DNA-binding domain. Science 245:635–637
52. Stoll R, Lee BM, Debler EW et al (2007) Structure of the Wilms tumor suppressor protein zinc finger domain bound to DNA. J Mol Biol 372:1227–1245
53. Cornilescu CC, Porter FW, Zhao KQ et al (2008) NMR structure of the mengovirus Leader protein zinc-finger domain. FEBS Lett 582:896–900
54. Kotaka M, Johnson C, Lamb HK et al (2008) Structural analysis of the recognition of the negative regulator NmrA and DNA by the zinc finger from the GATA-type transcription factor AreA. J Mol Biol 381:373–382
55. Chou CC, Lou YC, Tang TK et al (2010) Structure and DNA binding characteristics of the three-Cys(2)His(2) domain of mouse testis zinc finger protein. Proteins 78:2202–2212

56. Brown RS (2005) Zinc finger proteins: getting a grip on RNA. Curr Opin Struct Biol 15:94–98
57. Laity JH, Lee BM, Wright PE (2001) Zinc finger proteins: new insights into structural and functional diversity. Curr Opin Struct Biol 11:39–46
58. Wolfe SA, Nekludova L, Pabo CO (2000) DNA recognition by Cys2His2 zinc finger proteins. Annu Rev Biophys Biomol Struct 29:183–212
59. Lee S, Doddapaneni K, Hogue A et al (2010) Solution structure of Gfi-1 zinc domain bound to consensus DNA. J Mol Biol 397:1055–1066
60. Eustermann S, Brockmann C, Mehrotra PV et al (2010) Solution structures of the two PBZ domains from human APLF and their interaction with poly(ADP-ribose). Nat Struct Mol Biol 17:241–243
61. Isogai S, Kanno S, Ariyoshi M et al (2010) Solution structure of a zinc-finger domain that binds to poly-ADP-ribose. Genes Cells 15:101–110
62. Hudson BP, Martinez-Yamout MA, Dyson HJ et al (2004) Recognition of the mRNA AU-rich element by the zinc finger domain of TIS11d. Nat Struct Mol Biol 11:257–264
63. He Y, Imhoff R, Sahu A et al (2009) Solution structure of a novel zinc finger motif in the SAP30 polypeptide of the Sin3 corepressor complex and its potential role in nucleic acid recognition. Nucleic Acids Res 37:2142–2152
64. He F, Umehara T, Saito K et al (2010) Structural insight into the zinc finger CW domain as a histone modification reader. Structure 18:1127–1139
65. Watanabe S, Arai T, Matsumi R et al (2009) Crystal structure of HypA, a nickel-binding metallochaperone for [NiFe] hydrogenase maturation. J Mol Biol 394:448–459
66. Cavalli A, Salvatella X, Dobson CM et al (2007) Protein structure determination from NMR chemical shifts. Proc Natl Acad Sci USA 104:9615–9620
67. Shen Y, Lange O, Delaglio F et al (2008) Consistent blind protein structure generation from NMR chemical shift data. Proc Natl Acad Sci USA 105:4685–4690
68. Shen Y, Bryan PN, He Y et al (2010) De novo structure generation using chemical shifts for proteins with high-sequence identity but different folds. Protein Sci 19:349–356
69. Raman S, Huang YJ, Mao B et al (2010) Accurate automated protein NMR structure determination using unassigned NOESY data. J Am Chem Soc 132:202–207
70. Montalvao RW, Cavalli A, Salvatella X et al (2008) Structure determination of protein-protein complexes using NMR chemical shifts: case of an endonuclease colicin-immunity protein complex. J Am Chem Soc 130:15990–15996
71. Oz G, Pountney DL, Armitage IM (1998) NMR spectroscopic studies of $I = 1/2$ metal ions in biological systems. Biochem Cell Biol 76:223–234
72. Drakenberg T, Jaohansson C, Forsen S (1997) In: Reid DG (ed) Protein NMR techniques. Human Press, Totowa
73. Sun H (2002) Metallodrugs. In: Grant DM, Harris RK (eds) Encyclopedia of nuclear magnetic resonance: advances in NMR. Wiley, Chichester, pp 413–427
74. Blindauer CA, Harvey I, Bunyan KE et al (2009) Structure, properties, and engineering of the major zinc binding site on human albumin. J Biol Chem 284:23116–23124
75. Li H, Otvos JD (1996) ^{111}Cd NMR studies of the domain specificity of Ag^{+} and Cu^{+} binding to metallothionein. Biochemistry 35:13929–13936
76. Farrell RA, Thorvaldsen JL, Winge DR (1996) Identification of the Zn(II) site in the copper-responsive yeast transcription factor, AMT1: a conserved Zn module. Biochemistry 35:1571–1580
77. Kakalis LT, Kennedy M, Sikkink R et al (1995) Characterization of the calcium-binding sites of calcineurin B. FEBS Lett 362:55–58
78. Baleja JD, Marmorstein R, Harrison SC et al (1992) Solution structure of the DNA-binding domain of Cd2-GAL4 from S. cerevisiae. Nature 356:450–453
79. Pan T, Coleman JE (1990) GAL4 transcription factor is not a "zinc finger" but forms a Zn(II) 2Cys6 binuclear cluster. Proc Natl Acad Sci USA 87:2077–2081
80. Vasak M (1998) Application of ^{113}Cd NMR to metallothioneins. Biodegradation 9:501–512

81. Digilio G, Bracco C, Vergani L et al (2009) The cadmium binding domains in the metallothionein isoform Cd_7-MT10 from Mytilus galloprovincialis revealed by NMR spectroscopy. J Biol Inorg Chem 14:167–178
82. Serra-Batiste M, Cols N, Alcaraz LA et al (2010) The metal-binding properties of the blue crab copper specific CuMT-2: a crustacean metallothionein with two cysteine triplets. J Biol Inorg Chem 15:759–776
83. Daniels MJ, Turner-Cavet JS, Selkirk R et al (1998) Coordination of Zn^{2+} (and Cd^{2+}) by prokaryotic metallothionein. Involvement of his-imidazole. J Biol Chem 273:22957–22961
84. Stewart AJ, Blindauer CA, Berezenko S et al (2003) Interdomain zinc site on human albumin. Proc Natl Acad Sci USA 100:3701–3706
85. Narula SS, Mehra RK, Winge DR et al (1991) Establishment of the metal-to-cysteine connectivities in silver-substituted yeast metallothionein. J Am Chem Soc 113:9354–9358
86. Andersen RJ, diTargiani RC, Hancock RD et al (2006) Characterization of the first N2S (alkylthiolate)lead compound: a model for three-coordinate lead in biological systems. Inorg Chem 45:6574–6576
87. Claudio ES, ter Horst MA, Forde CE et al (2000) ^{207}Pb-^{1}H two-dimensional NMR spectroscopy: a useful new tool for probing lead(II) coordination chemistry. Inorg Chem 39:1391–1397
88. Aramini JM, Hiraoki T, Yazawa M et al (1996) Lead-207 NMR: a novel probe for the study of calcium-binding proteins. J Biol Inorg Chem 1:39–48
89. Neupane KP, Pecoraro VL (2010) Probing a homoleptic PbS3 coordination environment in a designed peptide using ^{207}Pb NMR spectroscopy: implications for understanding the molecular basis of lead toxicity. Angew Chem Int Ed Engl 49:8177–8180
90. Utschig LM, Bryson JW, O'Halloran TV (1995) Mercury-199 NMR of the metal receptor site in MerR and its protein-DNA complex. Science 268:380–385
91. DeSilva TM, Veglia G, Porcelli F et al (2002) Selectivity in heavy metal- binding to peptides and proteins. Biopolymers 64:189–197
92. Iranzo O, Thulstrup PW, Ryu SB et al (2007) The application of ^{199}Hg NMR and ^{199m}Hg perturbed angular correlation (PAC) spectroscopy to define the biological chemistry of Hg(II): a case study with designed two- and three-stranded coiled coils. Chemistry 13:9178–9190
93. Utschig LM, Wright JG, Dieckmann G et al (1995) The ^{199}Hg chemical-shift as a probe of coordination environments in blue copper proteins. Inorg Chem 34:2497–2498
94. Steele RA, Opella SJ (1997) Structures of the reduced and mercury-bound forms of MerP, the periplasmic protein from the bacterial mercury detoxification system. Biochemistry 36:6885–6895
95. Utschig LM, Baynard T, Strong C et al (1997) Probing copper-thioether coordination chemistry in rusticyanin and azurin by 2D ^{1}H-^{199}Hg NMR. Inorg Chem 36:2926–2927
96. Huffman DL, Utschig LM, O'Halloran TV (1997) Mercury-responsive gene regulation and mercury-199 as a probe of protein structure. Met Ions Biol Syst 34:503–526
97. Kornhaber GJ, Snyder D, Moseley HN et al (2006) Identification of zinc-ligated cysteine residues based on 13Calpha and 13Cbeta chemical shift data. J Biomol NMR 34:259–269
98. Kostic M, Matt T, Martinez-Yamout MA et al (2006) Solution structure of the Hdm2 C2H2C4 RING, a domain critical for ubiquitination of p53. J Mol Biol 363:433–450
99. Zuiderweg ER (2002) Mapping protein-protein interactions in solution by NMR spectroscopy. Biochemistry 41:1–7
100. Gao G, Williams JG, Campbell SL (2004) Protein-protein interaction analysis by nuclear magnetic resonance spectroscopy. Methods Mol Biol 261:79–92
101. Zeng YB, Zhang DM, Li H et al (2008) Binding of Ni^{2+} to a histidine- and glutamine-rich protein, Hpn-like. J Biol Inorg Chem 13:1121–1131
102. Syme CD, Viles JH (2006) Solution ^{1}H NMR investigation of Zn^{2+} and Cd^{2+} binding to amyloid-beta peptide (Abeta) of Alzheimer's disease. Biochim Biophys Acta 1764:246–256
103. Jones CE, Klewpatinond M, Abdelraheim SR et al (2005) Probing Cu^{2+} binding to the prion protein using diamagnetic Ni^{2+} and ^{1}H NMR: the unstructured N terminus facilitates the coordination of six Cu^{2+} ions at physiological concentrations. J Mol Biol 346:1393–1407

104. Jensen MR, Hass MA, Hansen DF et al (2007) Investigating metal-binding in proteins by nuclear magnetic resonance. Cell Mol Life Sci 64:1085–1104
105. Schumann FH, Riepl H, Maurer T et al (2007) Combined chemical shift changes and amino acid specific chemical shift mapping of protein-protein interactions. J Biomol NMR 39:275–289
106. Bertini I, Das Gupta S, Hu X et al (2009) Solution structure and dynamics of S100A5 in the apo and Ca^{2+}-bound states. J Biol Inorg Chem 14:1097–1107
107. Banci L, Bertini I, Ciofi-Baffoni S et al (2008) A structural-dynamical characterization of human Cox17. J Biol Chem 283:7912–7920
108. Banci L, Bertini I, Cantini F et al (2009) An NMR study of the interaction of the N-terminal cytoplasmic tail of the Wilson disease protein with copper(I)-HAH1. J Biol Chem 284:9354–9360
109. Pelton JG, Torchia DA, Meadow ND et al (1993) Tautomeric states of the active-site histidines of phosphorylated and unphosphorylated IIIGlc, a signal-transducing protein from Escherichia coli, using two-dimensional heteronuclear NMR techniques. Protein Sci 2:543–558
110. Lee BM, Buck-Koehntop BA, Martinez-Yamout MA et al (2007) Embryonic neural inducing factor Churchill is not a DNA-binding zinc finger protein: solution structure reveals a solvent-exposed beta-sheet and zinc binuclear cluster. J Mol Biol 371:1274–1289
111. Otting G (2010) Protein NMR using paramagnetic ions. Annu Rev Biophys 39:387–405
112. Arnesano F, Banci L, Piccioli M (2005) NMR structures of paramagnetic metalloproteins. Q Rev Biophys 38:167–219
113. Bertini I, Luchinat C, Rosato A (1996) The solution structure of paramagnetic metalloproteins. Prog Biophys Mol Biol 66:43–80
114. Otting G (2008) Prospects for lanthanides in structural biology by NMR. J Biomol NMR 42:1–9
115. Clore GM, Iwahara J (2009) Theory, practice, and applications of paramagnetic relaxation enhancement for the characterization of transient low-population states of biological macromolecules and their complexes. Chem Rev 109:4108–4139
116. Iwahara J, Tang C, Marius Clore G (2007) Practical aspects of ^{1}H transverse paramagnetic relaxation enhancement measurements on macromolecules. J Magn Reson 184:185–195
117. Bertini I, Ciurli S, Dikiy A et al (2001) The first solution structure of a paramagnetic copper (II) protein: the case of oxidized plastocyanin from the cyanobacterium Synechocystis PCC6803. J Am Chem Soc 123:2405–2413
118. Ubbink M, Worrall JA, Canters GW et al (2002) Paramagnetic resonance of biological metal centers. Annu Rev Biophys Biomol Struct 31:393–422
119. Hansen DF, Led JJ (2006) Determination of the geometric structure of the metal site in a blue copper protein by paramagnetic NMR. Proc Natl Acad Sci USA 103:1738–1743
120. Donaldson LW, Skrynnikov NR, Choy WY et al (2001) Structural characterization of proteins with an attached ATCUN motif by paramagnetic relaxation enhancement NMR spectroscopy. J Am Chem Soc 123:9843–9847
121. Iwahara J, Schwieters CD, Clore GM (2004) Ensemble approach for NMR structure refinement against ^{1}H paramagnetic relaxation enhancement data arising from a flexible paramagnetic group attached to a macromolecule. J Am Chem Soc 126:5879–5896
122. Yagi H, Loscha KV, Su XC et al (2010) Tunable paramagnetic relaxation enhancements by $[Gd(DPA)_3]^{3-}$ for protein structure analysis. J Biomol NMR 47:143–153
123. Su XC, Otting G (2010) Paramagnetic labelling of proteins and oligonucleotides for NMR. J Biomol NMR 46:101–112
124. Iwahara J, Schwieters CD, Clore GM (2004) Characterization of nonspecific protein-DNA interactions by ^{1}H paramagnetic relaxation enhancement. J Am Chem Soc 126:12800–12808
125. Fawzi NL, Doucleff M, Suh JY et al (2010) Mechanistic details of a protein-protein association pathway revealed by paramagnetic relaxation enhancement titration measurements. Proc Natl Acad Sci USA 107:1379–1384

126. Yu D, Volkov AN, Tang C (2009) Characterizing dynamic protein-protein interactions using differentially scaled paramagnetic relaxation enhancement. J Am Chem Soc 131:17291–17297
127. Liang B, Bushweller JH, Tamm LK (2006) Site-directed parallel spin-labeling and paramagnetic relaxation enhancement in structure determination of membrane proteins by solution NMR spectroscopy. J Am Chem Soc 128:4389–4397
128. Clore GM, Tang C, Iwahara J (2007) Elucidating transient macromolecular interactions using paramagnetic relaxation enhancement. Curr Opin Struct Biol 17:603–616
129. Tang C, Schwieters CD, Clore GM (2007) Open-to-closed transition in apo maltose-binding protein observed by paramagnetic NMR. Nature 449:1078–1082
130. Iwahara J, Clore GM (2006) Detecting transient intermediates in macromolecular binding by paramagnetic NMR. Nature 440:1227–1230
131. Volkov AN, Ubbink M, van Nuland NA (2010) Mapping the encounter state of a transient protein complex by PRE NMR spectroscopy. J Biomol NMR 48:225–236
132. Bertini I, Luchinat C, Parigi G (2002) Magnetic susceptibility in paramagnetic NMR. Prog Nucl Magn Reson Spectrosc 40:249–273
133. Biekofsky RR, Muskett FW, Schmidt JM et al (1999) NMR approaches for monitoring domain orientations in calcium-binding proteins in solution using partial replacement of Ca^{2+} by Tb^{3+}. FEBS Lett 460:519–526
134. Banci L, Bertini I, Bren KL et al (1996) The use of peudocontact shifts to refine solution structures of paramagnetic metalloproteins: Met80Ala cyano-cytochrome c as an example. J Biol Inorg Chem 1:117–126
135. Goodfellow BJ, Duarte IC, Macedo AL et al (2010) An NMR structural study of nickel-substituted rubredoxin. J Biol Inorg Chem 15:409–420
136. Arnesano F, Banci L, Bertini I et al (2003) A strategy for the NMR characterization of type II copper(II) proteins: the case of the copper trafficking protein CopC from Pseudomonas syringae. J Am Chem Soc 125:7200–7208
137. Bertini I, Kursula P, Luchinat C et al (2009) Accurate solution structures of proteins from X-ray data and a minimal set of NMR data: calmodulin-peptide complexes as examples. J Am Chem Soc 131:5134–5144
138. Pintacuda G, Park AY, Keniry MA et al (2006) Lanthanide labeling offers fast NMR approach to 3D structure determinations of protein-protein complexes. J Am Chem Soc 128:3696–3702
139. Pintacuda G, John M, Su XC et al (2007) NMR structure determination of protein-ligand complexes by lanthanide labeling. Acc Chem Res 40:206–212
140. Nguyen TH, Ozawa K, Stanton-Cook M et al (2010) Generation of pseudocontact shifts in protein NMR spectra with a genetically encoded cobalt(II)-binding amino acid. Angew Chem Int Ed Engl 50(3):692–694
141. Bertini I, Del Bianco C, Gelis I et al (2004) Experimentally exploring the conformational space sampled by domain reorientation in calmodulin. Proc Natl Acad Sci USA 101:6841–6846
142. Bertini I, Gupta YK, Luchinat C et al (2007) Paramagnetism-based NMR restraints provide maximum allowed probabilities for the different conformations of partially independent protein domains. J Am Chem Soc 129:12786–12794
143. Bertini I, Duma L, Felli IC et al (2004) A heteronuclear direct-detection NMR spectroscopy experiment for protein-backbone assignment. Angew Chem Int Ed 43:2257–2259
144. Bermel W, Bertini I, Felli IC et al (2006) C-13-detected protonless NMR spectroscopy of proteins in solution. Prog Nucl Magn Reson Spectrosc 48:25–45
145. Bertini I, Jimenez B, Pierattelli R et al (2008) Protonless ^{13}C direct detection NMR: characterization of the 37 kDa trimeric protein CutA1. Proteins 70:1196–1205
146. Machonkin TE, Westler WM, Markley JL (2002) $^{13}C\{^{13}C\}$ 2D NMR: a novel strategy for the study of paramagnetic proteins with slow electronic relaxation rates. J Am Chem Soc 124:3204–3205
147. Caillet-Saguy C, Delepierre M, Lecroisey A et al (2006) Direct-detected ^{13}C NMR to investigate the iron(III) hemophore HasA. J Am Chem Soc 128:150–158

148. Balayssac S, Bertini I, Luchinat C et al (2006) ^{13}C direct detected NMR increases the detectability of residual dipolar couplings. J Am Chem Soc 128:15042–15043
149. Bermel W, Bertini I, Felli IC et al (2003) ^{13}C direct detection experiments on the paramagnetic oxidized monomeric copper, zinc superoxide dismutase. J Am Chem Soc 125:16423–16429
150. Andersson P, Weigelt J, Otting G (1998) Spin-state selection filters for the measurement of heteronuclear one-bond coupling constants. J Biomol NMR 12:435–441
151. Ottiger M, Delaglio F, Bax A (1998) Measurement of J and dipolar couplings from simplified two-dimensional NMR spectra. J Magn Reson 131:373–378
152. Duma L, Hediger S, Lesage A et al (2003) Spin-state selection in solid-state NMR. J Magn Reson 164:187–195
153. Meissner A, Duus JO, Sørensen OW (1997) Integration of spin-state-selective excitation into 2D NMR correlation experiments with the heteronuclear ZQ/2Q pi rotations for $^{1}J_{XH}$-resolved E.COSY-type measurements of heteronuclear coupling constants in proteins. J Biomol NMR 10:89–94
154. Bermel W, Bertini I, Duma L et al (2005) Complete assignment of heteronuclear protein resonances by protonless NMR spectroscopy. Angew Chem Int Ed Engl 44:3089–3092
155. Bermel W, Bertini I, Felli IC et al (2006) Protonless NMR experiments for sequence-specific assignment of backbone nuclei in unfolded proteins. J Am Chem Soc 128:3918–3919
156. Bertini I, Felli IC, Kummerle R et al (2004) ^{13}C-^{13}C NOESY: an attractive alternative for studying large macromolecules. J Am Chem Soc 126:464–465
157. Bertini I, Felli IC, Kummerle R et al (2004) ^{13}C-^{13}C NOESY: a constructive use of ^{13}C-^{13}C spin-diffusion. J Biomol NMR 30:245–251
158. Arnesano F, Banci L, Bertini I et al (2003) A redox switch in CopC: an intriguing copper trafficking protein that binds copper(I) and copper(II) at different sites. Proc Natl Acad Sci USA 100:3814–3819
159. Madl T, Felli IC, Bertini I et al (2010) Structural analysis of protein interfaces from ^{13}C direct-detected paramagnetic relaxation enhancements. J Am Chem Soc 132:7285–7287
160. Babini E, Bertini I, Capozzi F et al (2004) Direct carbon detection in paramagnetic metalloproteins to further exploit pseudocontact shift restraints. J Am Chem Soc 126:10496–10497
161. Bermel W, Bertini I, Felli IC et al (2010) Exclusively heteronuclear NMR experiments to obtain structural and dynamic information on proteins. ChemPhysChem 11:689–695
162. Kostic M, Pochapsky SS, Pochapsky TC (2002) Rapid recycle $^{13}C'$, ^{15}N and ^{13}C, $^{13}C'$ heteronuclear and homonuclear multiple quantum coherence detection for resonance assignments in paramagnetic proteins: example of Ni^{2+} -containing acireductone dioxygenase. J Am Chem Soc 124:9054–9055
163. Bertini I, Jimenez B, Piccioli M et al (2005) Asymmetry in ^{13}C-^{13}C COSY spectra provides information on ligand geometry in paramagnetic proteins. J Am Chem Soc 127:12216–12217
164. Turano P, Lalli D, Felli IC et al (2010) NMR reveals pathway for ferric mineral precursors to the central cavity of ferritin. Proc Natl Acad Sci USA 107:545–550
165. Matzapetakis M, Turano P, Theil EC et al (2007) ^{13}C–^{13}C NOESY spectra of a 480 kDa protein: solution NMR of ferritin. J Biomol NMR 38:237–242
166. Williamson MP, Havel TF, Wuthrich K (1985) Solution conformation of proteinase inhibitor IIA from bull seminal plasma by ^{1}H nuclear magnetic resonance and distance geometry. J Mol Biol 182:295–315
167. Kupce E, Freeman R (2008) Fast multi-dimensional NMR by minimal sampling. J Magn Reson 191:164–168
168. Frueh DP, Sun ZY, Vosburg DA et al (2006) Non-uniformly sampled double-TROSY hNcaNH experiments for NMR sequential assignments of large proteins. J Am Chem Soc 128:5757–5763
169. Marion D (2005) Fast acquisition of NMR spectra using Fourier transform of non-equispaced data. J Biomol NMR 32:141–150

170. Szyperski T, Yeh DC, Sukumaran DK et al (2002) Reduced-dimensionality NMR spectroscopy for high-throughput protein resonance assignment. Proc Natl Acad Sci USA 99:8009–8014
171. Felli IC, Brutscher B (2009) Recent advances in solution NMR: fast methods and heteronuclear direct detection. ChemPhysChem 10:1356–1368
172. Kim S, Szyperski T (2003) GFT NMR, a new approach to rapidly obtain precise high-dimensional NMR spectral information. J Am Chem Soc 125:1385–1393
173. Freeman R, Kupce E (2003) New methods for fast multidimensional NMR. J Biomol NMR 27:101–113
174. Kupce E, Freeman R (2004) Projection-reconstruction technique for speeding up multi-dimensional NMR spectroscopy. J Am Chem Soc 126:6429–6440
175. Schanda P, Van Melckebeke H, Brutscher B (2006) Speeding up three-dimensional protein NMR experiments to a few minutes. J Am Chem Soc 128:9042–9043
176. Schanda P, Brutscher B (2005) Very fast two-dimensional NMR spectroscopy for real-time investigation of dynamic events in proteins on the time scale of seconds. J Am Chem Soc 127:8014–8015
177. Kupce E, Freeman R (2003) Fast multi-dimensional Hadamard spectroscopy. J Magn Reson 163:56–63
178. Hiller S, Wasmer C, Wider G et al (2007) Sequence-specific resonance assignment of soluble nonglobular proteins by 7D APSY-NMR spectroscopy. J Am Chem Soc 129:10823–10828
179. Shen Y, Atreya HS, Liu G et al (2005) G-matrix Fourier transform NOESY-based protocol for high-quality protein structure determination. J Am Chem Soc 127:9085–9099
180. Hiller S, Garces RG, Malia TJ et al (2008) Solution structure of the integral human membrane protein VDAC-1 in detergent micelles. Science 321:1206–1210
181. Tugarinov V, Kay LE, Ibraghimov I et al (2005) High-resolution four-dimensional ^{1}H-^{13}C NOE spectroscopy using methyl-TROSY, sparse data acquisition, and multidimensional decomposition. J Am Chem Soc 127:2767–2775
182. Baldwin AJ, Kay LE (2009) NMR spectroscopy brings invisible protein states into focus. Nat Chem Biol 5:808–814
183. Bermel W, Bertini I, Felli IC et al (2009) Speeding up ^{13}C direct detection biomolecular NMR spectroscopy. J Am Chem Soc 131:15339–15345
184. Lipton AS, Heck RW, Staeheli GR et al (2008) A QM/MM approach to interpreting ^{67}Zn solid-state NMR data in zinc proteins. J Am Chem Soc 130:6224–6230
185. Lipton AS, Heck RW, Primak S et al (2008) Characterization of Mg^{2+} binding to the DNA repair protein apurinic/apyrimidic endonuclease 1 via solid-state ^{25}Mg NMR spectroscopy. J Am Chem Soc 130:9332–9341

Top Curr Chem (2012) 326: 99–122
DOI: 10.1007/128_2011_212

Published online: 7 September 2011

Recent Developments in ^{15}N NMR Relaxation Studies that Probe Protein Backbone Dynamics

Rieko Ishima

Abstract Nuclear Magnetic Resonance (NMR) relaxation is a powerful technique that provides information about internal dynamics associated with configurational energetics in proteins, as well as site-specific information involved in conformational equilibria. In particular, ^{15}N relaxation is a useful probe to characterize overall and internal backbone dynamics of proteins because the relaxation mainly reflects reorientational motion of the N–H bond vector. Over the past 20 years, experiments and protocols for analysis of ^{15}N R_1, R_2, and the heteronuclear ^{15}N–{^{1}H} NOE data have been well established. The development of these methods has kept pace with the increase in the available static–magnetic field strength, providing dynamic parameters optimized from data fitting at multiple field strengths. Using these methodological advances, correlation times for global tumbling and order parameters and correlation times for internal motions of many proteins have been determined. More recently, transverse relaxation dispersion experiments have extended the range of NMR relaxation studies to the milli- to microsecond time scale, and have provided quantitative information about functional conformational exchange in proteins. Here, we present an overview of recent advances in ^{15}N relaxation experiments to characterize protein backbone dynamics.

Keywords Dispersion · Dynamics · NMR · Protein · Relaxation

Contents

1 NMR Relaxation to Detect Protein Dynamics 100
2 Recent Improvements in the ^{15}N R_1, R_2, and ^{15}N–{^{1}H} NOE Experiments 100
2.1 Practical Aspects in ^{15}N R_2 Experiment 101
2.2 Practical Aspects in ^{15}N R_1 Experiment 104
2.3 Practical Aspects in ^{15}N–{^{1}H} NOE Experiment 106

R. Ishima (✉)
Department of Structural Biology, University of Pittsburgh School of Medicine, Pittsburgh, PA 15260, USA
e-mail: ishima@pitt.edu

3 Extending Relaxation Measurements Beyond R_1, R_2, and ^{15}N–{^{1}H} NOE 107
4 Relaxation Dispersion Experiments 108
4.1 Relaxation Dispersion in General 108
4.2 ^{15}N CT-CPMG Relaxation Dispersion Experiment 109
4.3 ^{15}N Off-Resonance $R_{1\rho}$ Experiment 111
4.4 Bloch–McConnell Equation and Related Equations 112
4.5 Practical Aspects Parameter Optimization 112
References 115

1 NMR Relaxation to Detect Protein Dynamics

In this chapter, we describe mainly ^{15}N relaxation experiments to characterize protein backbone dynamics in solution. NMR spin-relaxation is a phenomenon in which perturbed magnetization is restored to statistical equilibrium by random fluctuations of local magnetic fields. The major local magnetic fields in diamagnetic proteins are generated by the amide ^{1}H–^{15}N dipolar interaction and ^{15}N chemical shift anisotropy (CSA). Interchange among different chemical shift environments by chemical exchange or conformational exchange also contributes to the spin relaxation. In this section, we review the types of ^{15}N relaxation experiments that are used to characterize protein backbone dynamics.

The model-free approach is the most frequently applied protocol to extract information about overall and fast (faster than overall) dynamics in proteins. In this approach the spectral density function characterizing the randomly fluctuating local fields is written in terms of correlation times for overall and internal motion and a generalized order parameter (the model free parameters). These parameters are obtained from the model-free analysis using ^{15}N longitudinal relaxation rate (R_1), transverse relaxation rate (R_2), and ^{15}N–{^{1}H} nuclear Overhauser effect (NOE) measurements. Measuring at least four relaxation rates, at two or more static magnetic field strengths, improves determination of the model-free parameters. Alternatively, ^{15}N–{^{1}H} NOE alone may be used to evaluate the high-frequency spectral density function, $J(\omega_H \pm \omega_N)$, instead of derivation of model-free parameters. However this approach is not straightforward at high magnetic field strength for proteins with significant mobility as described in Sect. 2.3.

Although relaxation of nuclei other than ^{15}N can be used to characterize backbone dynamics, ^{15}N relaxation experiments have been the most widely applied, and for this reason are the focuses of this chapter.

2 Recent Improvements in the ^{15}N R_1, R_2, and ^{15}N–{^{1}H} NOE Experiments

A set of ^{15}N R_1, R_2, and ^{15}N–{^{1}H} NOE observations is typically analyzed using model-free analysis to obtain the generalized-order parameter, S^2, that characterizes degree of internal motion [1, 2]. Since a set of ^{15}N R_1, R_2, and {^{1}H}–^{15}N NOE observables mainly reflects values of the spectral density functions at zero, ^{15}N, and ^{1}H frequencies, (i.e., $J(0)$, $J(\omega_N)$, and $J(\sim\omega_H)$, respectively), the model-free analysis using the three experimental data sets is suitable to characterize fast internal motion in proteins [3–5]. In the model-free analysis, a correlation time for internal motion, τ_i, is determined for each amide site in addition to S^2. Moreover, when the simple model-free spectral density function is unable to fit the data, an extended model that contains an order parameter for faster internal motion, S_f^2, an order parameter for slower internal motion, S_s^2, and a correlation time, τ_s, for the slower time scale motion is tested. A chemical exchange term, R_{ex}, is added to test for the presence of slow (milli- to microsecond) motions as well. These parameter optimizations are conducted by initially assuming a spherical rigid body rotation of the molecule, i.e., assuming a single rotational correlation time, τ_R. Subsequently, either an axially symmetric or fully asymmetric model of the molecular rotational diffusion may be tested. The principles, protocols, and verification of the parameterization derived by the model-free analysis have extensively been studied and described [6–34]. In this section we focus on the recent developments in applying the model-free approach.

2.1 *Practical Aspects in ^{15}N R_2 Experiment*

Since R_2 is the only observable that provides information about the $J(0)$ spectral density contribution, accurate measurement of this observable is of particular importance. Transverse relaxation rates are typically measured by either a spin-lock ($R_{1\rho}$) or a Carr–Purcell–Meiboom–Gill (CPMG) experiment. In the following, advantages and disadvantages of the two experiments are described with particular consideration of (1) limitations on the applied B_1 field strength in which $\omega_1 = \gamma_N B_1$ (γ_N is gyromagnetic ratio of ^{15}N), (2) off-resonance error, and (3) suppression of cross correlation by ^{1}H–^{15}N dipolar interaction (DD) and ^{15}N CSA.

2.1.1 Spin-Lock $R_{1\rho}$ Experiment

In the $R_{1\rho}$ experiment, in which relaxation is measured in a rotating reference frame, an rf field, B_1, is applied during the relaxation period, during which time the magnetization is "locked" almost parallel to B_1. $R_{1\rho}$ is a function of R_1 and R_2, given by

$$R_{1\rho} = R_2 \sin^2\theta + R_1 \cos^2\theta. \quad (1)$$

Here, θ is given by $\tan(\omega_1/(\omega_0 - \omega_{rot}))$, and ω_0 and ω_{rot} are the Larmor frequency of the signal and the angular frequency of the rotation frame, respectively. To obtain R_2 from $R_{1\rho}$ most accurately, it is advantageous to increase the R_2

contribution to $R_{1\rho}$. For this purpose, a strong B_1 field strength, which makes θ close to 90°, is employed.

Limitations of applicable B_1 field strength and the rf duty cycle depend on individual probes. In general, at ^{15}N resonance frequency at 61 MHz, we recommend that a spin lock field $\gamma_N B_1/2\pi > 2$ kHz be applied for 60–80 ms to determine ^{15}N $R_{1\rho}$. The $\gamma_N B_1/2\pi$ was calculated assuming that the entire chemical shift range for amide backbone ^{15}N signals in diamagnetic proteins is ±900 Hz (±15 ppm) at 61 MHz, which corresponds to sin $\theta > 0.9$. The spin lock duration was estimated based on approximate ^{15}N R_2 of a folded 10–20 kDa protein at room temperature. Although the highest measurement accuracy is obtained when data is recorded until the magnetization decays sufficiently (typically for a time $>=1/R_{1\rho}$), it may not be possible to satisfy this condition for ^{15}N sites which relax slowly (as in unfolded proteins or small peptides) without reducing B_1 which leads to a reduction in sin θ.

Errors in $R_{1\rho}$ resulting from off-resonance effects may be significant but can be corrected. It is an advantage of the $R_{1\rho}$ experiment that, even when a signal is off-resonance from the rf carrier frequency and for which sin θ is small, an accurate $R_{1\rho}$ value can be obtained using equation (1). Although the correction requires an R_1 value, this will be available when R_1 data is recorded to characterize fast backbone dynamics using model-free analysis. As described above, the accuracy of $R_{1\rho}$ measurements decreases for signals located far off-resonance. Compared to CPMG R_2 that is described in Sect. 2.1.2, $R_{1\rho}$ values do not need to be recorded at different carrier-frequencies without discarding any data.

Cross-correlation interference by ^{1}H–^{15}N dipolar interaction (DD) and ^{15}N CSA has to be suppressed to detect accurate ^{15}N transverse relaxation rates (Fig. 1). The cross term is suppressed by flipping the sign of the DD term by applying ^{1}H 180° pulses at a rate greater than the decay rate of the two ^{15}N–^{1}H J-coupled components [35–37]. However, in a weak B_1 field, the two J-coupled components undergo

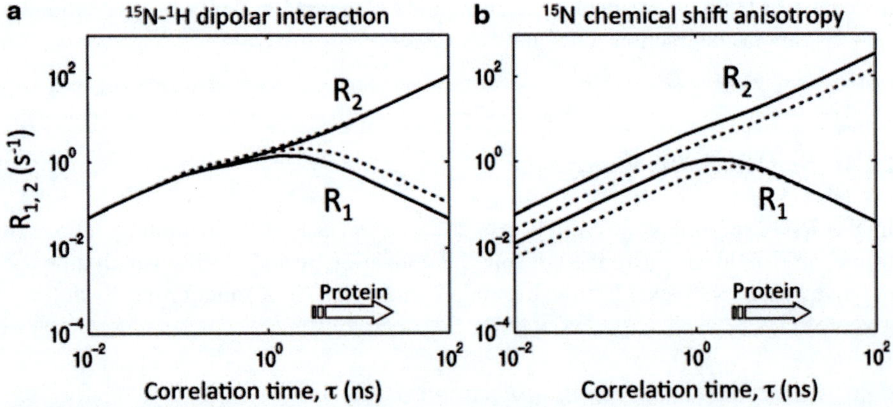

Fig. 1 (**a**) ^{15}N–H dipolar and (**b**) ^{15}N CSA contribution to longitudinal and transverse relaxation rates (R_1, and R_2) as a function of a correlation time. The rates were calculated assuming a simple Lorentzian spectral density function, $J(\omega) = \tau/(1 + \omega^2\tau^2)$. *Solid* and *dotted lines* indicate rates calculated assuming at 900 MHz and 600 MHz instruments, respectively

precessions at frequencies $\pm J/2$ relative to the rotating frame. Because this frame is significantly tilted from the rotating frame, application of frequent ^{1}H pulses results in incomplete cancelation of the cross term. Therefore, infrequent ^{1}H pulses are applied to suppress DD/CSA correlation when the off-resonance effect is significant [38, 39].

Other parameters that have to be specifically considered in the $R_{1\rho}$ experiment are the spatial homogeneity of the B_1 field strength and linearity of the power amplifier. Although B_1 homogeneity has been improved in recent NMR probes, estimation of the inhomogeneity remains important to confirm the accuracy of the obtained $R_{1\rho}$. For this purpose, measurement of the B_1 inhomogeneity using inverse detection is useful [40]. Amplifier linearity has also improved. However, since pulse power is switched for spin-lock, it is important to check that a phase shift accompanies a power change, and make a correction should one be needed.

2.1.2 Carr–Purcell–Meiboom–Gill R_2 Experiment

In CPMG R_2, *limitation of applicable B_1 field strength* is in general smaller than that of the $R_{1\rho}$ experiment. Since CPMG 180° pulses are applied with interpulse delays ($2\tau_{CP}$), CPMG pulses with much stronger $\gamma_N B_1/2\pi$ than that of spin-lock can be applied. However, in a protein, the RF field strength used in the ^{15}N CPMG R_2 experiment must be carefully considered because the interpulse delay is set short (typically, $2\tau_{CP}$ ~1 ms) to suppress generation of antiphase terms, $N_{X,Y}H_Z$, caused by ^{1}H–^{15}N J coupling. For example, if 6 kHz CPMG pulses are applied with $2\tau_{CP} = 1$ ms, the RF power delivered to the probe is nine times stronger than that of a 2 kHz spin lock. Given that the duty cycle is ca. 10% in the CPMG experiment, about the same amount of energy is deposited in the probe in both CPMG and $R_{1\rho}$ experiments.

In CPMG R_2, *off-resonance error* is negligible at low magnetic field strength but significant at high magnetic field strength [41, 42]. Although the stronger $\gamma_N B_1/2\pi$ for each CPMG pulse inverts magnetization more uniformly than the spin-lock, CPMG pulse train accumulates error caused by a combination of pulse imperfections and off-resonance effects. The CPMG error of a signal located at off-resonance frequency f_{off} (f_{off} is the difference between the signal and carrier frequencies) is estimated with a function of $2\tau_{CP}$ and B_1. Importantly, the off-resonance error is maximized at $2\tau_{CP}f_{off} = n$ (n is an integer): when $2\tau_{CP} = 1$ ms, the off-resonance error is significant at $f_{off} = 1$, 2, and 3 kHz [41]. Magnitude of the error depends on the B_1 field strength the 180° pulses. This relationship indicates that R_2 can be recorded without significant off-resonance error at 61 MHz for signals because the entire chemical shift range for amide backbone ^{15}N signals spans approximately less than $\pm$15 ppm. However, if the same $2\tau_{CP}$ is used at ^{15}N 91 MHz resonance frequency, signals located at $\pm$11 ppm off-resonance suffer from significant errors in measured R_2 values.

It is a disadvantage of CPMG R_2 experiment that there is no simple equation to correct for CPMG R_2 off-resonance effects. In practice, it is recommended to discard R_2 data obtained at $f_{off} = n/2\tau_{CP}$ frequency, and record the data at two different carrier frequencies. As an alternative, a phase cycle to average out the off-resonance effect in CPMG R_2 may be used [43, 44]. Using this method, R_2 is determined from the observed relaxation rate using known R_1, τ_{CP}, and the CPMG pulse width. Application of the sequence to relaxation dispersion has also been proposed [45]. Dissection of the relaxation contribution in the R_2 dispersion determined using the sequence has been described [46].

Cross-correlation of ^{1}H–^{15}N dipolar interaction (DD) and ^{15}N CSA in CPMG R_2 experiments has been suppressed by applying ^{1}H 180° pulses [35–37]. Since CSA relaxation increases as a function of the magnetic field strength (Fig. 1), relative contribution of DD/CSA cross-correlation in the ^{15}N transverse relaxation increases up to an external field strength.

2.1.3 Practical Relaxation Delay at High Magnetic Field Strength

Propagation of experimental noise, ΔI, to the uncertainty, ΔR, of a relaxation rate, R, in an experiment where signal is measured at only two relaxation time points is calculated using (2). Here, a two-point single exponential decay function, $I(t) = I^0\exp(-RT)$, is assumed [46, 47]:

$$\Delta R/R = (\Delta I/I^0)[1 + \exp(2RT)]^{1/2}/(RT). \tag{2}$$

Signal-to-noise ratio, $I^0/\Delta I$, is proportional to $(B_0)^{3/2}$ (the magnetic field strength is B_0) [48]. By taking the advantage of the gain of the signal-to-noise ratio, the time T can be reduced in the experiments performed at a higher magnetic field strength when similar fractional error, $\Delta R/R$, to that obtained at a lower magnetic field strength is desired. Assuming that R at 600 MHz is determined by a two-point exponential fitting at $T = 0$ and the optimal T (=$1/R$), the R_XT_X at X MHz is determined by the following function:

$$(B_0^X/B_0^{600})^{3/2} = 2.86R_XT_X/[1 + \exp(2R_XT_X)]^{1/2}. \tag{3}$$

This equation predicts how much T can be reduced to have R_X value with a similar uncertainty to that at 600 MHz (Fig. 2). For example, when R at 600 MHz is determined by a two-point exponential fitting at $T = 0$ and $T = 1/R$, the R_XT_X at $X = 900$ MHz can be reduced by a factor of 0.32 times the optimal value, $T_X = 1/R_X$. Since there are other factors that contribute to actual signal-to-noise ratio of relaxing magnetization and that determine sensitivity of two different magnetic fields, this calculation is a rough estimate.

2.2 *Practical Aspects in ^{15}N R_1 Experiment*

^{15}N R_1 depends upon $J(\omega_N)$ and $J(\omega_H)$, but not on either $J(0)$ or chemical exchange. The pulse sequence used to measure R_1 in proteins usually incorporates

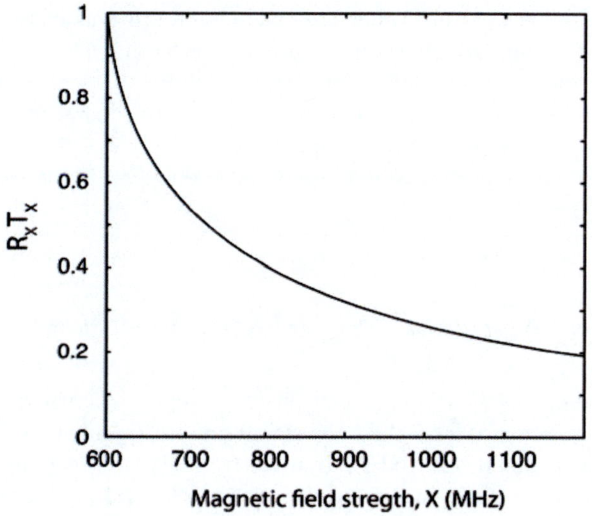

Fig. 2 Estimate of R_XT_X value at the magnetic field strength at X MHz to provide the uncertainty of the relaxation rate (R_X) equivalent to that obtained at 600 MHz. The graph was calculated using (3), in which ($I^0/\Delta I$) is proportional to $(B_0)^{3/2}$ and R at 600 MHz is determined by a two-point exponential fitting at $T = 0$ and $T = 1/R$. For example, at $X = 900$ MHz (*horizontal scale*), R_XT_X can be reduced to 0.32 (*vertical scale*) than the optimal $T_X = 1/R_X$ to obtain equivalent R_X uncertainty at 600 MHz

a Freeman–Hill phase cycle in which measured magnetization decays from an initial value of $I(0)$ to zero of magnetization starting from $I(0)$ to z [49]. Since this phase cycle avoids recovery from $-I(0)$ to $I(0)$, there is no need to record the magnetization recovery till equilibrium is attained. Otherwise, the total experimental time is enormously long [50]. In addition, because it is known that the $I(t)$ approaches zero at infinite time t, offset is not required as an unknown parameter in the exponential fitting. Typically, only two unknown parameters (R_1 and I^0) are optimized in the exponential fitting.

The pulse scheme to suppress DD/CSA cross-correlation interference is similar to that used to measure R_2 in which ^{1}H 180° pulses are applied during every 5–10 ms (at a rate greater than the decay rate of the faster-relaxing components of the ^{15}N–^{1}H J-coupled two components) [37]. As the magnetic field strength increases, the magnitude of the $\tau_R(1 + \tau_R^2\omega^2)$ term decreases (Fig. 1). However, in the slow molecular tumbling limit ($\tau_R\omega_N >> 1$), the R_1 contribution by the CSA term is field independent because $\sigma_{CSA}B_0$ increases cancels by the reduction of the $\tau_R(1 + \tau_R^2\omega^2)$ term. Thus, although the relative contribution of DD/CSA increases, the absolute cross-correlation effect on R_1 is not necessarily increased with increase in the magnetic field strength. The application rate of the ^{1}H 180° pulses may depend more on the apparent relaxation rates of the two components, i.e., the ^{1}H spin-flip rate, but not necessarily on the magnetic field strength.

To suppress DD/CSA cross-correlation in ^{15}N R_1 experiment, application of accurate ^{1}H 180° inversion pulses is important. In principle, it is best to invert only amide protons and not perturb water protons [51]. In this way, saturation of water proton magnetization, which reduces amide proton intensity by water-amide exchange at high-pH, is avoided. However, in practice this can be achieved only at high fields in which there is a sufficient chemical shift separation between amide and water proton chemical shifts. Otherwise, complete selective inversion is achieved at the cost of putting amide proton magnetization in the transverse plane for significant periods of time.

2.3 Practical Aspects in ^{15}N–{^{1}H} NOE Experiment

^{15}N–{^{1}H} NOE equals the ratio of steady state ^{15}N signal intensities recorded with/without ^{1}H saturation. Since the ^{15}N signal intensities have to be accurately encoded in the t_1 dimension, INEPT transfer from ^{1}H to ^{15}N is not used prior to the t_1 evolution. As a result, the sensitivity of the ^{15}N–{^{1}H} NOE experiment is ca. ten times lower than that of ^{15}N R_1 and R_2. To compensate for the low signal-to-noise ratio, experiment recording times are larger in the NOE experiment than in R_1 and R_2. In addition, a long magnetization recovery time (>3 s^{-1}) between each scan adds to the total time required to obtain data with adequate sensitivity [6, 52–54].

Sufficient magnetization recovery of ^{15}N (in the experiment with ^{1}H saturation) or both ^{1}H and ^{15}N (in the experiment without ^{1}H saturation) is crucial to determine NOE values accurately. In proteins where $\tau_R\omega >> 1$, dipolar longitudinal relaxation rates decrease as the resonance frequency, ω, increases. Therefore, a longer recovery time will be required for the ^{15}N–{^{1}H} NOE experiment at higher magnetic field strength. In a rigid protein with a rotational correlation time, τ_R, of 10 ns, ^{15}N R_1 is 1.1–1.4 s^{-1} at 61 MHz (in a 600 MHz NMR instrument) whereas ^{15}N R_1 is ca. 0.69–0.85 s^{-1} at 91 MHz (in a 900 MHz instrument). Thus, recovery times more than 3 and 5 s are required at 61 and 91 MHz, respectively. However, these are the recovery times estimated from ^{15}N R_1. ^{1}H recovery times often become longer when there is not much surrounding ^{1}H nuclei or less ^{1}H spin-flip (such as deuterated proteins, unfolded proteins, or in a loop region of a folded protein). In this case, insufficient ^{1}H Z-magnetization recovery is corrected using the equation derived by Bax and Grzesiek [52]. When the ^{15}N R_1 recovery is not sufficient, another correction equation that counts both ^{1}H and ^{15}N R_1 recovery is used [55] These corrections work reasonably once accurate ^{1}H and ^{15}N R_1 values are obtained.

When the amide proton magnetization does not recover in a single exponential manner, the correction equations do not give accurate results. In particular, when there is severe DD/CSA cross correlation in a deuterated protein in which the proton spin-flip rate is small at high-magnetic field strength, the decay of ^{1}H magnetization of one of the two ^{15}N coupled components becomes slow and nonexponential [56]. A simple solution will be to apply a sufficiently long recovery time. An alternative solution will be a pulse sequence that has recently introduced by Ferrage and coworkers [57, 58].

^{15}N–{^{1}H} NOE is thought to decrease monotonically as the rate of fast internal motion increases. However, the actual dependence of the NOE on correlation time is more complex. For example, when the model-free approach is used to express the spectral density using a correlation time for internal motion (τ_i) as well as an overall correlation time (τ_R), the NOE is a two-valued function of τ_R, attaining a maximum value at one value of τ_R (for example, see Fig. 19.10 in [59]).

3 Extending Relaxation Measurements Beyond R_1, R_2, and ^{15}N–{^{1}H} NOE

Although a set of ^{15}N R_1, R_2, and ^{15}N–{^{1}H} NOE is commonly used to characterize backbone protein dynamics, other relaxation experiments are also useful to characterize protein dynamics. Use of other than three experimental data allows an application of a more detail dynamics model than the conventional model-free model. For example, cross-correlated longitudinal (η_Z) and transverse (η_{XY}) rates between ^{1}H–^{15}N DD and ^{15}N CSA have provided useful information about protein backbone dynamics [14, 60, 61]. Since cross-correlated relaxation occurs together with auto-relaxation, the rate is obtained by multiple exponential fitting [36, 62–65]. As an alternative approach, the η_{XY} rate has been more accurately determined by taking intensity ratios of the inphase and antiphase magnetization [66].

In the analysis of cross-correlated relaxation rates to detect protein dynamics, the relative orientation between the ^{15}N–^{1}H dipole and the ^{15}N tensor ($\sigma_{\parallel}$) is required as an additional parameter for fitting the data [67, 68]. Numerous measurements of cross correlated relaxation have been used to estimate the ^{15}N CSA in protein backbone in solution [14, 15, 66, 67, 69–73]. According to these results, ^{15}N CSA is an axially symmetric CSA tensor, 169 $\pm$ 5 ppm, with a relative orientation about 21.4° $\pm$ 2.3° tilted against the N–H dipolar tensor [73]. Thus, a disadvantage of the use of η_{XY} is that this tilted angle from the N–H vector has to be included as a fixed parameter in the model-free analysis. Since the relative contribution of η_{XY} in S^2 depends on the degree of internal motion in each residue, use of η_{XY} may introduce an additional uncertainty in the model-free analysis. Therefore, it is important to clarify how much ^{15}N CSA varies site-specifically.

Recently, relaxation rate products ($2N_XH_X$, $2N_ZH_X$, $2N_XH_Z$, and $2N_ZH_Z$) have been measured to extract information about internal motion of proteins [74]. By addition and subtraction of these four terms there ideally remains only the relaxation rate, that contains the $J(0)$ term, obtained by dipolar coupling. Using this relaxation rate as well as R_1 and ^{15}N–{^{1}H} NOE values, S^2 values independent from the chemical exchange contribution were determined. The same set of data has also been applied to extract chemical exchange contribution [75].

Dynamics on a time scale much slower (i.e., ~10 ms) than can be measured by R_2 relaxation dispersion is often characterized by measuring the exchange of the longitudinal ^{1}H magnetizations among species undergoing chemical exchange

[76]. Detection of exchange crosspeaks in ^{1}H spectra is often difficult in large proteins because of interference from the many ^{1}H NOE cross peaks. In contrast, observation of such exchange through ^{15}N spectroscopy has the advantage that only exchange peaks are observed. The ^{15}N Z-exchange two-dimensional spectra are acquired using a pulse sequence similar to ^{15}N R_1 experiment but with t_1 chemical shift evolution period prior to the Z-mixing (relaxation) period [4, 77–79]. The measured rates of exchange provide information useful to characterize ligand–protein and protein–protein interactions.

4 Relaxation Dispersion Experiments

In contrast to ^{15}N R_1, R_2, and ^{15}N–{^{1}H} NOE experiments that characterize subnanosecond motions, CPMG and spin-lock relaxation experiments provide quantitative information about milli- to microsecond time scale motions. In this section, the relaxation dispersion is first defined, and, subsequently, CPMG and spin-lock R_2 dispersion experiments that have recently been developed for applications to proteins are reviewed.

4.1 Relaxation Dispersion in General

Although currently the term "relaxation dispersion" or "R_2 dispersion" often refers to CPMG or spin-lock (off-resonance or resonance $R_{1\rho}$) measurements, more generally the term refers to the relaxation rates measured as a function of magnetic field strength. Typically either the static field, B_0, provided by the spectrometer magnetic or the radio-frequency (RF) field, B_1, generated by the probe transmitter coil is varied over a wide range. B_0-dependent dispersion studies are also known as "NMR relaxometry," "field cycling," or "nuclear magnetic relaxation dispersion (NMRD)" in the literature. In these experiments R_1 of particular nucleus is measured as a function of B_0 [80–90]. In these relaxation dispersion applications, the spectral density function $J(\omega)$ is determined at numerous values of ω, allowing various dynamical features of macromolecules, such as paramagnetic interaction with proteins and residence times of water molecules in proteins to be obtained [84, 87]. The greatest advantage of the R_1 dispersion experiments is that the effective field strength varied is very wide. However, there are significant technical challenges to varying rapidly the static field strength of the samples and to increasing sensitivity [91–93]. In contrast, the dependence on R_2 on B_1 is readily measured, and has been used to study chemical exchange for a long time [94–101]. The R_2 dispersion experiment can detect such low field effects of the exchange in chemical shifts whereas the range of variable effective field strength in the R_2 dispersion is relatively small compared to that of the R_1 dispersion. R_2 dispersion experiments that are recently applied for biological systems will be described below.

4.2 ^{15}N CT-CPMG Relaxation Dispersion Experiment

The CPMG R_2 experiment has long been applied to detect chemical exchange phenomena [95, 96, 102]. A simple case is when the exchange rate is larger than the intrinsic relaxation rate, R_2^0 (determined by dipolar and CSA relaxation, details described below), and the observed R_2 is the sum of R_2^0 and the chemical exchange contribution, R_{ex}, [98, 103]:

$$R_2\tau_{CP} = R_2^0 + R_{ex}(\tau_{CP}). \tag{4}$$

Here, τ_{CP} is a half duration of CPMG interpulse delay, and the effective field strength is defined by $\nu_{CP} = 1/(4\tau_{CP})$. CPMG experiments have also been applied to characterize protein conformational equilibria and kinetics on the milli- to microsecond time scale [103–108]. The current form of the constant-time (CT) version of the experiment (Fig. 3) consists of the following two distinct steps.

One step averages the contributions to R_2^0 from the inphase and the antiphase (N_{XY} and $N_{XY}H_Z$, respectively) components by having tandem CPMG periods linked by an rc-INEPT [109]. Averaging is achieved because one CPMG period starts from the inphase coherence and the other from the antiphase coherence for all values of τ_{CP} [109]. This insures that R_2^0 remains the same at all values of τ_{CP}. Recently, use of a relatively strong ^{1}H CW field ($B_1 > 15$ kHz) that decouples the inphase and the antiphase terms has been introduced [110]. The approach is advantageous because the maximum value of τ_{CP} that can be achieved increases twofold, using the same transverse relaxation duration. However, application of a strong ^{1}H CW field may cause heating by dielectric or inductive losses for samples containing high salt at high magnet field strength [111–114].

The other step in the CPMG relaxation dispersion experiment is determination of R_2 values by two-point intensity measurement: one is the magnetization at the initial time and the other is the magnetization at time T_{CP} [115, 116]. Although

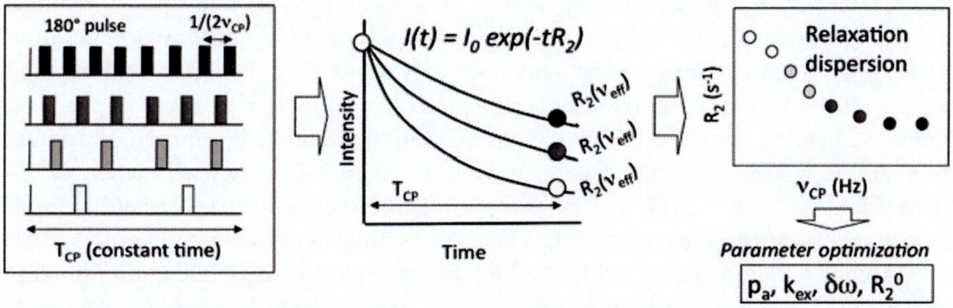

Fig. 3 Schematic flow chart of the experiment and analysis of constant-time Curr–Purcell–Meiboom–Gill (CT-CPMG) R_2 dispersion

such a two-point fitting has been studied [117, 118], it was not often applied for the conventional relaxation experiments for proteins, which record several data points to optimize the relaxation rate. For example, in the ^{15}N R_1 and R_2 experiments for the model-free analysis, it must be important to ensure that the signal decay is expressed as a single-exponential function so that the theoretical equations are applied for the analysis. This is not the case in the analysis of R_2 dispersion because R_2^0 is not dissected to extract parameters for internal motion. It will also be noteworthy that recent developments of the commercial NMR instrument have significantly increased signal-to-noise ratio, which enables two-point intensity measurements of ^{15}N R_2 of protein samples in more practical.

By using the rc-INEPT and by applying the two-point exponential fitting, the CPMG period is held constant in the CT-CPMG relaxation dispersion experiment. The template of this experiment was initially applied to probe side chain dynamics of Asn and Gln NH_2-sites, and then applied to detect backbone amide ^{15}N sites [115, 116]. There are three critical experimental parameters for the CT-CPMG experiment: 180° CPMG pulse width, p_{90}, the half duration between the CPMG pulses, τ_{CP}, and the total CPMG relaxation delay, T_{CP}. To obtain the relaxation dispersion profile, one reference spectrum without a CPMG period and a series of spectra with a fixed CPMG period (T_{CP}) but variable τ_{CP} (a half duration between 180° CPMG pulses) are recorded. Two-point exponential fitting to determine R_2 values is done for the entire set of CPMG spectra.

R_2 values are independent of ν_{CP} ($=1/(4\tau_{CP})$) when there is no chemical exchange ($R_{ex} = 0$) on the time scale similar to that of τ_{CP}. In contrast, R_2 typically decreases as ν_{CP} increases when there is chemical exchange. However, R_2 may not be independent of ν_{CP} even when there is no chemical exchange, if artifacts are introduced by off-resonance effects or CPMG pulse imperfections. To reduce these systematic errors, it is important to apply the strongest (shortest) 180° CPMG pulses possible within probe limits. Artifacts are maximized at $2\tau_{CP}f_{off} = n$ (here, the n is integer) as discussed in Sect. 2.1 [41, 119]. Typically, the author's group employs 90 μs or a shorter 180°-pulse at two different carrier frequencies, and records data up to 1 kHz ν_{CP}.

Sample heating is a more critical issue than regular CPMG R_2 experiment because τ_{CP} is shortened to achieve high ν_{CP}. In principle, heating of the samples by CPMG pulses should be avoided because heating is not uniform at varying τ_{CP} in the CT-CPMG R_2 dispersion experiment. However, one may insert a ^{15}N pulse scheme to compensate heating to perform experiments at uniform temperature [120, 121]. Such a compensation scheme for ^{1}H pulses has also been implemented in the CT-CPMG R_2 dispersion experiment with strong ^{1}H CW irradiation [110]. Once the heating is so severe that the compensation sequence is required, the actual temperatures during the experiments have to be recorded, particularly when the R_2 dispersion data are recorded at two or more static magnetic field strengths. From this aspect, it must be better to avoid heating as much as possible.

Another critical issue in ^{15}N CPMG R_2 dispersion is the magnitude of R_2^0 in large molecules. Since R_{ex} is extracted from the measured R_2, the accuracy of R_{ex} decreases when R_2^0 is large. In a pulse sequence that averages the inphase and antiphase components, ^{1}H R_1 contributes to R_2^0 as does ^{15}N R_2 [109, 116]. Thus, as the molecular tumbling becomes slower, both ^{15}N R_2 and ^{1}H R_1 increase (the later by proton spin-flip), resulting in increase in R_2^0. To overcome this problem, use of ^{1}H CW decoupling scheme decrease eliminates the ^{1}H R_1 term, resulting in R_2^0 determined only by ^{15}N R_2 [110]. Further reduction of R_2^0 is achieved by a pulse sequence in which there is an additional H_ZN_Z-relaxation period to set the ^{1}H R_1 term to be time-independent, called the "constant relaxation time" scheme [106, 122] with combination of TROSY [123, 124]. Alternatively, use of a deuterated ^{15}N protein is a simple way to reduce the ^{1}H R_1 contribution in the R_2^0 in ^{15}N CPMG R_2 dispersion.

4.3 ^{15}N Off-Resonance $R_{1\rho}$ Experiment

Characterization of milli- to microsecond motions in proteins based on chemical exchange can also be performed by the dispersion version of the ^{15}N off-resonance $R_{1\rho}$ experiment [39, 106, 125–128]. As described in the section on ^{15}N R_2 (Sect. 2.1), a simple equation, (1), is applied to determine $R_{1\rho}$ from R^{obs} and R_1. Although a strong B_1 field strength that satisfies $\omega_1 >> (\omega_0 - \omega_{rot})$ is applied to minimize the $R_1\cos^2\theta$ term in the standard (on-resonance) $R_{1\rho}$ experiment, the $(\omega_0 - \omega_{rot})$ term is significant and is varied in the off-resonance $R_{1\rho}$ experiment [40, 106]. $R_{1\rho}$ is plotted as a function of the off-resonance field strength, $\omega_e/2\pi$, given by $\omega_e = (\omega_0^2 + \omega_{rot}^2)^{1/2}$.

In theory, $R_{1\rho}$ is well-suited to record spectra at high $\omega_e/2\pi$ values because high $\omega_e/2\pi$ is achieved by increasing off-resonance field strength without increasing B_1 field strength. However, since R^{obs} decreases as $\omega_e/2\pi$ increases, a longer spin-lock period is required to record the reduced R^{obs} and this determines the limits of applicable $\omega_e/2\pi$. Although the sensitivity largely depend on sample concentration and intrinsic R_2^0 at each site, the data in the literatures cited above have been mostly acquired for 140–200 ms with $\omega_e/2\pi$ up to 2–4 kHz.

Most of the experimental parameters are the same as those of the on-resonance $R_{1\rho}$ experiment except for a couple of points. (1) Constant relaxation-time scheme is useful to subtract part of the R_1 component and to simplify the R^{obs} equation [106], and is also known as an "$R_{1\rho} - R_1$" sequence [126]. (2) A scheme to spin-lock most of the magnetization uniformly is needed. For this, a scheme with periods of evolution due to chemical shift offset [38, 39, 106] or adiabatic rotation of magnetization by amplitude and phase modulated pulses [125, 127] is applied. When there is a large chemical shift dispersion, such as a high magnetic field strength, the latter has been recommended [125]. (3) Suppression of DD/CSA cross correlation and the artifact caused by the antiphase component is required. As described in the section on ^{15}N R_2 (Sect. 2.1), sequences to take care of these effects have been used, in particular to record the data at weak off-resonance field strength [38, 39]. For application to large proteins, a TROSY-selected version has been proposed [128].

4.4 Bloch–McConnell Equation and Related Equations

Parameter optimization of CPMG R_2 dispersion data is carried out by minimizing chi squared as given by

$$x^2 = \sum_m \sum_i \left(\frac{R_{2a}^{i,exp} - R_{2a}^{i,cal}}{\sigma_{i,err}} \right) \tag{5}$$

Here, $R_{2a}{}^{i,exp}$ and $R_{2a}{}^{i,cal}$ are experimental and calculated R_2 values of *i*th σCP value, respectively, and is ex-perimental uncertainty of the *i* th' R_2 value. $R_{2a}{}^{i,exp}$ is calculated as described below. The number, m, indicates the number of residues to be analyzed. When fitting each residue, $m = 1$.$R_{2a}{}^{i,cal}$ is calculated by solving the Bloch-McConnell equation including the effects of 180° pulses iteratively or by using its analytical solutions [94, 95, 98, 103]. Typical analytical equations applied for CPMG R_2 dispersion are (1) Luz-Meiboom equation that is suitable to analyze fast exchange and easy to incorporate in optimization programs because $R_{ex}(\tau_{CP})$ is expressed by a single equation [95] or (2) Carver-Richards equation that is also suitable to analyze intermediate and fast exchange and when there are differences in $R_2{}^0$ in two sites [98, 103]. Violation of these equations in the slow limit has been well described in the literature [116]. In contrast to these analytical solutions, the Bloch-McConnell equation is applicable to any exchange regime and any relaxation rates [94]. However, since inten-sity is calculated step by step for each τ_{CP}, a relatively longer computation time is required.

When optimizing the parameters for on/off-resonance $R_{1\rho}$ experiments, the same principles apply in the minimization of χ_2 in (5). Since $R_{1\rho}$ is applied to investigate dynamics on the time scale faster than that of CPMG R_2, the fast exchange equation is often applied [39, 106]. Recently, the equation that is applicable for the slow time scale has become available, and has been applied to proteins undergoing slow conformational exchange [129–131].

4.5 Practical Aspects Parameter Optimization

Two steps are involved in optimizing the fitted exchange parameters: first, parameters are optimized for each residue ($m = 1$ in (5)) and, second, parameters are optimized for a group of residues ($m > 1$ in (5)). Overall flow is described as follows. First, the data of individual residues are fit to get verification that the R_2 dispersion data is in reasonable agreement with theory and to estimate the time scale of motions. This step may also select only the sites that exhibit significant dispersion profiles. Next, once this is done, one fits the data of a group of residues to determine global exchange parameters. When only one R_2 dispersion is fit, the maximum number of unknown parameters in the two-site exchange model is four:

p_a, k_{ex}, $\delta\omega/2\pi$, and R_2^0. In the group fits, the p_a and k_{ex} are assumed to be uniform in the group, while $\delta\omega/2\pi$ and R_2^0 are assumed to be residue specific. The total number of parameters in a group fit is therefore equal to 2 + 2*m* (*m* is the number of residues, as described in (5)). As described below, there are variations in the number of parameters depending on the models and the kinds of experimental data applied for the analysis.

In the individual fit, there are a couple of practical points for better optimization. To explain it, the following simplified fast-exchange (6) and slow-exchange (7) [97, 98, 116] are useful (Fig. 4):

$$R_2 = R_2^0 + p_a p_b (\delta\omega)^2 k_{ex} / (k_{ex}^2 + (2\pi\nu_{CP})^2) \tag{6}$$

$$R_2^a = R_2^{a0} + p_b k_{ex} - p_a k_{ex} \{\sin(\delta\omega/4\nu_{CP})/(\delta\omega/4\nu_{CP})\}. \tag{7}$$

First, as seen in (6) and (7), the parameters are not independent of each other. For example, in the fast-exchange regime (6), p_a and $\delta\omega$ are not independently determined from one relaxation dispersion profile (Fig. 4a). In such a fit for each R_2 dispersion profile, the $p_a p_b (\delta\omega)^2$ term is given as a single term, Φ_{ex}, and the k_{ex} is determined [132, 133]. In particular, these parameters are extracted from an analysis of $R_{1\rho}$, in which a strong B_1 field strength is applied and exchange is therefore assumed to be in the fast limit. In the slow exchange, ($\delta\omega/4\nu_{CP}$) term is not separated (Fig. 4b). Moreover, since the dispersion profiles in the fast and slow exchanges are similar [134], the simplified equations for each time scale may be used when the time scale of exchange has been estimated by other experiments. Second, since the parameters are not such

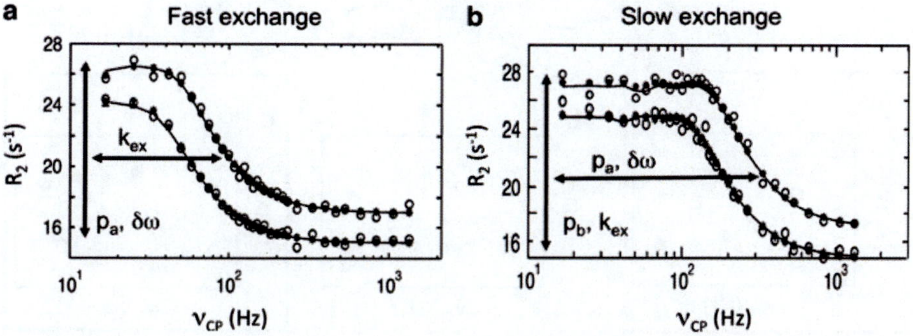

Fig. 4 Generated (**a**) fast-exchange and (**b**) slow-exchange R_2 dispersion data (*open circles*) with 1% intensity noise, and their data points obtained by fit (*asterisks*), and the fit curve (*solid lines*). In (**a**), the noisy data points were generated assuming a two-site exchange with the following parameters: $p_a = 0.9$, $k_{ex} = 100\ \text{s}^{-1}$, $\delta\omega/2\pi = 60$ Hz, and $R_2^0 = 15\ \text{s}^{-1}$ at 61 MHz. In (**b**), the noisy data points were generated assuming a two-site exchange with the following parameters: $p_a = 0.9$, $k_{ex} = 100\ \text{s}^{-1}$, $\delta\omega/2\pi = 200$ Hz, and $R_2^0 = 15\ \text{s}^{-1}$ at 61 MHz. In both cases, the two sets of data with the noise were generated assuming 61 and 81 MHz. The fit data points were slightly different from the fit curves because of the noise. The *vertical* and *horizontal bars* and parameters besides indicate the parameters that determine magnitudes of R_{ex} and the effective field strength, ν_{CP}, respectively

independent, it is safe to acquire multiple (>2) R_2 dispersion data sets at different B_0 field strength. When two sets of R_2 dispersion data recorded at two magnetic field strengths are analyzed for each residue, the unknown parameters for optimization are p_a, k_{ex}, $\delta\omega/2\pi$, and R_2^0 at one B_0 field, and R_2^0 at another B_0 field strength. Third, even when R_2 data recorded at two magnetic field strengths are available, the parameters may not be well optimized because the experimentally varied ν_{CP} range may not be sufficiently wide. Thus, it is beneficial if R_2^0 value(s) are determined by independent experiments so that the number of unknown parameters is decreased. Several reports of R_2^0 determinations have been made [14, 74, 122, 135]. However, such determination of R_2^0 by other methods may not be necessarily performed because of the added measurement time and when the group fit will be performed subsequently.

The group fit is useful to extract p_a and k_{ex} of the group of residues [116, 136–138]. As described above, p_a and k_{ex} may not be accurately determined by the individual fit.

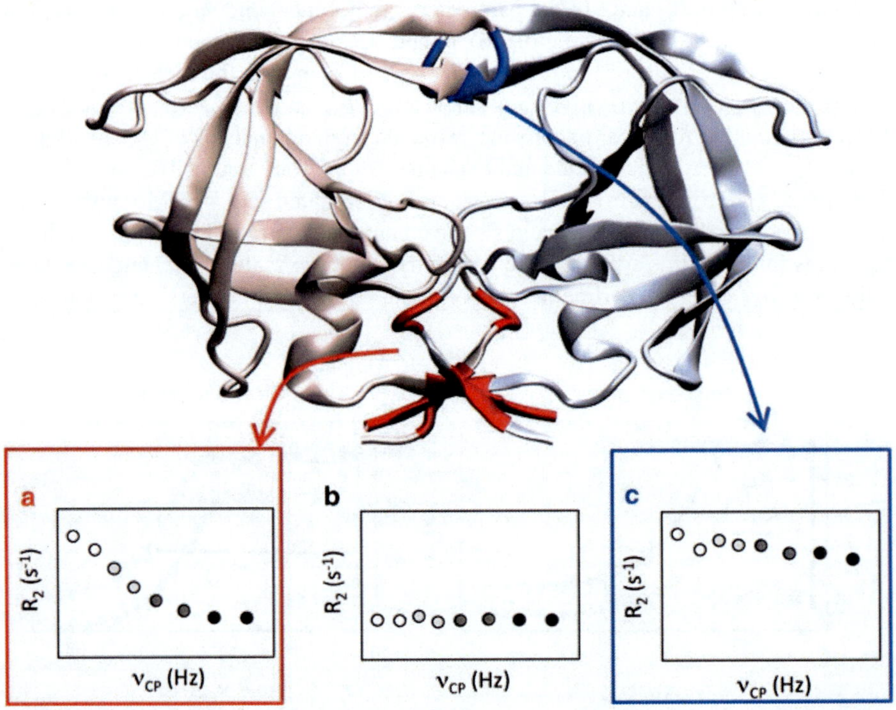

Fig. 5 Ribbon presentation of HIV-1 protease and the overview of the regions CPMG R_2 dispersion profiles for (**a**, *red*) the terminal β-sheet region, (**b**, *gray*) the core of the protein, and (**c**, *blue*) the flap region. The terminal β-sheet residues exhibited significant chemical exchange (p_a ~ 0.94 and k_{ex} ~ 650 s^{-1}) by the CPMG R_2 dispersion experiments [137]. In contrast, the flap region exhibited too large R_2 values in the CPMG R_2 dispersion experiments to be analyzed [137]. However, the flap region had been found to undergo conformational exchange by the model-free analysis previously [141]. To prevent misinterpretation of data, since CPMG R_2 dispersion experiments can detect exchange in a limited time scale, inspection of R_2 is important as well as evaluation of the optimized parameters

The group fit may be performed for (1) a group of residues and/or (2) for a set of R_2 dispersion data recorded by observing different nuclei [47, 139, 140]. The fit using a group of residues will be advantageous to identify regions that undergo conformational equilibrium in the same time scale and population, presumably indicating cooperative dynamics. Selection of the group of residues may not be straightforward because uncertainty of the optimized parameters at each residue is not necessarily Gaussian distributed [47]. In the fits of data of different types of nuclei, there may be systematic errors in individual experiments so that contour maps of the χ_2 values (5) consistent with data acquired for all types of nuclei may not be obtained [139].

Finally, a practical aspect of evaluation of regions that undergo conformational exchange is described. General technical limitation of spectroscopies, such as NMR relaxation, is that information of only a limited frequency range is obtained. R_2 dispersion experiment detects chemical (conformational) exchange on or around the ν_{CP} range. In the CT-CPMG R_2 dispersion studies of Human Immunodeficiency Virus-1 (HIV-1) protease (as depicted in Fig. 5), significant R_2 dispersion profiles were detected to optimize the exchange parameters only in the terminal β-sheet region [137]. Although the flap region exhibited very high R_2 values, the data did not have the sensitivity to be fit, presumably because the time scale of motion is much faster than the studied ν_{CP} range. Such fast dynamics (comparing to the ν_{CP}) of the HIV-1 protease has previously been characterized by the model-free analysis and amide ^{1}H $R_{1\rho}$ experiments [141, 142]. Thus, application of the R_2 dispersion with other experiments will be important to avoid misinterpretation of data.

Acknowledgments This study was supported by grants from the National Institutes of Health (AI077424 and GM066524), the National Science Foundation (MCB 0814905), and funds from the University of Pittsburgh. We thank Dennis Torchia and Robert Bryant for discussion and critical reading.

References

1. Lipari G, Szabo A (1982) Model-free approach to the interpretation of nuclear magnetic resonance relaxation in macromolecules. 1. Theory and range of validity. J Am Chem Soc 104:4546–4559
2. Lipari G, Szabo A (1982) Model-free approach to the interpretation of nuclear magnetic resonance relaxation in macromolecules. 2. Analysis of experimental results. J Am Chem Soc 104:4559–4570
3. Peng JW, Wagner G (1995) Frequency spectrum of NH bonds in eglin c from spectral density mapping at multiple fields. Biochemistry 34:16733–16752
4. Farrow NA, Zhang O, Forman-Kay JD, Kay LE (1995) Comparison of the backbone dynamics of a folded and an unfolded SH3 domain existing in equilibrium in aqueous buffer. Biochemistry 34:868–878
5. Ishima R, Nagayama K (1995) Protein backbone dynamics revealed by quasi spectral density function analysis of amide N-15 nuclei. Biochemistry 34:3162–3171
6. Kay LE, Torchia DA, Bax A (1989) Backbone dynamics of proteins as studied by nitrogen-15 inverse detected heteronuclear NMR spectroscopy: application to staphylococcal nuclease. Biochemistry 28:8972–8979

7. Clore GM, Driscoll PC, Wingfield PT, Gronenborn AM (1990) Analysis of the backbone dynamics of interleukin-1 beta using two-dimensional inverse detected heteronuclear ^{15}N-1H NMR spectroscopy. Biochemistry 29:7387–7401
8. Palmer AG, Rance M, Wright PE (1991) Intramolecular motions of a zinc finger DNA-binding domain from Xfin characterized by proton-detected natural abundance 13 C heteronuclear NMR spectroscopy. J Am Chem Soc 113:4371–4380
9. Schneider DM, Dellwo MJ, Wand AJ (1992) Fast internal main-chain dynamics of human ubiquitin. Biochemistry 31:3645–3652
10. Mandel AM, Akke M, Palmer AG (1995) Backbone dynamics of *Escherichia-coli* ribonuclease Hi – correlations with structure and function in an active enzyme. J Mol Biol 246:144–163
11. Tjandra N, Kuboniwa H, Ren H, Bax A (1995) Rotational dynamics of calcium-free calmodulin studied by ^{15}N-NMR relaxation measurements. Eur J Biochem 230:1014–1024
12. Dayie KT, Wagner G, Lefevre JF (1996) Theory and practice of nuclear spin relaxation in proteins. Annu Rev Phys Chem 47:243–282
13. Lee LK, Rance M, Chazin WJ, Palmer AGr (1997) Rotational diffusion anisotropy of proteins from simultaneous analysis of 15N and 13C alpha nuclear spin relaxation. J Biomol NMR 9:287–298
14. Kroenke CD, Loria JP, Lee LK, Rance M, Palmer AG (1998) Longitudinal and transverse H-1-N-15 dipolar N-15 chemical shift anisotropy relaxation interference: unambiguous determination of rotational diffusion tensors and chemical exchange effects in biological macromolecules. J Am Chem Soc 120:7905–7915
15. Fushman D, Tjandra N, Cowburn D (1999) An approach to direct determination of protein dynamics from N-15 NMR relaxation at multiple fields, independent of variable N-15 chemical shift anisotropy and chemical exchange contributions. J Am Chem Soc 121:8577–8582
16. Andrec M, Montelione GT, Levy RM (1999) Estimation of dynamic parameters from NMR relaxation data using the Lipari-Szabo model-free approach and Bayesian statistical methods. J Magn Reson 139:408–421
17. Lee AL, Wand AJ (1999) Assessing potential bias in the determination of rotational correlations times of proteins by NMR. J Biomol NMR 13:101–112
18. Campbell AP, Spyracopoulos L, Irvin RT, Sykes BD (2000) Backbone dynamics of a bacterially expressed peptide from the receptor binding domain of Pseudomonas aeruginosa pilin strain PAK from heteronuclear ^{1}H-15N NMR spectroscopy. J Biomol NMR 17:239–255
19. Ishima R, Torchia DA (2000) Protein dynamics from NMR. Nat Struct Biol 7:740–743
20. Fushman D, Cowburn D (2001) Nuclear magnetic resonance relaxation in determination of residue-specific N-15 chemical shift tensors in proteins in solution: protein dynamics, structure, and applications of transverse relaxation optimized spectroscopy. Methods Enzymol Nucl Magn Reson Biol Macromol Pt B 339:109–126
21. Palmer AG 3rd (2001) NMR probes of molecular dynamics: overview and comparison with other. Annu Rev Biophys Biomol Struct 30:129–155
22. Bruschweiler R (2003) New approaches to the dynamic interpretation and prediction of NMR relaxation data from proteins. Curr Opin Struct Biol 13:175–183
23. d'Auvergne EJ, Gooley PR (2003) The use of model selection in the model-free analysis of protein dynamics. J Biomol NMR 25:25–39
24. Pelupessy P, Ravindranathan S, Bodenhausen G (2003) Correlated motions of successive amide N-H bonds in proteins. J Biomol NMR 25:265–280
25. Idiyatullin D, Daragan VA, Mayo KH (2003) (NH)-N-15 backbone dynamics of protein GB1: comparison of order parameters and correlation times derived using various "model-free" approaches. J Phys Chem B 107:2602–2609
26. Redfield C (2004) Using nuclear magnetic resonance spectroscopy to study molten globule states of proteins. Methods Mol Biol 34:121–132
27. Chen J, Brooks CLI, Wright PE (2004) Model-free analysis of protein dynamics: assessment of accuracy and model selection protocols based on molecular dynamics simulation. J Biomol NMR 29:243–257

28. Korchuganov DS, Gagnidze IE, Tkach EN, Schulga AA, Kirpichnikov MP, Arseniev AS (2004) Determination of protein rotational correlation time from NMR relaxation data at various solvent viscosities. J Biomol NMR 30:431–442
29. Kay LE (2005) NMR studies of protein structure and dynamics. J Magn Reson 173:193–207
30. Jarymowycz VA, Stone MJ (2006) Fast time scale dynamics of protein backbones: NMR relaxation methods, applications, and functional consequences. Chem Rev 106:1624–1671
31. Igumenova TI, Frederick KK, Wand AJ (2006) Characterization of the fast dynamics of protein amino acid side chains using NMR relaxation in solution. Chem Rev 106:1672–1699
32. Spyracopoulos L (2006) A suite of Mathematica notebooks for the analysis of protein main chain (15)N NMR relaxation data. J Biomol NMR 36:215–224
33. Frederick KK, Sharp KA, Warischalk N, Wand AJ (2008) Re-evaluation of the model-free analysis of fast internal motion in proteins using NMR relaxation. J Phys Chem B 112: 12095–12103
34. Reddy T, Rainey JK (2010) Interpretation of biomolecular NMR spin relaxation parameters. Biochem Cell Biol 88:131–142
35. Peng JW, Thanabal V, Wagner G (1991) Improved accuracy of heteronuclear transverse relaxation time measurements in macromolecules. elimination of antiphase contributions. J Magn Reson 95:421–427
36. Palmer AG, Skelton NJ, Chazin WJ, Wright PE, Rance M (1992) Suppression of the effects of cross-correlation between dipolar and anisotropic chemical-shift relaxation mechanisms in the measurement of spin relaxation rates. Mol Phys 75:699–711
37. Kay LE, Nicholson LK, Delaglio F, Bax A, Torchia DA (1992) Pulse sequences for removal of the effects of cross correlation between dipolar and chemical-shift anisotropy relaxation mechanisms on the measurement of heteronuclear T1 and T2 values in proteins. J Magn Reson 97:359–375
38. Korzhnev DM, Skrynnikov NR, Millet O, Torchia DA, Kay LE (2002) An NMR experiment for the accurate measurement of heteronuclear spin-lock relaxation rates. J Am Chem Soc 124:10743–10753
39. Massi F, Johnson E, Wang C, Rance M, Palmer AGr (2004) NMR R1 rho rotating-frame relaxation with weak radio frequency fields. J Am Chem Soc 126:2247–2256
40. Guenneugues M, Berthault P, Desvaux H (1999) A method for determining B1 field inhomogeneity. Are the biases assumed in heteronuclear relaxation experiments usually underestimated? J Magn Reson 136:118–126
41. Ross A, Czisch M, King GC (1997) Systematic errors associated with the CPMG pulse sequence and their effect on motional analysis of biomolecules. J Magn Reson 124:355–365
42. Korzhnev DM, Tischenko EV, Arseniev AS (2000) Off-resonance effects in N-15 T-2 CPMG measurements. J Biomol NMR 17:231–237
43. Gullion T, Baker DB, Conradi MS (1990) New, compensated Carr-Purcell sequences. J Magn Reson 89:479–484
44. Yip GN, Zuiderweg ER (2004) A phase cycle scheme that significantly suppresses offset-dependent artifacts in the R2-CPMG 15 N relaxation experiment. J Magn Reson 171:25–36
45. Long D, Liu M, Yang D (2008) Accurately probing slow motions on millisecond timescales with a robust NMR relaxation experiment. J Am Chem Soc 130:2432–2433
46. Myint W, Gong Q, Ishima R (2009) Practical aspects of 15 N CPMG transverse relaxation experiments for proteins in solution. Concepts Magn Reson 34A:63–75
47. Ishima R, Torchia DA (2003) Extending the range of amide proton relaxation dispersion experiments in proteins using a constant-time relaxation-compensated CPMG approach. J Biomol NMR 25:243–248
48. Abragam A (1961) Principles of nuclear magnetism. Oxford University Press, Oxford
49. Freeman R, Hill HDW (1971) Fourier transform study of NMR spin-lattice relaxation by "progressive saturation". J Chem Phys 54:3367–3377
50. Markley JL, Horsley WJ, Klein MP (1971) Spin–lattice relaxation measurements in slowly relaxing complex spectra. J Chem Phys 55:3604–3605

51. Chill JH, Louis JM, Baber JL, Bax A (2006) Measurement of 15N relaxation in the detergent-solubilized tetrameric KcsA potassium channel. J Biomol NMR 36:123–136
52. Grzesiek S, Bax A (1993) The importance of not saturating H_2O in protein NMR – application to sensitivity enhancement and NOE measurements. J Am Chem Soc 115:12593
53. Li YC, Montelione GT (1994) Overcoming solvent saturation-transfer artifacts in protein NMR at neutral pH – application of pulsed-field gradients in measurements of H-1 N-15 Overhauser effects. J Magn Reson Ser B 105:45–51
54. Renner C, Schleicher M, Moroder L, Holak TA (2002) Practical aspects of the 2D N-15-{H-1}-NOE experiment. J Biomol NMR 23:23–33
55. Freedberg DI, Ishima R, Jacob J, Wang YX, Kustanovich I, Louis JM, Torchia DA (2002) Rapid structural fluctuations of the free HIV protease flaps in solution. Protein Sci 11:221–232
56. Gong Q, Ishima R (2007) 15 N-{^{1}H} NOE experiment at high magnetic field strengths. J Biomol NMR 37:147–157
57. Ferrage F, Piserchio A, Cowburn D, Ghose R (2008) On the measurement of 15 N-{^{1}H} nuclear Overhauser effects. J Magn Reson 192:302–313
58. Ferrage F, Cowburn D, Ghose R (2009) Accurate sampling of high-frequency motions in proteins by steady-state (15)N-{(1)H} nuclear Overhauser effect measurements in the presence of cross-correlated relaxation. J Am Chem Soc 131:6048–6049
59. Rule GS, Hichens KT (2006) Fundamentals of protein NMR spectroscopy, vol 5. Springer, Dordrecht
60. Vugmeyster L, Raleigh DP, Palmer AGr, Vugmeister BE (2003) Beyond the decoupling approximation in the model free approach for the interpretation of NMR relaxation of macromolecules in solution. J Am Chem Soc 125:8400–8404
61. Vugmeyster L, McKnight CJ (2008) Slow motions in Chicken Villin headpiece subdomain probed by cross-correlated NMR relaxation of amide NH bonds in successive residues. Biophys J 95:5941–5950
62. Goldman M (1984) Interference effects in the relaxation of a pair of unlike spin-1/2 nuclei. J Magn Reson 60:437–452
63. Boyd J, Hommel U, Campbell ID (1990) Influence of cross-correlation between dipolar and anisotropic chemical shift relaxation mechanisms upon longitudinal relaxation rates of 15 N in macromolecules. Chem Phys Lett 175:477–482
64. Wang L, Kurochkin AV, Zuiderweg ER (2000) An iterative fitting procedure for the determination of longitudinal NMR cross-correlation rates. J Magn Reson 144:175–185
65. Bouguet-Bonnet S, Mutzenhardt P, Canet D (2004) Measurement of 15 N csa/dipolar cross-correlation rates by means of spin state selective experiments. J Biomol NMR 30:133–142
66. Tjandra N, Szabo A, Bax A (1996) Protein backbone dynamics and N-15 chemical shift anisotropy from quantitative measurement of relaxation interference effects. J Am Chem Soc 118:6986–6991
67. Fushman D, Cowburn D (1998) Model-independent analysis of 15 N chemical shift anisotropy from NMR relaxation data. Ubiquitin as a test example. J Am Chem Soc 120:7109–7110
68. Fushman D, Tjandra N, Cowburn D (1998) Direct measurement of N-15 chemical shift anisotropy in solution. J Am Chem Soc 120:10947–10952
69. Kroenke CD, Rance M, Palmer AGr (1999) Variability of the 15N chemical shift anisotropy in *Escherichia coli* Ribonuclease H in Solution. J Am Chem Soc 121:10119–10125
70. Boyed J, Redfield C (1999) Characterization of 15 N chemical shift anisotropy from orientation-dependent changes to 15 N chemical shifts in dilute bicelle solutions. J Am Chem Soc 121:7441–7442
71. Kurita J, Shimahara H, Utsunomiya-Tate N, Tate S (2003) Measurement of 15N chemical shift anisotropy in a protein dissolved in a dilute liquid crystalline medium with the application of magic angle sample spinning. J Magn Reson 163:163–173
72. Lipsitz RS, Tjandra N (2003) 15N chemical shift anisotropy in protein structure refinement and comparison with NH residual dipolar couplings. J Magn Reson 164:171–176

73. Damberg P, Jarvet J, Gräslund A (2005) Limited variations in 15N CSA magnitudes and orientations in ubiquitin are revealed by joint analysis of longitudinal and transverse NMR relaxation. J Am Chem Soc 127:1995–2005
74. Hansen DF, Yang D, Feng H, Zhou Z, Wiesner S, Bai Y, Kay LE (2007) An exchange-free measure of 15N transverse relaxation: an NMR spectroscopy application to the study of a folding intermediate with pervasive chemical exchange. J Am Chem Soc 129:11468–11478
75. Hansen DF, Feng H, Zhou Z, Bai Y, Kay LE (2009) Selective characterization of microsecond motions in proteins by NMR relaxation. J Am Chem Soc 131:16257–16265
76. Jeener J, Meier BH, Bachmann P, Ernst RR (1979) Investigation of exchange processes by two-dimensional NMR spectroscopy. J Chem Phys 71:4546–4553
77. Sahu D, Clore GM, Iwahara J (2007) TROSY-based z-exchange spectroscopy: application to the determination of the activation energy for intermolecular protein translocation between specific sites on different DNA molecules. J Am Chem Soc 129:13232–13237
78. Li Y, Palmer AG (2009) TROSY-selected ZZ-exchange experiment for characterizing slow chemical exchange in large proteins. J Biomol NMR 45:357–360
79. Robson SA, Peterson R, Bouchard LS, Villareal VA, Clubb RT (2010) A heteronuclear zero quantum coherence Nz-exchange experiment that resolves resonance overlap and its application to measure the rates of heme binding to the IsdC protein. J Am Chem Soc 132:9522–9523
80. Fite W II, Redfield AG (1967) Nuclear spin relaxation in superconduting mixed-state vanadium. Phys Rev 162:358–367
81. Koenig SH, Schillinger WE (1969) Nuclear magnetic relaxation dispersion in protein solutions. J Biol Chem 244:3283–3289
82. Kimmich R (1979) Field cycling in NMR relaxation spectroscopy: applications in biological, chemical and polymer physics. Bull Magn Reson 1:195–218
83. Noack F (1986) NMR field-cycling spectroscopy: principles and applications. Prog NMR Spectrosc 18:171–276
84. Bertini I, Briganti F, Xia ZC, Luchinat C (1993) Nuclear magnetic relaxation dispersion studies of hexaaquo Mn(II) ions in water-glycerol mixtures. J Magn Reson A 101:198–201
85. Hodges MW, Cafiso DS, Polnaszek CF, Lester CC, Bryant RG (1997) Water translational motion at the bilayer interface: an NMR relaxation dispersion measurement. Biophys J 75: 2575–2579
86. Koenig SH, Brown RD (1990) Field-cycling relaxometry of protein solutions and tissue: implications for MRI. Prog NMR Spectrosc 22:487–567
87. Halle B, Denisov VP (1995) A new view of water dynamics in immobilized proteins. Biophys J 69:242–249
88. Roberts MF, Redfield AG (2004) Phospholipid bilayer surface configuration probed quantitatively by P-31 field-cycling NMR. Proc Natl Acad Sci USA 101:17066–17071
89. Kimmich R, Anoardo E (2004) Field-cycling NMR relaxometry. Prog NMR Spectrosc 44: 257–320
90. Diakova G, Goddard YA, Korb JP, Bryant RG (2010) Water and backbone dynamics in a hydrated protein. Biophys J 98:138–146
91. Wagner S, Dinesen TRJ, Rayner T, Bryant RG (1999) High-resolution magnetic relaxation dispersion measurements of solute spin probes using a dual-magnet system. J Magn Reson 140:172–178
92. Redfield AG (2003) Shuttling device for high-resolution measurements of relaxation and related phenomena in solution at low field, using a shared commercial 500 MHz NMR instrument. Magn Reson Chem 41:753–768
93. Victor K, Kavolius V, Bryant RG (2004) Magnetic relaxation dispersion probe. J Magn Reson 171:253–257
94. McConnell HM (1958) Reaction rates by nuclear magnetic resonance. J Chem Phys 28: 430–431

95. Luz Z, Meiboom S (1963) Nuclear magnetic resonance study of the protolysis of trimethylammonium ion in aqueous solution – order of the reaction with respect to solvent. J Chem Phys 39:366–370
96. Allerhand A, Gutowsky HS (1964) Spin-echo NMR studies of chemical exchange. 1. Some general aspects. J Chem Phys 41:2115
97. Gutowsky HS, Vold RL, Wells EJ (1965) Theory of chemical exchange effects in magnetic resonance. J Chem Phys 43:4107–4125
98. Carver JP, Richards RE (1972) General 2-site solution for chemical exchange produced dependence of T2 upon Carr-Purcell pulse separation. J Magn Reson 6:89–105
99. Jen J (1978) Chemical exchange and NMR T2 relaxation – multisite case. J Magn Reson 30: 111–128
100. Bleich HE, Day AR, Freer RJ, Glasel JA (1979) NMR rotating frame relaxation studies of intramolecular motion in peptides. Tyrosine ring motion in methionine-enkephalin. Biochem Biophys Res Commun 87:1146–1153
101. Snyder GH, Rowan R III, Karplus S, Sykes BD (1979) NMR rotating frame relaxation studies of intramolecular motion in peptides. Tyrosine ring motion in methionine-enkephalin. Biochem Biophys Res Commun 87:1146–1153
102. Bloom M, Reeves LW, Wells EJ (1965) Spin echoes and chemical exchange. J Chem Phys 42:1615–1624
103. Davis DG, Perlman ME, London RE (1994) Direct measurements of the dissociation-rate constant for inhibitor-enzyme complexes via the T-1-Rho and T-2 (Cpmg) methods. J Magn Reson Ser B 104:266–275
104. Szyperski S, Luginbühl P, Otting G, Güntert P, Wüthrich K (1993) Protein dynamics studied by rotating frame. J Biomol NMR 3:151–164
105. Orekhov VY, Pervushin KV, Arseniev AS (1994) Backbone dynamics of (1–71)bacterioopsin studied by two-dimensional ^{1}H-15N NMR spectroscopy. Eur J Biochem 219:887–896
106. Akke M, Palmer AGr (1996) Monitoring macromolecular motions on microsecond to millisecond time scales by $R1\rho-R1$ constant relaxation time NMR spectroscopy. J Am Chem Soc 118:911–912
107. Zinn-Justin S, Berthault P, Guenneugues M, Desvaus H (1997) Off-resonance rf fields in heteronuclear NMR. Application to the study of slow motions. J Biomol NMR 10:363–372
108. Mulder FA, van Tilborg PJ, Kaptein R, Boelens R (1999) Microsecond time scale dynamics in the RXR DNA-binding domain from a combination of spin-echo and off-resonance rotating frame relaxation measurements. J Biomol NMR 13:275–288
109. Loria JP, Rance M, Palmer AG (1999) A relaxation-compensated Carr-Purcell-Meiboom-Gill sequence for characterizing chemical exchange by NMR spectroscopy. J Am Chem Soc 121:2331–2332
110. Hansen DF, Vallurupalli P, Kay LE (2007) An improved (15)N relaxation dispersion experiment for the measurement of millisecond time-scale dynamics in proteins. J Phys Chem B 112(19):5898–5904
111. Gadian DG, Robinson FNH (1979) Radiofrequency losses in NMR experiments on electrically conducting samples. J Magn Reson 34:449–455
112. Hoult DI, Lauterbur PC (1979) The sensitivity of the zeugmatographic experiment involving human samples. J Magn Reson 34:425–433
113. Kelly AE, Ou HD, Withers R, Dotsch V (2002) Low-conductivity buffers for high-sensitivity NMR measurements. J Am Chem Soc 124:12013–12019
114. Horiuchi T, Takahashi M, Kikuchi J, Yokoyama S, Maeda H (2005) Effect of dielectric properties of solvents on the quality factor for a beyond 900 MHz cryogenic probe model. J Magn Reson 174:34–42
115. Mulder FA, Skrynnikov NR, Hon B, Dahlquist FW, Kay LE (2001) Measurement of slow (micros-ms) time scale dynamics in protein side chains by (15)N relaxation dispersion NMR spectroscopy: application to Asn and Gln residues in a cavity mutant of T4 lysozyme. J Am Chem Soc 123:967–975

116. Tollinger M, Skrynnikov NR, Mulder FA, Forman-Kay JD, Kay LE (2001) Slow dynamics in folded and unfolded states of an SH3 domain. J Am Chem Soc 123:11341–11352
117. Jones JA, Hodgkinson P, Barker AL, Hore PJ (1996) Optimal sampling strategies for the measurement of spin-spin relaxation times. J Magn Reson Seres B 113:25–34
118. Jones JA (1997) Optimal sampling strategies for the measurement of relaxation times in proteins. J Magn Reson 126:283–286
119. Czisch M, King GC, Ross A (1997) Removal of systematic errors associated with off-resonance oscillations in T2 measurements. J magn Reson 126:154–157
120. Wang AC, Bax A (1993) Minimizing the effects of radio-frequency heating in multidimensional NMR experiments. J Biomol NMR 3:715–720
121. Yip GN, Zuiderweg ER (2005) Improvement of duty-cycle heating compensation in NMR spin relaxation experiments. J Magn Reson 176:171–178
122. Wang C, Grey MJ, Palmer AG 3rd (2001) CPMG sequences with enhanced sensitivity to chemical exchange. J Biomol NMR 21:361–366
123. Loria JP, Rance M, Palmer AG (1999) A TROSY CPMG sequence for characterizing chemical exchange in large proteins. J Biomol NMR 15:151–155
124. Wang CY, Palmer AG (2003) Solution NMR methods for quantitative identification of chemical exchange in N-15-labeled proteins. Magn Reson Chem 41:866–876
125. Mulder FAA, de Graaf RA, Kaptein R, Boelens R (1998) An off-resonance rotating frame relaxation experiment for the investigation of macromolecular dynamics using adiabatic rotations. J Magn Reson 131:351–357
126. Kempf JG, Jung JY, Sampson NS, Loria JP (2003) Off-resonance TROSY (R1 rho – R1) for quantitation of fast exchange processes in large proteins. J Am Chem Soc 125:12064–12065
127. Kim S, Baum J (2004) An on/off resonance rotating frame relaxation experiment to monitor millisecond to microsecond timescale dynamics. J Biomol NMR 30:195–204
128. Igumenova TI, Palmer AG (2006) Off-resonance TROSY-selected R-1p experiment with improved sensitivity for medium- and high-molecular-weight proteins. J Am Chem Soc 128: 8110–8111
129. Trott O, Palmer AGr (2002) R1rho relaxation outside of the fast-exchange limit. J Magn Reson 154:157–160
130. Korzhnev DM, Orekhov VY, Dahlquist FW, Kay LE (2003) Off-resonance R1rho relaxation outside of the fast exchange limit: an experimental study of a cavity mutant of T4 lysozyme. J Biomol NMR 26:39–48
131. Miloushev VZ, Palmer AGr (2005) R(1rho) relaxation for two-site chemical exchange: general approximations and some exact solutions. J Magn Reson 177:221–227
132. Vugmeyster L, Kroenke CD, Picart F, Palmer AG, Raleigh DP (2000) N-15 R-1 rho measurements allow the determination of ultrafast protein folding rates. J Am Chem Soc 122: 5387–5388
133. Massi F, Grey MJ, Palmer AGr (2005) Microsecond timescale backbone conformational dynamics in ubiquitin studied with NMR R1rho relaxation experiments. Protein Sci 14: 735–742
134. Ishima R, Torchia DA (1999) Estimating the time scale of chemical exchange of proteins from. J Biomol NMR 14:369–372
135. Gardino AK, Kern D (2007) Functional dynamics of response regulators using NMR relaxation techniques. Methods Enzymol 423:195–205
136. Korzhnev DM, Salvatella X, Vendruscolo M, Di Nardo AA, Davidson AR, Dobson CM, Kay LE (2004) Low-populated folding intermediates of Fyn SH3 characterized by relaxation. Nature 430:586–590
137. Ishima R, Torchia DA (2005) Error estimation and global fitting in transverse-relaxation dispersion. J Biomol NMR 32:41–54
138. Kovrigin EL, Kempf JG, Grey MJ, Loria JP (2006) Faithful estimation of dynamics parameters from CPMG relaxation dispersion measurements. J Magn Reson 180:83–104

139. Hansen DF, Vallurupalli P, Lundström P, Neudecker P, Kay LE (2008) Probing chemical shifts of invisible states of proteins with relaxation dispersion NMR spectroscopy: how well can we do? J Am Chem Soc 130:2667–2675
140. Lundström P, Vallurupalli P, Hansen DF, Kay LE (2009) Isotope labeling methods for studies of excited protein states by relaxation dispersion NMR spectroscopy. Nat Protoc 4: 1641–1648
141. Nicholson LK, Yamazaki T, Torchia DA, Grzesiek S, Bax A, Stahl SJ, Kaufman JD, Wingfield PT, Lam PY, Jadhav PK, Hodge CN, Domaille PJ, Chong-Hwan C (1995) Flexibility and function in HIV-1 protease. Nat Struct Biol 2:274–280
142. Ishima R, Wingfield PT, Stahl SJ, Kaufman JD, Torchia DA (1998) Using amide H-1 and N-15 transverse relaxation to detect millisecond time-scale motions in perdeuterated proteins: application to HIV-1 protease. J Am Chem Soc 120:10534–10542

Top Curr Chem (2012) 326: 123–186
DOI: 10.1007/128_2011_306

Published online: 13 December 2011

Contemporary Methods in Structure Determination of Membrane Proteins by Solution NMR

Tabussom Qureshi and Natalie K. Goto

Abstract Integral membrane proteins are vital to life, being responsible for information and material exchange between a cell and its environment. Although high-resolution structural information is needed to understand how these functions are achieved, membrane proteins remain an under-represented subset of the protein structure databank. Solution NMR is increasingly demonstrating its ability to help address this knowledge shortfall, with the development of a diverse array of techniques to counter the challenges presented by membrane proteins. Here we document the advances that are helping to define solution NMR as an effective tool for membrane protein structure determination. Developments introduced over the last decade in the production of isotope-labeled samples, reconstitution of these samples into the growing selection of NMR-compatible membrane-mimetic systems, and the approaches used for the acquisition and application of structural restraints from these complexes are reviewed.

Keywords Detergents · Global folds · Isotope labeling · NMR structure restraints · Protein expression

Contents

1 Introduction 125
2 Production of Membrane Protein Samples for Solution NMR 126
  2.1 Factors Affecting Choice of Expression System 126
  2.2 Membrane Protein Expression in *E. Coli* 126
  2.3 Cell-Free Expression Approaches 127

T. Qureshi and N.K. Goto (✉)
Department of Chemistry and Department of Biochemistry, Microbiology and Immunology, University of Ottawa, Ottawa, ON K1N 6N5, Canada
e-mail: ngoto@uottawa.ca

3 Membrane Mimetic Systems for Solution NMR 129
3.1 Micelles 129
3.2 Bicelles 140
3.3 New Developments in Membrane-Mimetic Solvents for Solution NMR 142
3.4 Approaches for Sample Screening 146
4 Methods for Acquisition of Solution NMR Spectra of Membrane Proteins 147
4.1 Isotope Labeling Schemes for Large Protein Complexes 148
4.2 Relaxation Optimization 152
5 Strategies for Membrane Protein Structure Determination by Solution NMR 154
5.1 NOE-Based Methods of Structure Determination 154
5.2 Non-uniform Sampling 155
5.3 Global Folds in the Absence of Long-Range NOEs 157
6 Concluding Remarks 164
References 165

Abbreviations

β-OG	β-Octyl glucoside
14-*O*-PC	1,2-Di-*O*-tetradecyl-*sn*-glycero-3-phosphocholine
6-*O*-PC	1,2-Di-*O*-hexyl-*sn*-glycero-3-phosphocholine
C_6-DHPC	1,2-Dihexanoyl-*sn*-glycero-3-phosphocholine
C_7-DHPC	1,2-Diheptanoyl-*sn*-glycero-3-phosphocholine
CF	Cell-free
CHAPS	3-[(3-Cholamidopropyl) dimethylammonio]-1-propanesulfonate
COSY	Correlation spectroscopy
CSA	Chemical shift anisotropy
CTAB	Cetyl-trimethylammonium bromide
DAGK	Diacylglycerol kinase
DDM	Dodecylmaltoside
DHAP	Dihexadecyldimethylammonium bromide
DPC	Dodecylphosphocholine
DTAB	Dodecyl-trimethylammonium bromide
EPR	Electron paramagnetic resonance
gA	Gramicidin A
GpA	Glycophorin A
GPCR	G-protein coupled receptor
HSQC	Heteronuclear single quantum coherence
LDAO	Lauryldimethylamine-oxide
LMPC	Lyso-myristoyl phosphatidylcholine
LMPG	Lyso-myristoyl phosphatidylglycerol
LPPG	Lyso-palmitoyl phosphatidylglycerol
MDD	Multidimensional decomposition
MSP	Membrane scaffold protein
MTSL	1-Oxyl-2,2,5,5-tetramethyl-3-pyrroline-3-methyl-methanethiosulfonate
NMR	Nuclear magnetic resonance

NOE	Nuclear Overhauser enhancement
NOESY	Nuclear Overhauser spectroscopy
NUS	Non-uniform sampling
Omp	Outer membrane protein
PRE	Paramagnetic relaxation enhancement
pSRII	Photosensitive rhodopsin II
RCSA	Residual chemical shift anisotropy
RDC	Residual dipolar coupling
SDS	Sodium dodecyl sulfate
TM	Transmembrane
TROSY	Transverse relaxation spectroscopy
UNC2	Uncoupling protein 2
VDAC	Voltage dependent anion channel

1 Introduction

Membrane proteins confer a remarkable array of functionalities to the membranes that define cellular boundaries [1, 2]. They are responsible for the controlled transport of nutrients, electrolytes, signaling agents, and toxins across an otherwise inert lipid bilayer, and also make it possible for a cell to sense and communicate with its environment, a vital process for a wide range of biological events. The fact that alterations in membrane protein function are linked to a number of disease states; e.g., cystic fibrosis, Alzheimer's disease, and long QT syndrome [3–6], and that 50% of known drug targets are membrane proteins [7, 8], has made this class of proteins an attractive target in drug discovery efforts. Consequently there is a high level of interest in understanding how membrane proteins function at the atomic level, and in finding ways in which these functions can be disrupted or enhanced. High-resolution structures greatly facilitate efforts to address these issues, yet at present there are only ~300 unique membrane proteins for which structures have been determined (http://blanco.biomol.uci.edu/mpstruc/listAll/list). Although a large number of these have been provided by X-ray crystallography, relatively recent developments in the study of large protein complexes by solution NMR have greatly increased the ability of this approach to provide important insights into membrane protein structure and function.

Solution NMR has unique capabilities to provide structural insights for proteins that are refractory to crystallization [9–11], and to characterize functionally relevant dynamic processes at atomic resolution [12–15]. However, the hydrophobic nature of membrane proteins greatly complicates handling and biophysical analyses in general. This gives rise to significant challenges, particularly for solution NMR of membrane proteins, in (1) development of cost-effective strategies to produce isotopically labeled membrane protein samples, (2) identification of detergent or lipid solutions that can maintain the protein in a folded, soluble state with a complex size that would be compatible with solution NMR, and (3) acquisition of the NMR data required to determine their structures. We will outline current approaches being used to address these challenges and review the progress being made to increase the range of membrane protein systems that can be studied by solution NMR. It should be noted that solid-state NMR also continues to make impressive advances in the study of membrane protein structure and dynamics, as interested readers can refer to in these authoritative reviews [16–19].

2 Production of Membrane Protein Samples for Solution NMR

2.1 *Factors Affecting Choice of Expression System*

Solution NMR of membrane proteins usually requires that uniformly ^{15}N, ^{13}C-labeled samples be produced to facilitate chemical shift and NOE assignments. Samples with uniform or site-specific incorporation of ^{2}H atoms are also often needed to reduce the number of unfavorable relaxation pathways that can significantly attenuate the NMR signal in large protein–detergent complexes [20]. Although the price of these isotopes has come down over the last decade, cost-effective production of multiple NMR samples still requires the use of expression systems that can produce a high yield of the target protein from the simplest possible metabolic precursors. *Escherichia coli* is widely regarded as an ideal expression host for this purpose since it is simple to use, can produce high levels of protein, and has available a large variety of expression plasmids and strains for this purpose [21–23]. In addition, condensed phase approaches that allow large volumes of bacterial culture in unlabeled media to be resuspended in reduced volumes of labeled media for protein expression can further reduce associated costs [24–28]. Although yeast and mammalian cell expression systems are also being developed as alternate sources of isotope-labeled protein [29–33], the convenience of the bacterial expression system has preserved its dominance as a host for NMR sample production. Notably, almost all membrane protein structures that have been determined by solution NMR to date were produced using proteins expressed in *E. coli*, or peptides produced through solid-phase chemical synthesis (nicely summarized at http://www.drorlist.com/nmr/MPNMR.html).

2.2 *Membrane Protein Expression in* E. Coli

Special considerations for membrane protein expression in *E. coli* include issues of targeting; ideally the expressed protein can be incorporated into the bacterial cell membrane, allowing extraction of folded samples from detergent-solubilized cell membranes [34–37]. This has been the case for the small number of polytopic helical membrane protein structures that were successfully determined by solution NMR, namely diacylglycerol kinase (DAGK) [38], the disulfide bond isomerase DsbB [39], sensory rhodopsin II (pSRII) [40], and the mitochondrial uncoupling protein 2 (UCP2) [41]. However insertion of large amounts of expressed protein in the cell membrane is often not well tolerated by the host [42, 43]. Compromised cell viability may arise from changes in lipid bilayer fluidity [44], or by overwhelming the cytoplasmic protein translocation machinery, fundamentally altering the composition of both the cell envelope and cytoplasmic proteome [45–47]. Special strains have been developed that better tolerate the stresses of toxic protein expression [48] that have proven useful for some membrane proteins [49, 50]. However, even in these systems, many membrane proteins are produced at levels that are too low to facilitate structural studies by solution NMR.

One strategy that has been used to circumvent these toxicity issues is to express the membrane protein as inclusion bodies, thereby avoiding insertion into the membrane [51–53]. These insoluble aggregates of misfolded proteins are usually non-toxic to the host cell [52]. Since they are also typically resistant to proteolytic cleavage, expression levels as high as 25% of the total cell protein has been attained through this approach [53]. While overexpression of some membrane proteins spontaneously gives rise to inclusion body formation, most notably with β-barrels missing their signal sequences, and mitochondrial carrier proteins [54, 55], for other proteins it is possible to use fusion tags that target them to inclusion bodies [56, 57]. For example, expression of a trp operon L gene that has been modified to allow translation through its native stop codon generates a polypeptide of 105 residues called trpΔLE [58, 59]. This polypeptide has a strong tendency to form inclusion bodies either when expressed on its own or when expressed as an N-terminal fusion to smaller membrane proteins (i.e., one to two transmembrane (TM) helices [60–65]).

Many solution NMR structures of polypeptides comprised of a single TM helix have been produced by inclusion body targeting [62, 63, 65–67]. In the case of larger proteins that span the membrane multiple times, the development of high-yielding refolding protocols can present a significant impediment. Nonetheless this strategy has proven to be particularly useful for the production of β-barrel structures, e.g., the bacterial palmitoyltransferase PagP [68], outer membrane proteins OmpA [69, 70] and OmpX [71], the pH-sensitive OmpG porin [72], and the mitochondrial voltage-dependent anion channel VDAC-1 [73]. In addition, there are a number of examples of polytopic helical membrane proteins that have been expressed and refolded from inclusion bodies, e.g., a mammalian G-protein coupled receptor (GPCR) [29, 56, 74], the Y-2 receptor [75] bacteriorhodopsin [76], and a range of mitochondrial carriers [55, 77, 78]. These examples suggest some potential for inclusion body targeted expression of larger proteins, although no membrane protein NMR structure has yet been produced by this approach for any protein containing more than two TM segments.

2.3 Cell-Free Expression Approaches

A promising approach that circumvents complications arising from membrane protein toxicity or refolding is the cell free expression system comprised of purified extracts from bacteria [79, 80] or wheat-germs [81, 82] added to a mixture of tRNAs, amino acids, nucleotides, enzymes, and cofactors [83–87]. The increasing availability of pure isotope-labeled amino acids has allowed a growing number of groups to use this approach to produce specifically labeled samples for NMR [87–97]. In the case of membrane proteins, lipids or detergents can be included in the reaction mix to maintain them in a soluble state during the synthesis [91, 98, 99]. Using these methods it is possible for several milligrams of the target be synthesized from only a few milliliters of reaction mix.

An exciting development in the cell-free approach to membrane protein production is mounting evidence suggesting that additives normally intended to maintain expressed membrane proteins in solution (e.g., lipids, detergents) might not be necessary. In the absence of added lipids most of the expressed membrane protein tends to precipitate into an insoluble fraction; however, it has been shown that this aggregate can be resolubilized in mild detergents [96, 100–103]. The validity of this method has been supported by functional assays on some of these resolubilized aggregates [100, 101, 103], and in one case by highly similar NMR spectra for samples produced from cell-free precipitates vs conventional *E. coli*-based production systems [88]. Meanwhile the purity of these precipitates tends to be high, potentially eliminating the requirement for subsequent purification steps. Consequently, in some cases it is possible to directly resolubilize the cell-free expression pellet in the desired volume and composition of buffer to allow immediate acquisition of NMR data. Elimination of chromatography, dialysis, fusion protein cleavage, and protein concentration steps provides significant time savings, and also reduces material costs, particularly in the consumption of expensive detergents and lipids [88].

In addition to the advantages in sample purity offered by cell-free membrane protein expression, unique possibilities for amino acid specific labeling are also introduced with this method [88, 92]. In the case of bacterial expression systems, metabolic processes can break down amino acids added to the expression media, leading to label dilution and non-specific incorporation. Use of auxotrophic strains or a limited subset of amino acids can help to get around these issues [104, 105], but neither approach has the flexibility of specific amino acid labeling that is characteristic of cell-free expression systems. This feature has been particularly useful for the development of selective labeling approaches to facilitate backbone assignment of membrane proteins with poor spectral dispersion. Most rely on combinatorial labeling approaches, with samples having specific combinations of ^{15}N-labeled and/or ^{13}C-labeled amino acids. This allows inter-residue heteronuclear ^{15}N–^{13}C coupling, such as that used in the HNCO experiment, to rapidly identify adjacent pairs of amino acids [106] (example described in Fig. 1). There is a range of combinatorial strategies that have been developed to maximize the information that can be obtained from a limited number of differently labeled samples [102, 107, 108]. Since membrane proteins have unique sequence and spectral characteristics compared to their water-soluble counterparts, a Monte Carlo method was developed to determine the optimal labeling strategy for a specific protein sequence [88]. Using this approach it was possible to elucidate structures for three different membrane protein structures in an impressively short 8-month period.

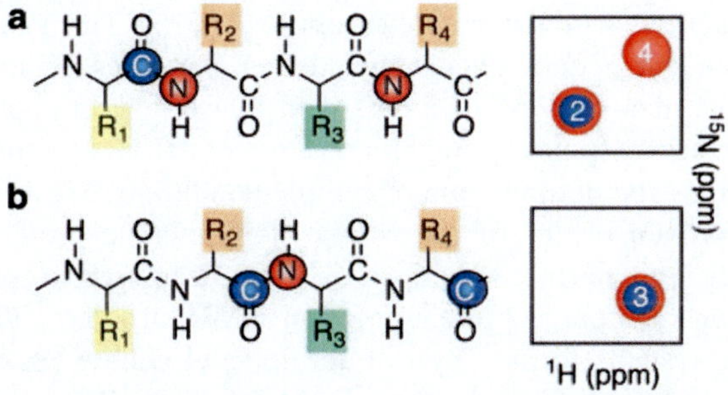

Fig. 1 Example of a combinatorial isotope labeling scheme that identifies adjacent amino acids in the sequence [106]. In this simplified example, backbone structures for the same segment with two differently labeled samples are shown on the *left* in **a** and **b**, with the corresponding ^{1}H–^{15}N 2D projection of the HNCO spectrum (*smaller blue peaks*) superimposed on the ^{1}H–^{15}N HSQC spectrum from the same sample (*red peaks*). Peaks are labeled with residue numbers for this tetrapeptide segment. In sample **a**, two of the amino acids with the same side chain (residues 2 and 4, *orange side chains*) are labeled with ^{15}N (highlighted in *red* in the backbone structure), and one of the amino acids (*yellow*) is labeled with ^{13}C. The appearance of a peak in the HNCO projection for one of the ^{15}N-labeled amino acids allows the identification of the preceding amino acid type for that residue (i.e., the sequence is *yellow-orange*). In sample **b** residues 2 and 4 are now labeled with carbonyl ^{13}C (*blue*), with ^{15}N-labeling of 'green' amino acids. The single peak that would be seen in the HSQC spectrum for this sample would also be observed in the HNCO projection, allowing the sequence to be extended to *yellow-orange-green* in this example. The number of samples and specific amino acids that would need to be labeled for a full assignment will depend on the method used to find the combination that gives the maximum number of inter-residue correlations with a minimum number of samples [102, 107, 108]

Meanwhile, a combinatorial optimization method has subsequently been developed that allows a wider range of auxiliary conditions to be factored into the design of protein sequence-specific labeling protocols [109]. Overall, these tools should help to make cell-free expression and selective labeling increasingly accessible to a wider range of laboratories in membrane protein structural biology.

3 Membrane Mimetic Systems for Solution NMR

3.1 Micelles

One of the challenges for solution NMR of membrane proteins is the identification of conditions that can mimic the native lipid bilayer environment while maintaining the sample in a stable, folded state with a total complex size of ~100 kDa or less [110–112]. This has most commonly been achieved through the use of detergents that form smaller micelles (~10–30 kDa) with a relatively high critical micelle concentration (cmc). However, the ability of some detergents to maintain membrane proteins in a water-soluble state sometimes works against structural studies, essentially solubilizing the protein so well that the native contacts are disrupted by competing interactions with solvent detergent [113–115]. This can be exacerbated by the frequent need to acquire data at elevated temperatures (35 to >40 °C) [112]. Consequently, the identification of a detergent that offers an appropriate balance between solubilization efficiency and preservation of native structure often requires extensive screening and optimization of sample conditions. While a wide range of detergents has proven useful for the study of membrane protein structure and function, a surprisingly narrow selection has so far been used to solve the majority of high-resolution membrane protein solution NMR structures (Table 1). These include micellar systems formed by sodium dodecyl sulfate (SDS), dodecylphosphocholine (DPC), or short-chain phosphatidylcholines (C_6- or C_7-DHPC), and small bicelles with C_6-DHPC/DMPC (Table 1).

3.1.1 SDS

SDS was one of the first detergents to be used for solution NMR of membrane proteins [116, 117], and has continued to be used for this purpose, particularly for the study of smaller membrane proteins containing only a single TM or amphipathic helix. SDS micelles have the ability to maintain a high concentration of membrane proteins in solution while forming protein–detergent complexes with size and dynamic properties that are often favorable for NMR spectroscopy. Although SDS is generally classified as a denaturing detergent due to its tendency to unfold water-soluble proteins [114, 118], hydrophobic peptide segments that interact with the micelle usually adopt a helical structure that minimizes exposure of polar backbone atoms to the hydrophobic interior of the micelle in the same way that would occur in lipid bilayers [119–121]. This has made this detergent a convenient solvent for the study of isolated TM segments [122, 123] and amphipathic membrane surface binding helices (e.g., [124–127]).

In spite of the generally accepted classification of SDS as a denaturing detergent, it has been widely used for the study of intermolecular interactions between isolated TM helices [123, 128, 129] or multi-spanning membrane proteins connected by short turn sequences [130, 131]. The suitability of this detergent for the study of TM-helix dimer structures has also been substantiated in a number of systems by comparison of mutational effects on dimer affinity in SDS and bilayer environments [129, 132–134] although small discrepancies do occur in some cases. However, it should be noted that examples of interactions identified in SDS exist that did not appear in the structure of the full-length protein (e.g., DAGK [135]). Moreover, there is a potential for disruptive interactions to occur between SDS and longer inter-helical loops or globular domains (e.g., the Y2 receptor [75]), making it necessary to adopt a cautious approach when interpreting structural information for membrane proteins in SDS. To date only a handful of larger (i.e., more than one TM segment) membrane proteins have been studied by solution NMR in this detergent, although in all cases the structural data was well validated by comparison with complimentary functional or structural data [136–138].

Table 1 Properties of detergents that have been used for membrane protein structure determination by solution NMR

Charge	Detergent	Structure	MW (Da)	CMC (mM)	Aggregation number	Micelle MW (kDa)	Ref
Zwitterionic	DPC	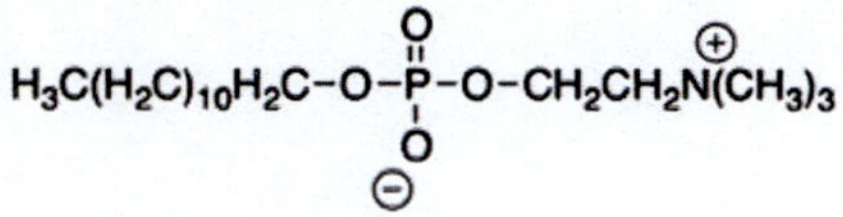	352	1.5	50–60	19	[142, 380]
	TDC	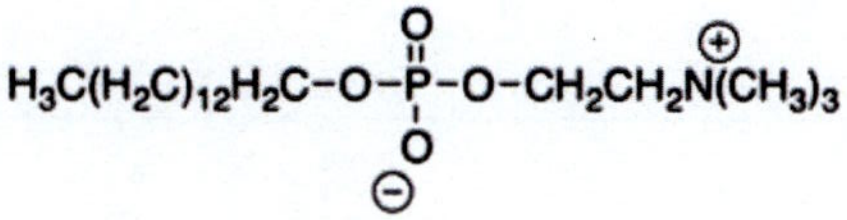	380	0.12	108	47	[246][a]
	C_6-DHPC	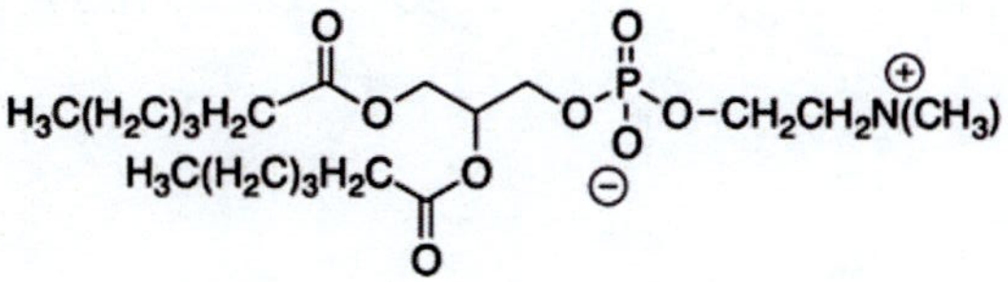	453	14	19–25	16	[180, 388]
	C_7-DHPC	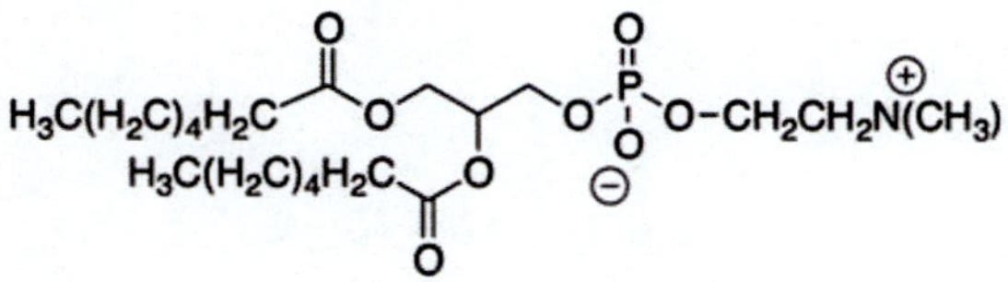	482	1.1	42–200	–	[388, 389]
	LDAO	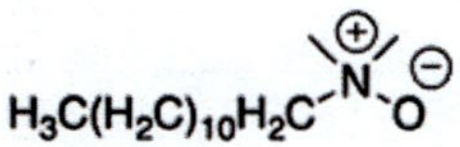	229	2	69–73	17	[389]

(continued)

Table 1 (continued)

Charge	Detergent	Structure	MW (Da)	CMC (mM)	Aggregation number	Micelle MW (kDa)	Ref
Anionic	SDS	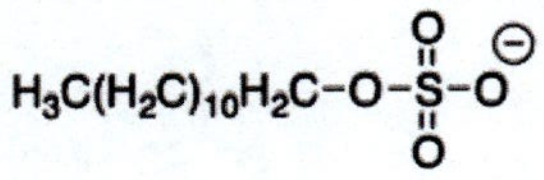	288	8.2	62	18	[389]
	LMPG	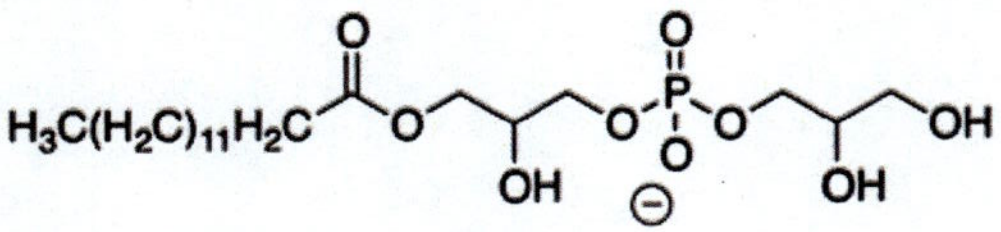	478	0.2–0.3	55	26	[390]
	LPPG	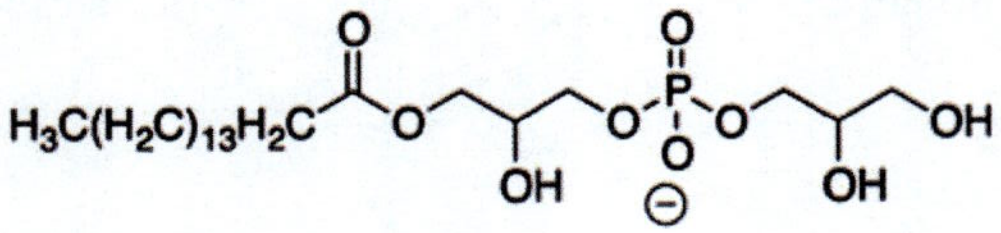	506	0.02–0.6	125	65	[180, 390]

[a]Cmc, aggregation number from Affymetrix/Anatrace

3.1.2 Dodecylphosphocholine

Dodecylphosphocholine (DPC) has a zwitterionic headgroup (Table 1) with a reduced tendency to participate in interactions that would disrupt protein structure [139]. It has proven to be particularly useful for NMR structure elucidation of TM helix interactions, as was first demonstrated with the GpA dimer [140]. It has since been used for some of the largest multi-spanning helical membrane protein systems determined to date (e.g., DAGK [38], phospholamban [141], and UCP2 [41]). The ready availability of deuterated DPC, its stable lipid-like headgroup structure, and its relatively small micelle aggregation number and narrow micelle size distribution over a range of conditions [142] are all favorable features contributing to the general utility of this detergent. For these reasons DPC has become one of the most popular detergents for solution NMR of membrane proteins, with more than one-third of the integral membrane protein NMR structures being determined in this detergent (Table 2).

3.1.3 Short-Chain Phosphatidylcholines

For those looking for closer mimics of a native phospholipid bilayer, short chain phosphatidylcholines, namely 1,2-diheptanoyl- and 1,2-dihexanoyl-*sn*-glycero-3-phosphocholine (C_7-DHPC and C_6-DHPC, respectively), have provided an attractive alternative [143]. These phospholipids differ from those found in biological membranes in the length of the acyl chains, giving rise to a tendency to form micelles instead of bilayer structures. Meanwhile the motionally averaged conformational properties of these lipids resemble those of the long-chain phosphatidylcholine bilayers at higher temperatures [143]. Particularly useful for solution NMR is the low polydispersity and stable size of the C_6-DHPC micelle over a range of concentrations [144]. However, the best compromise between sample stability and spectral quality was actually found with C_7-DHPC for the seven TM-helix GPCR pSRII [145]. Although the molecular mass and polydispersity of protein-free C_7-DHPC micelles is highly dependent on its concentration [144], the complexes formed with pSRII were spectrally homogeneous, illustrating the strong influence of the protein on the properties of the detergent–protein complex as has been previously observed for other proteins [146].

3.1.4 Lysolipids

Lysolipids have also been gaining attention as effective, relatively mild solubilization agents [88, 147–150]. The polar glycerol spacer between the headgroup and alkyl chain provides a gentler transition between hydrophobic and charge-containing phases of the micelle, reducing potentially denaturing effects of the headgroup charges. This was illustrated with DAGK, which was shown to retain full activity in lyso-myristoyl phosphatidylglycerol (LMPG) micelles while

Table 2 Summary of unique integral membrane protein structures determined by solution NMR

Method[a]	Structure class	Protein	Size, kDa[b]	# of TM α (β)[c]	Detergent	Detergent: protein molar ratio	Temp °C	PDB ID	Ref
(I) Conventional[d]	Single TM helix[e]	Na^+/H^+ exchanger TM6	3	1	DPC	75	30	2L0E	[373]
		Integrin β3	9	1	DHPC/POPC (3:1) DPC	750	40 (bicelle) 25 (DPC)	2RMZ 2RN0	[162]
	TM helix dimer	Glycophorin A	4	1 × 2	DPC	60	40	1AFO	[140, 391]
		BNip3	10	1 × 2	DHPC/DMPC (4:1)	40	40	2J5D	[188]
		Erb B2	9	1 × 2	DHPC/DMPC (4:1)	40	40	2JWA	[287]
		Eph A1	8	1 × 2	DHPC/DMPC (4:1)	40	40	2K1K 2K1L	[189]
		Eph A2	9	1 × 2	DHPC/DMPC (4:1)	40	40	2K9Y	[392]
		Erb B3	9	1 × 2	DPC	45	40	2L9U	[393]
	TM helix trimer	DAP12-NKG2C	11	1 + 1 × 2	TDC/SDS (10:1)	>450	30	2L35	[65]
(II) I + TROSY + fractional ^2H-labeling	TM helix oligomer	Phospholamban[f]	30	1 × 5	DPC	200	30	1ZLL	[338]
		Influenza A M2 tetramer	20	1 × 4	C_6-DHPC	400	30	2RLF	[166]
		Influenza B M2 tetramer	14	1 × 4	C_6-DHPC	270	32	2KIX	[326]

(III) TROSY, uniform ^{2}H-labeling, $^{1}H^{N}$ NOEs	β-Barrel	OmpA	19 (46)	(8)	DPC	600	50	1G90	[69]
		PagP	20	(8)	DPC	570	45	1MM4	[68]
					βOG	200	45	1MM5	
		OmpG	28	(14)	DPC	<300	40	2JQY	[72]
(IV) III + methyl NOEs		KpOmpA	22	(8)	DHPC	300	40	2K0L	[70]
		OmpX	16	(8)	DHPC	100	30	1Q9F	[71]
		VDAC	32 (75–90)	(19)	LDAO	>300	30	2K4T	[73]
(V) IV + non-methyl NOEs	TM helix heterodimer	Integrin β3/αII$_b$	9	1 + 1	DHPC/POPC/ POPS (9:2:1)	>400	30	2K9J	[184]
	Multi-TM helix	pSRII	26 (50–70)	7	C_7-DHPC	–	50	2KSY	[40]
		Ste2p TM 1-2	9	2	LPPG	>500	47	2K9P	[64]
(VI) IV + PREs + RDCs		DsbB	18 (40)	4	DPC	–	40	2K73	[39]
(VII) Short-range NOEs + PREs/RDCs	Single TM + amphipathic helix	Pf1 coat protein	5	1	DHPC	100	50	2KLV	[323]
		Stannin	14	1	SDS	>860	60	1ZZA	[394]
		FXYD1	8	1	SDS	>500	40	2JO1	[366]
		FXYD4	7	1	SDS	>500	40	2JP3	[395]
		CD4 TM fragment[g]	8	1	DPC	200	45	2KLU	[396]
		Phospholamban monomer[g]	6	1	DPC	750	50	1N7L	[397]
		KCNE1	16	1	LMPG	80	40	2K21	[67]
	Multi-TM helix	MerF	6	2	SDS	700	60	1WAZ	[137]
		AcrB	13	2	LMPG	200	45.37	2KSD	[88]

(continued)

Table 2 (continued)

Method[a]	Structure class	Protein	Size, kDa[b]	# of TM α (β)[c]	Detergent	Detergent: protein molar ratio	Temp °C	PDB ID	Ref
		QseC[h]	21	2	LMPG	330	45.37	2KSE	[88]
		KdpD[h]	11	4	LMPG	330	45.37	2KSF	[88]
		Presenilin catalytic fragment	19	3	SDS	>140	40	2KR6	[136]
		DAGK trimer[i]	43	3 × 3	DPC	–	45	2KDC	[38]
		UCP2[j]	33	6	DPC/DMPC/CL (150:2:1)	190	33	2LCK	[41]

[a]Method used to determine structures are divided into seven different classes, depending on the type of data and experiments that were used for the structure calculation
[b]Total molecular weight for all protein subunits in the complex
[c]Number of β-strands spanning the membrane are denoted in parentheses
[d]Standard NOE-based structure determination protocols typically applied to small water-soluble globular proteins
[e]Selected representatives
[f]Positioning of amphipathic helices in this structure represents ensemble average on the timescale of RDC and NOE restraints
[g]PREs were used to localized solvent-exposed or micelle-bound residues
[h]No NOE restraints directly used, but hydrogen bond restraints were imposed for the TM helices
[i]Data from disulfide bond mapping was used to create intermolecular restraints
[j]No NOE data used

yielding spectra of comparable quality to those obtained in the inactivating DPC conditions used to solve its structure [148]. A potential explanation for this arose from the observation of NOEs between the glycerol spacer of lyso-myristoyl phosphatidylcholine (LMPC) and Trp indole side chains at the peripheral regions of the TM helices; in DPC these same side chains showed NOEs to the phosphocholine headgroup. Based on these observations it was suggested that these indole-phosphocholine interactions were destabilizing, and that the glycerol spacer provided a more appropriate environment for interfacial side chains. The potential utility of detergents with polar spacers to improve the stability of membrane proteins has also been highlighted by synthetic variants of DPC with modifications that mimic the properties of the polar spacer [151]. In this study superior stabilization of OmpX for folding and solution NMR was obtained for DPC with a β-hydroxylated, or ethyl-amide-linked alkyl chain.

Lysolipids have also proven useful for direct resolubilization of membrane proteins from cell-free expression pellets for solution NMR structure determination, as was illustrated for three bacterial membrane proteins [88]. However, these structures were all characterized by loose helical packing, which may reflect a destabilizing influence of LMPG on these proteins. Alternatively, the loose structures could have been a consequence of the type and number of structural restraints used, leaving unanswered the question of possible denaturing effects of this detergent on these structures.

3.1.5 Mixed Micelles

For some proteins an appropriate balance between solubilization and stabilization could not be provided by a single detergent system, but was attained using a mixture of detergents. For example, the TM helix ζζ dimer could only be solubilized in SDS after its final purification step, but could subsequently be transferred into a less denaturing system by the addition of a fivefold molar excess of DPC over SDS, with samples undergoing aggregation if the ratio of DPC:SDS exceeded ~10:1 [152]. Similarly, NMR spectra of the KvAP voltage sensing domain were found to be optimal in 2:1 DPC/LDAO (lauryldimethylamine-oxide), with DPC alone giving rise to exchange-broadened spectra, while LDAO yielded favorable spectral properties but short sample lifetimes [153]. An N-terminal fragment from the Y4 GPCR containing two TM segments also benefited from the use of a detergent mixture (i.e., DPC and lyso-palmitoyl phosphatidylglycerol (LPPG)), since DPC could solubilize the sample but yielded spectra with broad lineshapes, while LPPG was a poor solubilization agent for this sample [154]. The common theme arising from these studies is that detergent mixtures can offer an improved capacity to provide the best compromise between sample solubility and stability. Also, in some cases it may be necessary to include small amounts of phospholipids normally found in the membrane environment, as was required for the mitochondrial uncoupling protein 2 (UCP2) [41].

Aside from mixtures that alter the charge of micelle surfaces, it is also possible to choose detergent combinations that alter the mean size of the micelle hydrophobic core. Analysis of small angle X-ray scattering data showed that the smaller hydrophobic core in short-chain detergent micelles can be increased by the addition of longer-chain detergents [155, 156]. A linear relationship between micelle size and detergent long- to short-chain molar ratios was observed, suggesting that micelle dimensions could be tuned in a straightforward manner to match the size of a protein hydrophobic domain. For a model 2-TM helix system, detergent mixtures that optimized hydrophobic matching between micelle and protein gave rise to the highest quality spectra, and also promoted more compact protein structures [155]. Although the applicability of these trends for other membrane proteins remains to be established, this study has identified an additional parameter that can be explored when optimizing protein–detergent complexes for solution NMR.

3.1.6 Potential Drawbacks to the Use of Micellar Membrane-Mimetics

While most solution NMR structures of larger membrane proteins have been elucidated in non-SDS detergent micelles, there are some caveats to keep in mind whenever any detergent is used to study membrane protein structure (Fig. 2). For example, with most NMR-friendly detergents having high cmcs, there is a significant concentration of monomeric detergent in the solvent that can potentially bind to solvent-exposed regions of the protein that do not normally interact with lipids. X-ray crystal structures have provided some examples of this, with detergent being found in the active site cavity of the β-barrel PagP [157], and inside the M2 channel [158]. For smaller membrane proteins this is particularly significant since there are fewer intramolecular interactions that stabilize the protein fold compared to the relatively large number of detergent–protein interactions. This larger proportion of residues that are exposed to solvent detergent increases the potential for

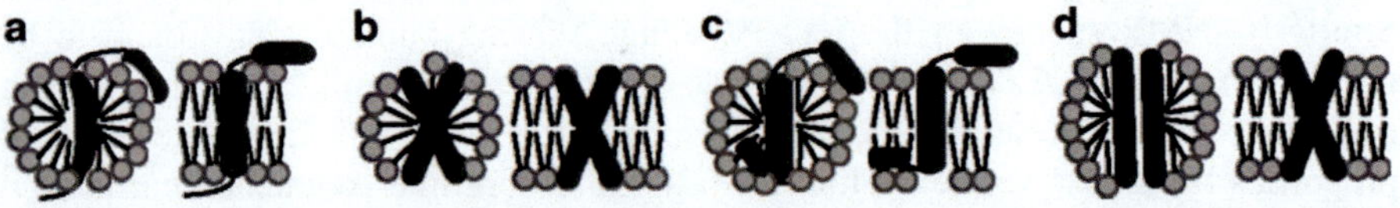

Fig. 2 Schematic representation of potential changes in integral membrane protein structure that could be imposed by a micellar environment (left hand side of each panel), compared to the native structure in bilayers (*right*). Possible distortions include; (**a**) micelle-induced curvature in the TM helix or amphipathic helix; (**b**) monomeric detergent molecules bound to a solvent-exposed region, in this case an aqueous cavity close to the micelle surface; (**c**) altered relative orientations of amphipathic vs TM helices; (**d**) loss of tilt relative to other TM segments. In this scenario hydrophobic mismatch between the TM helix and micelle are minimized by distortions in micelle structure that allow hydrophobic protein surfaces to remain in the hydrophobic phase. In the bilayer environment hydrophobic mismatch induces tilt, favoring a non-zero inter-helical crossing angles

detergent–protein interactions to alter the structure. This has been suggested to be a factor in the differences observed between some solid-state structures obtained in lipid membranes and solution-state detergent-solubilized forms [159].

Another potential complication arising from the use of micelles is the significant curvature of the micelle surface, a particularly relevant concern for surface-binding elements. This is capable of inducing curvature in amphipathic helices to maximize burial of the hydrophobic side of the helix in the hydrophobic phase of the micelle [160]. Structural elements with an affinity for the polar regions of the native bilayer could also follow the curved dimension of the headgroup phase, potentially leading to distortions of adjacent parts of the structure (e.g., the voltage dependent potassium channel KvAP [161]). Similarly, mismatch between the size of the hydrophobic core of the micelle and TM segment length can induce curvature in the TM helix to minimize its exposure to the aqueous phase [162]. In addition, the less densely packed, more dynamic state of the detergent monomers makes the hydrophobic phase of the micelle easier to access compared to that of a lipid bilayer [163]. This can promote detergent–protein interactions that may not occur in native lipid membranes, potentially distorting and/or destabilizing other parts of the protein [164, 165].

An illustrative example of how detergents can influence structure is provided by the controversy centered about the influenza A M2 channel. In C_6-DHPC micelles M2 was found in a tightly packed tetrameric state that bound the inhibitor rimantidine at four allosteric sites in the lipid-facing part of the channel exterior, inhibiting the channel by stabilizing the closed state [166]. In contrast, X-ray crystallographic [158] as well as solid-state NMR data acquired in lipid bilayers [167] on a shorter M2 construct clearly showed a single inhibitor-binding site inside the channel, with no interactions at the exterior allosteric sites. Based on these differences it was suggested that the helix tilt of the M2 TM segment in lipid bilayers could be important in maintaining a more open, active channel conformation [168]. According to this idea the malleability of loosely packed detergent molecules could allow coverage of a larger hydrophobic surface (Fig. 2d), eliminating the environment-induced tilt required to promote larger helix crossing angles. However, functional evidence can be cited that supports the biological relevance of both forms M2 channel [158, 166, 169], suggesting that the two structure types represent different functional states for this protein. This hypothesis is supported by the conformational dynamics observed for M2 in C_6-DHPC [166] and the structure of a drug-resistant mutant that had a reduced affinity for rimantidine at the allosteric site [170]. Moreover, subsequent experiments have shown that DPC micelles can support a tetrameric state that binds rimantidine at a single site within the channel [171]. This confirms that a micellar environment is capable of capturing the open state of this channel, although the physical properties of the micelle that promote one form of the channel over the other remain one of the interesting questions to be addressed.

3.2 Bicelles

Avoiding some of the potential problems associated with micellar detergents, small isotropic bicelles have increasingly been used for solution NMR of membrane proteins (reviewed in [172–174]). These are formed by mixtures of up to a fivefold molar excess of short-chain lipid (e.g., C_6-DHPC) or detergent (e.g., CHAPS) over long-chain lipid (e.g., DMPC, DOPC) [175, 176]. The composition of a bicelle is most accurately described by q, the molar ratio of lipid and *bicellar* detergent (i.e., $[\text{detergent}]_{\text{bicelle}} = [\text{detergent}]_{\text{total}} - \text{cmc}$) [112], which for small isotropic bicelles is typically in the range of 0.25–0.5. It has been shown that bicelles with this range of q values have a disk-shaped morphology, containing a distinct lipid bilayer phase and edges coated by a more mobile detergent phase (Fig. 3) [177–179]. The size of the protein-free small bicelle depends on q, and can be comparable to the size of commonly used detergent micelles (e.g., ~22 kDa for $q = 0.15$ [180]). However, spectra obtained for proteins reconstituted in bicelles have generally shown broader peaks than those for the same sample in a micelle [40, 137, 181, 182], with the larger complex size for the bicelle–protein complex contributing to this broadening. Yet in some cases only a bicelle environment could uniquely confer a functionally folded, spectrally homogeneous sample (e.g., the small multidrug resistance pump (Smr) [183]). A bicelle formulation mimicking physiological membrane compositions was also found to be instrumental for structure determination of the weakly interacting integrin αIIbβ3 TM-heterodimer [184], a complex that was not supported by DPC micelles [185].

In contrast with micelles, the introduction of proteins into bicelles may require an additional optimization step, since there are a few different approaches that can be used [186, 187]. However, most membrane proteins that have been reconstituted into small isotropic bicelles for solution NMR could be prepared in solvent-free

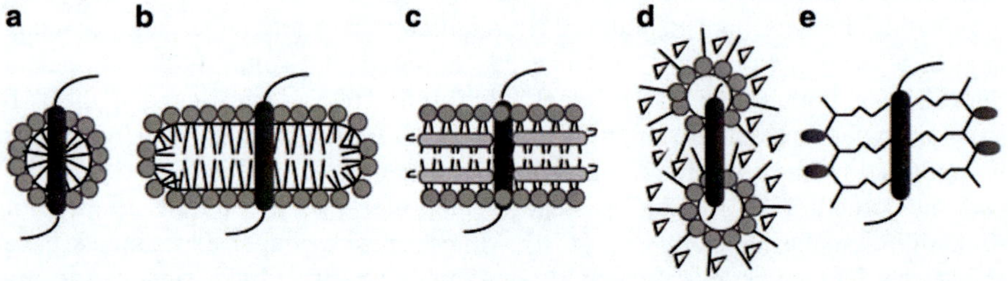

Fig. 3 Schematic diagram showing the general structure of various membrane mimetic systems used for solution NMR studies of membrane proteins. One TM segment is shown embedded in (**a**) a micelle, (**b**) a bicelle, (**c**) a nanodisc, (**d**) reverse micelles, and (**e**) amphipols. Polar detergent or lipid headgroups are represented by *spheres*, with hydrophobic acyl chains as *straight lines*. The MSPs surrounding the periphery of the nanodisc are shown as *gray rods*, and co-surfactants or co-solvents that stabilize reverse micelles are shown as *triangles*. Two amphipols are shown surrounding the TM segment in (**e**), with the polar blocks (*gray*) connected to hydrophobic blocks (*lines*) that interact with the protein

lyophilized (e.g., BNip3 [188], EphA1 [189], integrin αIIbβ3 [184]) or acetone-precipitated (the potassium channel voltage sensing domain [153]) states, allowing direct resolubilization into bicelle solutions. Similarly, proteins that are purified into organic solvents can be directly added to an organic solution of bicelle-forming lipids. As shown for Smr, evaporation of the organic solvent followed by resuspension in aqueous buffer can generate a functional bicelle-solubilized receptor [183].

For proteins that cannot be refolded directly into bicelle solutions, it may be possible to transfer a micelle-solubilized sample into a bicelle solution through on-column exchange, as was done for KCNE3 [190]. Alternatively, for proteins that cannot be folded in bicelle-compatible micelles, it may be necessary first to reconstitute into lipid vesicles [186]. Once in this state, sample concentration and buffer exchange can be achieved by centrifugation, followed by resuspension of the pellet in the bicelle-forming detergent solution. This approach was found to be required for the incorporation of the G protein-coupled receptor CXCR1 into small bicelles [191]. However, only extramembraneous N- and C-terminal regions of the sample could be observed in ^{1}H–^{15}N HSQC spectra, and only for very low q ratio bicelles (0.1).

When working with bicelles, the stability of the bicelle itself may require special attention since hydrolysis of lipid carboxy–ester bonds changes the composition of the bicelle over time, leading to phase separation [192]. This reaction is accelerated under acidic or basic conditions, with the more water-accessible state of the short-chain phase being particularly susceptible to hydrolysis. For this reason, the hydrolysis-resistant ether-linked analog 6-*O*-PC (1,2-di-*O*-hexyl-*sn*-glycero-3-phosphocholine) is being increasingly used to extend bicelle sample lifetimes [186]. It is also possible to use ether-linked long-chain lipids (e.g., 14-*O*-PC or 16-*O*-PC), although this change in the lipid headgroup can alter the structure of embedded proteins [193]. Bicelle stability can also be improved by spiking the solution with charged amphiphiles to increase charge repulsion between bicelles [194, 195]. Both stabilization strategies were used for Smr, with a 3:1 mixture of short- and long-chain ether lipids (6-*O*-PC and 14-*O*-PC), doped with 10% of a 3:1 mixture of short- and long-chain phosphatidylserines [183]. Even with all these modifications to the basic bicelle mix, the sample half-life was typically on the order of ~1 week, illustrating the inherent difficulties of maintaining some membrane proteins in a folded functional state.

One aspect of any bicelle solution that should not be neglected is the significant concentration of monomeric detergent that exists in equilibrium with the bicelle-bound state. These small amphipathic molecules may be capable of binding to any exposed hydrophobic patches on the protein that protrude from the bicelle [196], potentially destabilizing the protein [165]. Under these circumstances the slightly larger hydrophobicity of ether lipids may confer an additional advantage since this gives rise to a small decrease in monomeric detergent concentrations [197]. More dramatic reductions in monomeric detergent concentrations can be realized by making bicelles with lower-cmc detergents (e.g., C_7-DHPC; [198]), although the utility of this bicelle type for solution NMR applications has yet to be demonstrated.

3.3 New Developments in Membrane-Mimetic Solvents for Solution NMR

While micelles and bicelles have served as the membrane-mimetic solvent for the majority of solution NMR studies on integral membrane proteins, the identification of suitable NMR-compatible solvents remains a formidable challenge. With <30 unique multi-spanning membrane protein structures determined by solution NMR, there is a clear need to expand the current library of solubilization agents for this purpose. In response to this problem, creative approaches are currently being developed that may open the door to solution NMR for a wider range of membrane proteins in the future. Some of the systems that have started to yield promising results include nanodiscs, amphipols, and reverse micelles.

3.3.1 Nanodiscs

Nanodiscs are comprised of a bilayer containing ~130–160 lipids, maintained in a discrete, water-soluble state by the association of two copies of the membrane scaffold protein (MSP) from apolipoprotein A-I wrapped around the hydrophobic rim of the bilayer [199–202] (Fig. 3). Originally developed for the solubilization of functionally active integral membrane proteins, they have since been used for solid-state [203], and more recently, solution NMR applications [153, 204–207].

Incorporation of an integral membrane protein in a nanodisc is usually achieved by incubation of a detergent-solubilized sample with detergent-solubilized lipids, along with MSP. Detergent removal, often with adsorbent beads, can promote the spontaneous formation of nanodisc-reconstituted samples that remain stable for many months [202, 208]. However, a number of variables must be optimized when formulating a reconstitution procedure, with the choice of detergent for membrane protein solubilization being identified as a particularly critical one [200]. Even when solubilized in detergents, a membrane protein may undergo irreversible aggregation that cannot be rescued by reincorporation into a native lipid bilayer. Consequently, high yield reconstitution requires a detergent that can minimize the formation of these aggregates, often with the use of excess nanodisc-forming components.

Once reconstituted into a nanodisc, the membrane protein is embedded in a bilayer phase that shares some similarity to that in the bicelle. However, the absence of detergent in the final nanodisc preparation can prevent the folding/stability problems that may be introduced by the significant concentrations of monomeric detergent in bicelle and micelle solutions. The nanodiscs themselves are also more stable than bicelles, and can be subjected to the same manipulations as soluble proteins (e.g., lyophilization, chromatography, concentration, etc.) [200]. However, there is a significant trade-off in that nanodisc sizes tend to be on the large side for NMR applications, at ~200 kDa. While this remains within the accessible range for relaxation-optimized NMR experiments (see Sect. 4.2), most solution NMR spectra for nanodisc-embedded membrane proteins have shown broad ^{1}H line widths [153, 204, 207]. Rapid axial rotation of smaller membrane proteins within the bilayer can significantly reduce these line widths [205]. Unfortunately this is unlikely to occur for the larger membrane proteins that are more prone to require reconstitution in a detergent-free lipid environment. Consequently, the most common application for nanodisc preparations in solution NMR to date has been to provide a reference 2D correlation spectrum of a bilayer-reconstituted sample to allow comparisons with results obtained in less bilayer-like solvents [153, 205, 209]. The GPCR CC-chemokine receptor 5 (CCR5) has also been reconstituted in nanodiscs, allowing its interaction with its chemokine ligand to be studied by solution NMR [210]. Nanodiscs have additionally proven useful for the study of interactions at membrane surfaces (e.g., peripheral phosphoinositide-binding proteins [206]). Meanwhile further developments in nanodisc formulations and protein labeling may help to expand the utility of these complexes for solution NMR in the future.

3.3.2 Amphipols

A distinct class of surfactant that is also being developed for a range of membrane protein applications is the amphipathic polymer, otherwise known as the amphipol [211]. These molecules are typically comprised of an amphipathic "backbone" with hydrophobic branches interspersed with polar or charged groups (reviewed in [212–214]). The result is a polymer that could be thought of as a unimolecular micelle with covalent bonds linking polar headgroups (Fig. 3). The equilibrium between monomeric and micellar states that characterizes unlinked detergent solutions is avoided by the covalent linkages, and only a few amphipol molecules are required to envelop and solubilize the protein target.

Transfer of a membrane protein into amphipols can be achieved by sub-cmc dilution of a detergent-solubilized sample into an amphipol solution. The increase in entropy that results from the release of multiple detergent molecules upon amphipol binding makes the exchange highly favorable [215], allowing functional reconstitution of a wide range of membrane proteins [211, 216–218]. Particularly impressive is the ability of amphipols to refold SDS-solubilized GPCRs to greater yields than could detergent/lipid mixtures that had been identified from extensive screening experiments [219]. This allowed the structure of a GPCR ligand in its receptor-bound state to be determined by solution NMR [220]. Meanwhile, feasibility for solution NMR of the solubilized protein itself has also been demonstrated with the transmembrane domain from OmpA [221]. While a small increase in ^{1}H line widths was observed in the amphipol relative to the C_6-DHPC spectrum, this could be attributed to the slightly larger size of the amphipol-OmpA complex, along with chemical exchange processes. These results with a first-generation amphipol provide an encouraging indication of the potential that future amphipols could have for membrane protein structural biology. The development of amphipols that can remain soluble under more acidic conditions that decrease rates of amide proton exchange with solvent, along with methods that improve the homogeneity of prepared samples will be particularly important for this application.

3.3.3 Reverse Micelles

A fundamental restriction for solution NMR of solubilized membrane proteins is imposed by the large size of the complexes they make with detergents and lipids, since they usually exhibit slow molecular reorientation rates, giving rise to significant line broadening and decreased coherence transfer efficiencies. To reduce the influence of molecular mass on rotational correlation times, and hence NMR spectral quality, the Wand group has pioneered the use of reverse micelles to encapsulate water-soluble proteins [222–224]. When dissolved in an organic solvent of low viscosity, molecular tumbling rates of these complexes are significantly enhanced compared to those in water. When the increase in molecular reorientation rates exceeds the decrease that comes from the added mass of the reverse micelle, significant improvements in size-sensitive NMR spectral properties can result [225]. For a reverse micelle in pentane this benefit may not appear for a water-soluble protein until its size exceeds ~50 kDa; in contrast, membrane proteins already require detergents for solubilization. Moreover, the benefits of reverse micelle encapsulation can be dramatically improved by the use of solvents with lower viscosities such as butane and propane, which require only slightly elevated pressures to maintain the liquid state in regular NMR tubes [222]. Even larger advantages can be realized by the use of liquid ethane under elevated pressures (~4000 psi) prepared through the use of a special apparatus [226]. In this solvent a >100-kDa protein can be conferred with molecular tumbling properties rivaling those of a 10-kDa protein in aqueous solution.

Although reverse micelle systems were first developed for the study of water-soluble proteins, their application to integral membrane proteins has since been demonstrated with gramicidin A (gA) in a dioctyl sulfosuccinate/pentane system [227]. The pattern of NOEs obtained for the gA peptide dimer in this medium was consistent with its native $\beta^{6.3}$-helix. In addition, the intermolecular NOEs showed preservation of native N- to N-terminal intersubunit hydrogen-bonding interactions within the hydrophobic phase. Introduction of gA into this medium was relatively straightforward, since this peptide can be isolated in a lipid and/or detergent-free form and then directly introduced into the reverse micelle system. This was a critical advantage, since the standard reverse micelle systems formulated for water-soluble samples are not compatible with the detergents normally used to purify larger membrane proteins [225].

To extend the utility of reverse micelles to those samples that require detergent or lipid to prevent aggregation and promote folding, reverse micelle formulations have recently been introduced that use detergents capable of forming regular aqueous-phase micelles (e.g., LDAO, dodecyl- or cetyl-trimethylammonium bromide (DTAB or CTAB, respectively)) [228]. These can be used to solubilize membrane proteins in the aqueous micelle conditions usually used during purification.

Dehydration, followed by addition of the organic phase, allows micelles to "flip" out into a reverse micelle configuration while maintaining the hydrophobic domain of the membrane protein in a membrane-mimetic environment. In this complex, each membrane protein is thought to be associated with two reverse micelles that converge around the hydrophobic domain of the protein (Fig. 3) [225, 227, 229]. In some cases co-surfactants (e.g., dihexadecyldimethylammonium bromide (DHAB)) and/or co-solvents (e.g., hexanol) are also required to stabilize this structure. This approach was shown to be successful using the tetrameric KcsA channel as a test system [228]. Reconstitution in reverse micelles was achieved by purification in CTAB, followed by lyophilization and addition of DHAB, hexanol, pentane, and water. The result was a well-resolved ^{1}H–^{15}N HSQC spectrum for KscA, with significantly enhanced transverse relaxation times relative to those in water (80 ms vs 20 ms). Moreover, potassium ion-dependent chemical shifts in the channel selectivity filter could be observed (Fig. 4), confirming a functional state for this sample. Although protein concentrations in reverse micelles tend to be lower than what can be obtained in

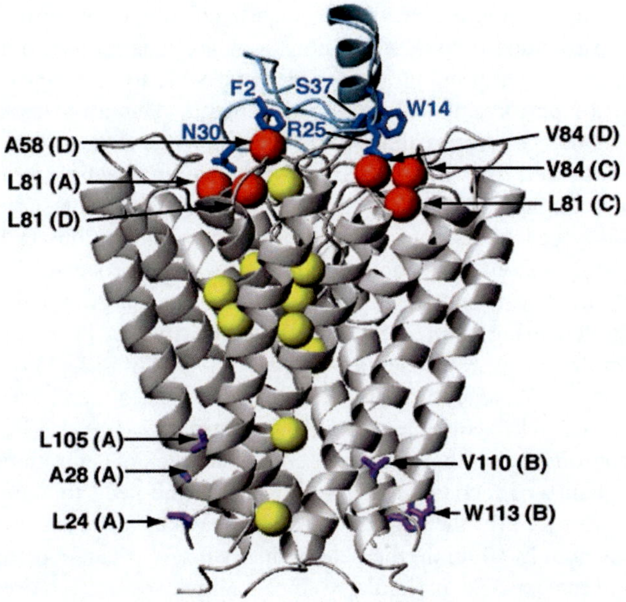

Fig. 4 Novel insights into KcsA potassium channel structure and function revealed by solution NMR. A ribbon diagram representation of the complex determined for the KcsA tetramer (*gray*) bound to charybdotoxin (*blue*) [298]. Highlighted are KcsA methyl groups (*red balls*) that gave rise to intermolecular NOEs with charybdotoxin side chains (*sticks*), showing the importance of methyl groups in defining this complex. Selectivity filter residues that were observed to give rise to potassium-dependent chemical shifts are shown in *purple* [228, 273]. In addition, methyl groups of side chains showing the largest chemical shift differences between active and resting states are shown in *yellow* for one subunit (residues L24, L40, L59, V70, V76, V91, V95, I100, and L105), indicating widespread changes throughout the TM region between the two states [273]. Residue labels include KcsA subunits A, B, C, and D in parentheses. All ribbon structure figures were made with MOLMOL [398]

aqueous systems (e.g., ~0.2 mM for KcsA), this is offset by the increase in cryogenic probe sensitivity (via an increase in probe quality factor Q) that comes from low sample conductance [230]. In general, these results provide an encouraging indication of future benefits that reverse micelles may bring, particularly for the elimination of the requirement for deuterium isotope labeling for some large complexes (described in Sect. 4.1.1).

3.4 Approaches for Sample Screening

A consistent theme that has emerged in membrane protein structural biology is that the selection of an appropriate membrane-mimetic solvent is a largely empirical process. This was well illustrated for the pSRII GPCR, which required over 20 different sets of conditions to be systematically explored before C_7-DHPC was identified as the most appropriate solvent [40]. In this case screening was greatly facilitated by the ability to use the unique absorbance properties of folded pSRII to identify promising conditions rapidly. For those proteins with folding states that cannot be so conveniently monitored it is still possible to screen a wide range of small samples by performing stability measurements. This involves incubation of each sample under conditions that would simulate those of a typical 3D NMR experiment, followed by SDS-PAGE of the soluble fraction separated from aggregated precipitates [37]. Identification of factors that preserve membrane protein solubility can be used to narrow the range of conditions to be examined by more informative but time- and sample-intensive approaches. These include experiments that estimate size (e.g., size exclusion chromatography with or without dynamic light scattering) [40, 138], translational diffusion [146, 231], effective rotational correlation time measurements [70, 138, 207, 232, 233], small angle X-ray scattering [37], analytical ultracentrifugation [234], and spectral quality (e.g., 1D ^{1}H or 2D ^{1}H–^{15}N HSQC (TROSY) spectra, [37, 149, 159]).

While cryoprobe technology has reduced the sample concentrations required to evaluate the quality of reconstituted samples [235], the need for sample volumes that exceed ~275 μL limits the number of conditions that can be tested from a single protein preparation. In addition, the detergents themselves can constitute a significant fraction of the expense, particularly since a large excess is required to ensure a 1:1 micelle:protein ratio [111, 236]. However, with the advent of smaller-diameter NMR probes it is now possible to reduce the amount of sample used in each experiment, since the increase in the probe coil length-to-diameter ratio confers an increase in probe mass sensitivity [237–239]. Although acquisition times for biomolecules are lengthened due to the reduction in total sample quantities, it is nonetheless still possible to use sample concentrations similar to those used in conventional probes to obtain 2D spectra within a few hours [240–242]. For example, a 1-mm Bruker TXI microcoil probe with a ~10 μL sample volume was used to evaluate refolding efficiencies of the *E. coli* outer membrane protein OmpX [151]. In this case it took ~9 h to record a ^{1}H–^{15}N TROSY with ~1.2 mM detergent solubilized OmpX. The small quantities involved made it possible to evaluate the NMR spectral properties of refolded samples in a series of novel detergents that were available in a very limited supply. Meanwhile, even faster screening can be achieved if the first set of conditions is evaluated using 1D experiments, with 2D spectra only being acquired for the most promising samples [147]. Also exciting is the recent availability of the cryogenically cooled 1.7-mm microprobe, with an increase in mass sensitivity that compensates for the tenfold reduction in sample volume (30 μL vs 300 μL in a conventional 5-mm probe). With spectral acquisition times and sample concentration requirements being comparable to those of the 5-mm room temperature probe, it is in fact now possible to use these reduced sample volumes for acquisition of all NMR data types required [147, 232].

One aspect of detergent screening that is gaining increasing attention is the concentration of the detergent itself. While it has long been recognized that there is a minimum concentration required for solubilization, samples that exceed this minimum can also exhibit poor spectral quality [232]. Screening experiments with a 1.7-mm microprobe showed that the degradation in OmpX spectral quality associated with higher decylphoshocholine concentrations was caused by an increase in the solution viscosity due to the presence of excess protein-free micelles. Based on these results it could be shown that the optimal detergent concentration for this system approximately corresponded to the number of detergent molecules bound in the protein–detergent complex multiplied by the concentration of OmpX, added to the cmc of the detergent. High detergent-to-protein ratios can also complicate the study of oligomeric states of membrane proteins, with excess micelles reducing the effective concentration of TM segments that can interact [243–245]. However, the practice of concentrating samples in ultracentrifugation devices can increase the concentration of detergent in the final sample, even when the molecular weight cut-off is smaller than the size of the protein-free micelle [234, 246]. This can be particularly problematic for detergents that tend toward greater polydispersity and concentration dependence of micelle size. Consequently, it is important to choose ultrafiltration units with molecular weight cut-offs that can maximize retention of the target protein while minimizing undesired increases in detergent concentrations, particularly for any samples that require concentration or detergent exchange as a final preparation step. It is useful to monitor detergent concentrations during these procedures, with detergent peak intensities in 1D ^{1}H NMR spectra providing a convenient read-out of concentration for many of the commonly used detergents for solution NMR [234].

4 Methods for Acquisition of Solution NMR Spectra of Membrane Proteins

As discussed above, most of the systems used to reconstitute a membrane protein sample for solution NMR create large, slowly tumbling complexes that suffer from rapid transverse relaxation processes. Size estimations of some of these protein–detergent complexes show that the detergent can add as much as 60 kDa to the protein molecular weight (e.g., DAGK [247], VDAC-1 [73]). In order to counter the loss of NMR signal sensitivity that is characteristic of these large complexes, it is usually necessary to use approaches originally developed for study of large water-soluble proteins. Those that have found the most widespread utility are deuterium isotope sample labeling and relaxation-optimized pulse sequences.

4.1 Isotope Labeling Schemes for Large Protein Complexes

4.1.1 Deuteration

To compensate for the unfavorable relaxation properties of protons in large protein complexes, all the non-exchangeable carbon-bound protons can be replaced with deuterium (reviewed in [20, 248, 249]). As the gyromagnetic ratio of ^{2}H is 6.7-fold lower than that for ^{1}H, many of the relaxation pathways that would otherwise be present in fully protonated samples are greatly attenuated in a deuterated sample. This leads to an increase in T_2 transverse relaxation times, and more effective preservation of signal over the numerous coherence transfer elements that occur in a typical multidimensional NMR pulse sequence. Significant sensitivity gains have been demonstrated for uniformly deuterated proteins in many triple resonance applications with both water-soluble [20] and membrane protein samples [111]. As a result, uniformly deuterated samples are now routinely used to obtain backbone assignments for large membrane proteins.

Since backbone assignment experiments use amide proton magnetization to generate and detect the NMR signal, deuterated backbone amides must first undergo complete exchange with solvent $^{1}H_2O$ protons [20]. Although the extended exposure to $^{1}H_2O$ solutions during purification can be sufficient to re-introduce protons at all sites, many of the larger membrane proteins that exhibit good spectral properties are highly stable and therefore resistant to this process [112]. In these cases re-introduction of protons at exchangeable sites can be extremely slow in the core, making it impossible to achieve significant exchange without the exposure to destabilizing conditions. This can be particularly challenging for proteins that have been optimized to maximize stability, potentially necessitating the development of high-yielding unfolding/refolding protocols. This was done for the 4-TM DsbB, which had to be solubilized in a DPC/SDS mixture for solvent exchange before reconstitution back into DPC [39].

One way to get around this problem is to treat the protein as a two-domain system, since non-exchangeable sites tend to be clustered together for several contiguous residues along the protein sequence [250]. As shown for the tetrameric potassium channel protein KcsA in SDS, a complementary pair of samples can be generated; one that retains amide protons at the solvent-exposed sites, and the other with protons at non-exchangeable sites [138]. The former sample is generated using conventional protocols for perdeuteration followed by amide-proton exchange with solvent ^1H_2O. However, for the latter sample, expression is done in a ^1H_2O-based medium that is deficient in glucose and instead contains deuterated amino acids from an algal lysate. Subsequent exchange of the purified sample into D_2O allows simplification of the amide spectrum, with only protected site remaining visible. Since the NMR spectral properties of micelle-embedded core regions of membrane protein structure tend to differ from solvent-exposed regions, use of these two labeled samples allows assignment strategies to be specifically tailored to each region.

4.1.2 Methyl Protonation

Despite the utility of perdeuteration in the assignment of backbone chemical shifts, the elimination of all but the exchangeable protons impedes structural studies that rely on conventional NOE-based approaches. Although in some cases it is possible to use only amide proton NOEs to obtain a protein global fold, the accuracy of these structures tends to be low due to the small proportion of distance restraints between protons from non-sequential residues (e.g., 5–8 Å backbone pairwise rmsd to target structure [251, 252]). Therefore to increase the number of protons in the protein core while maintaining the benefits of extensive deuteration, a number of methods have been developed to retain protons at specific non-exchangeable sites using a "reverse isotope" labeling approach [20, 248, 249]. Methyl groups have been the principal targets for selective protonation since they are enriched in protein hydrophobic cores [253], making them structurally informative sources of NOE-based restraints [251, 254, 255]. In addition, rapid rotation about the methyl symmetry axis causes its three protons to give rise to a single peak with narrow ^1H line widths that are additionally narrowed for methyls that terminate flexible amino acid side chains [256]. In the context of a slowly tumbling macromolecule, this rapid methyl rotation also creates ideal conditions for the optimization of relaxation in HMQC-type experiments (methyl-TROSY, described in Sect. 4.2).

A widely used strategy for the selective introduction of methyl protons into deuterated proteins is known as the ILV method, since targeted methyl groups reside in the amino acids Ile, Leu, and Val (reviewed in [256–258]). In this method, deuterated α-keto acid precursors retaining protons at methyl sites are added to a bacterial expression culture growing in D_2O minimal media approximately 1 h prior to induction of protein expression (Fig. 5) [259, 260]. To maintain a high background of deuterium incorporation, uniformly deuterated glucose is included in the growth as the only other source of carbon. After a relatively short induction period (~4–6 h) designed to maximize incorporation and minimize metabolic scrambling [259], methyl protons from α-ketobutyrate and α-ketoisovalerate will be incorporated into the Ile(δ1), and Leu/Val methyl groups, respectively. There is a large variety of isotope label combinations available for these α-keto acids [258]; those shown in Fig. 5 have been the most commonly used for the study of large proteins [257].

The ILV approach has been applied to structure determination of membrane proteins from the β-barrel family, as first shown for OmpX in DHPC [261].

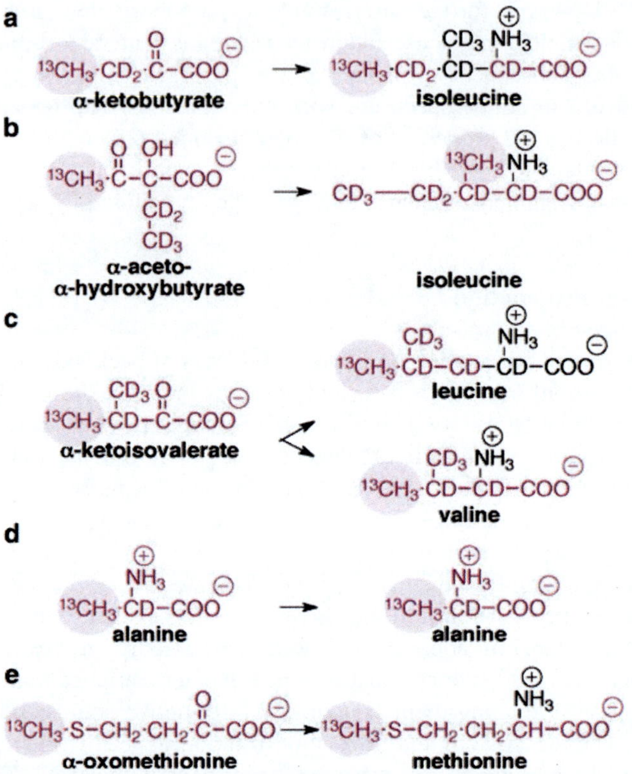

Fig. 5 Metabolic precursors added to expression media prior to induction of expression for the selective incorporation of ^{13}C and ^{1}H into methyl groups of ^{12}C, ^{2}H-labeled (**a**) Ile (Cδ1-labeled), (**b**) Ile (Cγ2-labeled), (**c**) Leu and Val, (**d**) Ala, and (**e**) ^{13}C, ^{1}H-Met. The parts of the amino acid that originate from each precursor are shown in *purple*, with the ^{13}C, ^{1}H-labeled methyl group highlighted. In all cases the expression medium must include ^{13}C, ^{2}H-labeled glucose as the carbon source. Selective methyl labeling of Ala also requires media supplementation with deuterated succinate, α-ketoisovalerate and isoleucine to suppress isotope scrambling. In this procedure glucose and glycerol together are used as the carbon sources prior to addition of methyl labeling agents [263]

NOEs involving methyl protons yielded a fourfold increase in the number of distance restraints over ^{1}H^{N}-^{1}H^{N} NOEs alone [71], and a twofold increase in NOE restraints for the 210 residue KpOmpA β-barrel [70]. Smaller gains were obtained for the 283 residue VDAC-1, with 324 methyl-associated NOEs adding to the 288 amide proton NOEs [73]. However, in all these cases, the overall fold of the β-barrel is already well-defined by NOEs between amide protons, and hence the impact of methyl NOEs on structure quality tends to be modest, albeit significant, for these folds (e.g., OmpX NOEs from methyls decreased the backbone rmsd to the mean from 2.13 Å to 1.42 Å, [71]).

While the utility of methyl protonation for structure determination of β-barrel folds has been demonstrated, the full benefits of this strategy for structure determination of helical integral membrane proteins has been more difficult to realize. Even in the case of helical water-soluble proteins it was noted that fewer long-range methyl-based restraints are available relative to other fold types [251]. Similar studies focusing on all-helical membrane protein folds substantiated this observation, with structure accuracies in the 5 Å region even when complete assignment of methyl-associated NOEs was assumed to be possible [262]. In reality the low spectral dispersion that is characteristic of helical membrane proteins can significantly impede chemical shift and NOE assignment for these methyl groups, as was the case for the DAGK trimer [111]. On the other hand, relatively high spectral quality for the pSRII GPCR allowed ~50% of expected inter-helical NOEs involving these methyl groups to be assigned, with the remainder being either absent from the spectrum or buried under strong diagonal signals [40]. Yet this level of NOE assignment was still not sufficient to generate high-resolution structures, making it was necessary to acquire a greater number of long-range restraints involving Ala, Thr, Ile(γ2), and Met methyl group NOEs. In the case of pSRII it was possible to use the ILV methyl assignments to help extend the chemical shift assignments to these methyl groups using a fully protonated sample. This afforded a ~2.5-fold increase in the number of inter-helical NOEs that could be assigned and helped to increase the quality of the resulting structures.

When the spectral quality of a fully protonated sample is not sufficient to help increase the number of methyl proton assignments, other methyl protonation strategies are also available. For example, methyl protonated Ala can be directly incorporated into deuterated proteins so long as a trio of deuterated precursors (Fig. 5) that suppresses metabolic scrambling to undesired sites is also added to the media. [263]. The direct bond between the Ala methyl group and Cα backbone atoms make these methyl groups excellent sources of information for backbone structure and dynamics [264]. Importantly, the ILV and Ala-methyl labeling strategies are complementary, allowing protons to be simultaneously introduced to ILV and Ala methyl groups in a single sample. Ile(δ1) and Ala methyl groups can also be simultaneously labeled in an alternate strategy that uses deuterated rich media supplemented with appropriately labeled Ala and α-ketobutyrate [265]. Methionine methyls can similarly be targeted by the inclusion of protonated Met [266] or a selectively protonated α-keto acid derivative [267] into the D_2O minimal expression medium (Fig. 5). The Cγ2 proton of Ile can also be targeted by including α-aceto-α-hydroxybutyrate in the minimal media [268]. Meanwhile, selective incorporation of protons beyond the methyl groups has been demonstrated in the stereo-array isotope labeling (SAIL) strategy [269, 270]. This technique uses cell free expression to incorporate a complete suite of synthetically prepared stereo- and regiospecifically ^{2}H-labeled amino acids, producing a sample with reduced ^{1}H density that retains a larger number of structurally informative protons. Overall, the general utility of these various strategies will depend on the characteristics of each sample, as well as cost-effective availability of labeled precursors. However, they should prove to be increasingly useful as the number of large membrane proteins being studied by solution NMR continues to increase.

4.1.3 Exploring New Frontiers Using Methyl Protonated Samples

Although methyl protons have been widely used to assist structure determination of large protein complexes, their high sensitivity and amenability to TROSY-type sequences (described in Sect. 4.2) has provided the foundation for exciting applications that probe large systems in dynamic equilibrium [256, 257]. This was spectacularly demonstrated for the 670-kDa 20S core-particle proteasome, with the identification of two interchanging structural states for gating residues using $^{13}CH_3$-labeled Met in U-^{2}H, ^{12}C-labeled samples [271], along with interactions inside the proteasome antechamber that maintained ILV-labeled substrate proteins in an unfolded state [272]. Meanwhile, applications with membrane protein samples are also beginning to appear, as shown with the DDM-solubilized KcsA channel [273]. In this study, methyl TROSY spectra from U-^{2}H, Leu/Val-[$^{13}CH_3$, $^{12}CD_3$], Tyr-^{1}H KcsA showed pH-dependent chemical shifts reflecting interconversion between three different functional states, and coupling between gating residues in two distinct regions of the structure (Fig. 4). Selectively ILV-labeled samples have also been used to characterize the spectroscopically-invisible vesicle-bound states of α-synuclein by transfer-NOE type experiments involving Leu methyl protons that were detected in the ^{1}H–^{15}N HSQC spectrum of the free state [274].

In an even more ambitious application, methyl group signals allowed ligand-induced changes in the extracellular surface of the 365-residue human β2 adrenergic receptor in DDM to be investigated without the assistance of deuterium labeling [275]. In this study ^{13}C-methyl groups were introduced through reductive methylation of solvent-exposed Lys ε-NH_2 groups. The presence of a salt-bridge interaction on the extracellular side could be confirmed, with changes in this interaction being detected in the ^{1}H–^{13}C HMQC spectrum upon agonist binding. This data provided new insights into the structure of the active state, a form that had not yet been captured by X-ray crystallography. The relatively large signal-to-noise ratio provided by methyl groups at the ends of the long flexible Lys side chain was instrumental for the success of this experiment, particularly given the low concentrations of samples that were available (60–200 μM). These experiments illustrate the great potential for methyl protons to provide a window into previously inaccessible dynamic states of large membrane proteins.

4.2 *Relaxation Optimization*

Given the large size of typical membrane protein samples, selective labeling approaches must usually be accompanied by relaxation-optimized NMR experiments. Just over a decade ago, rapid transverse relaxation rates associated with large, slowly tumbling molecules defined a ~30–40 kDa molecular weight ceiling for protein NMR. Beyond this limit, broad peaks and magnetization losses during coherence transfer thwarted attempts to acquire standard spectra required for structure determination. The introduction of transverse relaxation optimized spectroscopy (TROSY) by Wuthrich and co-workers [276] opened up a wealth of new opportunities for solution NMR on large protein systems, including detergent-solubilized membrane proteins.

First demonstrated for amide ^{1}H–^{15}N [276] followed by aromatic ^{1}H–^{13}C groups [277], the basic idea exploits interference effects between the two dominant mechanisms of relaxation, namely dipole–dipole coupling and chemical shift anisotropy (CSA). The effect of this interference can be seen in an uncoupled COSY spectrum, which for a directly bonded ^{1}H–^{15}N pair gives rise to a doublet in both the ^{1}H and ^{15}N dimensions, each separated by the one-bond coupling constant $^{1}J_{HN}$. In this quartet, one peak will exhibit a narrower line width relative to the other peaks as a result of partial cancellation of dipole-dipole and CSA relaxation pathways. This effect is most pronounced at higher magnetic field strengths, with maximal cancellation occurring at ~1 GHz proton frequency for both ^{1}H and ^{15}N magnetization. In TROSY-type NMR pulse sequences, the coherence that gives rise to the slow-relaxing component of the quartet is preserved while the other components are either eliminated [276] or allowed to decay to negligible levels during the course of the experiment [278, 279]. Although all four coherences would make contributions to the single ^{1}H–^{15}N peak in the decoupled correlation spectrum that is normally run for smaller proteins, interconversion between states that undergo very fast vs slow relaxation give rise to magnetization losses that are dominated by the rapidly decaying coherence. Consequently, for larger proteins (~50 kDa and higher) the signal lost by discarding three parts of the quartet is compensated by the increase in resolution and sensitivity that comes from avoiding the more rapidly relaxing states during acquisition.

Numerous multidimensional triple-resonance NMR experiments have since been designed that isolate this slow-relaxing component during evolution and acquisition periods (reviewed in [280]). Additional strategies that take advantage of the unique relaxation properties of slowly tumbling proteins have since been described that provide additional sensitivity gains for large systems (e.g., polarization transfer schemes that use cross-correlated relaxation [281, 282], or longitudinal relaxation optimization [283]). This has allowed backbone assignments to be obtained for many proteins exceeding 40 kDa, including a number of large membrane protein–detergent complexes [110, 284]. This has usually required the use of deuterium-labeled samples to eliminate ^{1}H–^{1}H dipole dipole interactions that also contribute significantly to transverse relaxation rates. In some cases it can be possible to skip deuterium labeling, particularly when only 2D ^{1}H–^{15}N correlation spectra are required, as was the case when screening sample conditions for DAGK [285]. However, the ability to obtain backbone resonance assignments for typical membrane protein samples requires that the TROSY effect be maximized through the use of uniform deuterium incorporation and high spectrometer field strengths (i.e., 700 MHz and greater). Even when these conditions are fulfilled, peak intensities in the TROSY spectrum may be attenuated by microsecond to millisecond timescale exchange processes. This has been commonly observed for membrane proteins in detergent micelles [37, 70, 111, 209, 285–287], making the optimization of sample conditions that can minimize exchange on this timescale a particularly important consideration for membrane proteins [155].

Relaxation optimization strategies have since been extended to side chain aliphatic groups with the important recognition by Kay and coworkers that the coherence transfer pathways available to a rapidly rotating, isolated $^{13}C^{1}H_3$ spin system in a slowly tumbling macromolecule undergoes different rates of relaxation [288, 289]. Most significantly, they established that the HMQC is a TROSY type of experiment for isolated $^{13}C^{1}H_3$ systems since, unlike the HSQC experiment, there is no interconversion of slowly- and rapidly-relaxing coherences. The benefits of this effect are optimal for ^{2}H, ^{12}C-labeled proteins with specific incorporation of a single $^{13}C^{1}H_3$ group in a subset of amino acid types (e.g., ILV [290]). Deuterium spin relaxation measurements in $^{13}CHD_2$-labeled methyl groups can also take advantage of this relaxation optimization strategy [291], opening the door to the study of dynamic structural states for extremely large protein complexes, as demonstrated for the 670-kDa 20S proteasome core particle [271].

Although unassigned methyl group resonances can be used as reporters of structural transitions in large proteins [257, 292], any requirement for site-specific information requires that these methyl peaks be assigned. This is straightforward when backbone assignments are available, since TROSY-type sequences can be used to correlate methyl proton or carbon shifts to amide ^{1}H and ^{15}N chemical shifts [293–295]. Alternatively, in some cases it has been possible to transfer assignments made on smaller fragments or isolated subunits to these same components in the intact complex [257, 266, 295]. Meanwhile, efforts are also being made to automate the process of methyl shift assignment, with one protocol using X-ray structure-based chemical shift predictions along with methyl-NOESY data to correctly assign 99% of a 300-kDa ILV-labeled proteasome [296]. This approach may become important for large single-chain membrane protein–detergent complexes, particularly since these samples would not easily be adapted to the fragment-based approach, and ^{1}H–^{15}N spectral quality may not be sufficient for through-bond correlations with methyl groups. However, in the event that backbone assignments are not available or methyl correlation spectra do not allow unambiguous assignment, methyl shift assignments can still be made by using spectra of mutants [266], sometimes paired with NOEs, or paramagnetic relaxation enhancement patterns [297].

5 Strategies for Membrane Protein Structure Determination by Solution NMR

5.1 NOE-Based Methods of Structure Determination

The approach used for structure determination of a membrane protein varies significantly depending on the size of the complex, the protein fold, and the quality of spectra that can be attained. When high quality spectra of relatively small detergent–protein complexes are available, conventional NOE-based approaches can be applied in a straightforward manner without the need for deuterium labeling or TROSY type experiments. This tends to be the case for structures of single TM helices in monomeric or dimeric forms, examples of which are shown in Table 2, Method I, Fig. 6a. Structures of larger oligomeric states formed by single TM helices have also been determined in this way, although some ^{2}H-labeling and TROSY-based backbone assignment experiments were required to deal with the larger size of these complexes (Table 2, Method II, Fig. 6b).

When the length of the polypeptide exceeds that of these single TM-helix constructs, TROSY-based experiments and uniform deuteration are usually used to obtain backbone assignments. In cases where spectral quality or complexity prevents straightforward assignment of backbone resonances from uniformly ^{15}N, ^{13}C, ^{2}H-labeled samples, it is possible to increase the number of assignments with samples selectively ^{15}N- or ^{15}N, ^{13}C-labeled with a single amino acid. For example, ~20 different selectively labeled samples were used to help assign backbone atoms in the human mitochondrial VDAC β-barrel [73]. Alternatively, if the complex being studied is comprised of more than one polypeptide chain, backbone assignment can be assisted by combining labeled and unlabeled subunits, as was done for the heterotrimeric natural killer cell-activating complex [65]. Additional assignments are also sometimes accessible through the use of a range of temperatures [184] or pH conditions [40] that allow observation of different subsets of resonances. As a last resort it is also possible to use mutagenesis [273, 297, 298], although this approach is not practical for the assignment of a large number of residues.

In cases where spectral quality allows a high level of backbone assignments to be made, it is usually possible to acquire assignable NOEs between amide protons. For β-barrel folds these NOEs can be used with backbone torsion angle input derived from secondary backbone shift data [299], and hydrogen bond restraints inferred from amide solvent exchange to construct an informative global fold (Table 2, Method III, Fig. 6c) [68, 69, 72]. Selectively methyl labeled ILV has also been used for this fold type as an additional source of NOEs in some cases (Table 2, Method IV, Fig. 6c) [70, 71, 73]. For helical membrane proteins, more NOEs are required to obtain a structure of comparable precision (as described in Sect. 4.1.2), usually by extending chemical shift and NOE assignments to other side chain atoms. This requires that 3D ^{1}H–^{1}H correlation spectra on partially or fully protonated samples are of sufficient quality for assignment, an achievement that that has only been realized in a very small number of cases (see Table 2, Method V).

5.2 *Non-uniform Sampling*

One of the most challenging membrane protein structures determined to a high resolution via an NOE-driven approach is the pSRII GPCR (Fig. 6d), which required additional sensitivity-enhancing approaches to be employed during data

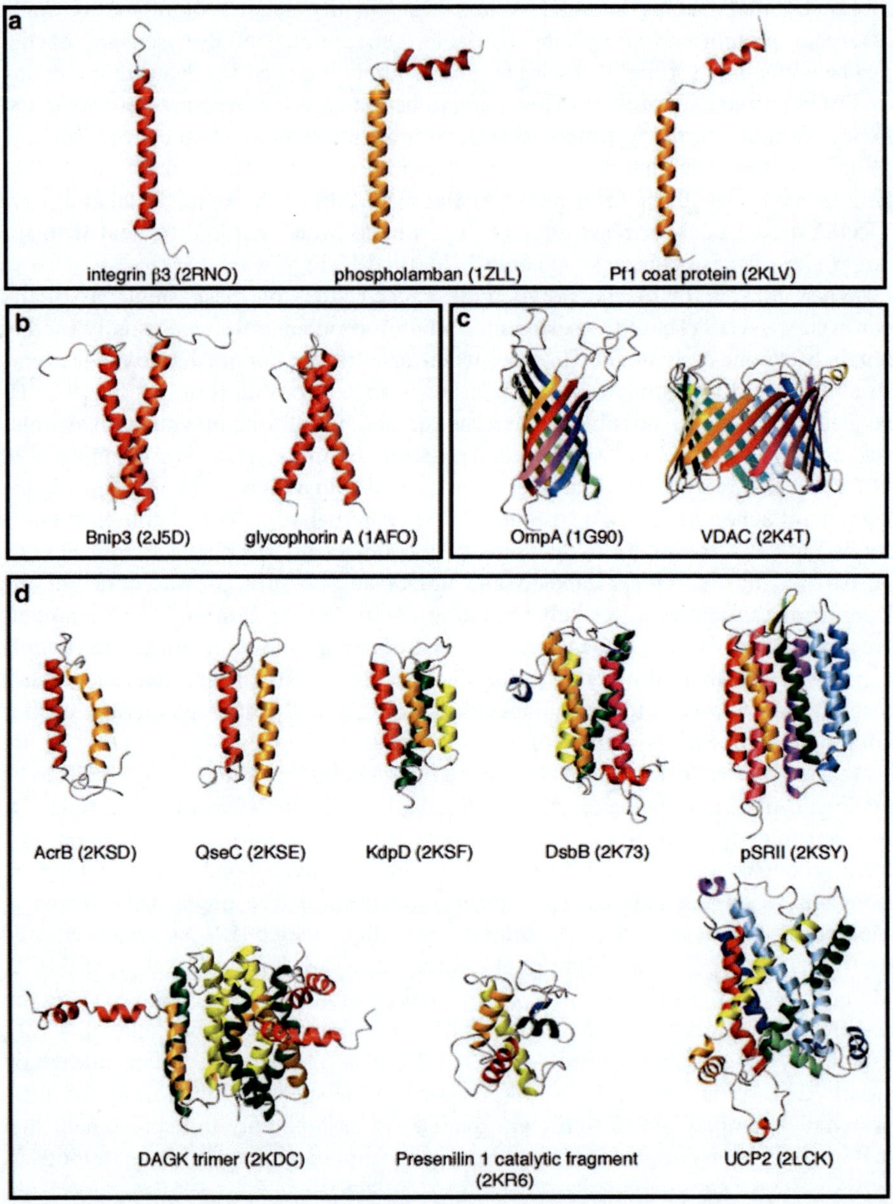

Fig. 6 Representative samples of integral membrane protein structures solved using solution NMR, comprised of either (**a**) a single TM helix, (**b**) a TM helix dimer, (**c**) a β-barrel, or (**d**) multiple TM helices from a single chain (with the exception of DAGK, which is a trimer of 3-TM helix subunits). Coloring from the N terminus to C goes from *red* to *orange, yellow, green, blue, purple,* and then *pink*. All PDB accession numbers are indicated in parentheses for each structure

acquisition [40, 145]. Specifically, many of the 3D spectra were acquired using non-uniform sampling in all indirectly detected dimensions [300]. In this method, data is only acquired for a subset of the incremented evolution periods, with more frequent sampling in regions of short evolution times that have larger signal-to-noise ratios [301]. The emphasis on data acquisition in parts of the time domain having the greatest signal intensity leads to increases in sensitivity per unit time over conventional uniformly sampled experiments [302]. Maximum entropy reconstruction was used to process this sparsely sampled dataset, producing multidimensional spectra that retained the resolution of the conventionally sampled experiment in spite of the reduced number of indirect points [303, 304]. In the case of pSRII, the signal-to-noise ratio was further enhanced by adding multiple datasets acquired on one or more samples [40], a useful approach when sample stability limits the time that can be used to acquire a single dataset.

Non-uniform sampling (NUS) also opens the door to experiments of increased dimensionality that might normally require several months of acquisition time to obtain adequate resolution in the indirectly detected domains [302, 305]. This strategy was used to address the problem of chemical shift degeneracy in methyl NOESY data for the 283-residue VDAC β-barrel (Fig. 6d) [73]. Four-dimensional spectra were acquired on this sample to correlate ^{1}H shifts with the shift of the directly attached heteroatom on both sides of the NOE interaction [73, 306]. These spectra were processed using multidimensional decomposition (MDD) [307, 308], an approach that is particularly well-suited for the accurate reproduction of signal intensities in spectra having a wide range of peak intensities, as is seen in NOESY spectra [309]. Moreover, significant sensitivity increases for weak peaks were realized by combining multiple datasets during MDD. This technique, called coupled MDD, was used to process a ^{13}C–^{13}C separated 4D [^{1}H–^{1}H] NUS-NOESY by co-processing with a ^{1}H–^{13}C HMQC template spectrum [310]. Coupled MDD allowed NOEs involving the 15% of VDAC methyl groups that did not appear in the MDD-processed spectrum to be detected and assigned, many of which were critical for determination of the global fold.

Overall, the application of non-linear sampling and other random sampling techniques to membrane proteins is still in its early days, and does not reflect the diversity of sparse sampling techniques that have been developed to date (reviewed in [302, 305]). Nonetheless these studies demonstrate the significant advantages that can be gained by applying these data acquisition strategies to challenging membrane protein samples, which should motivate an increasing number of research groups to adopt this technology.

5.3 Global Folds in the Absence of Long-Range NOEs

Even when sample labeling and NMR experiments are optimized to minimize unfavorable relaxation processes and increase sensitivity, it is not unusual for difficulties in side chain and NOE assignment to preclude an NOE-driven approach to structure determination. In addition, some smaller membrane proteins are comprised of structural elements that only weakly interact with each other, if at all, and would therefore not be expected to give rise to long-range NOEs. Regardless of whether the absence of assignable long-range NOEs is due to structural or spectroscopic causes, in both cases distance restraints are accessible using approaches that do not require assignment of side chain chemical shifts or long-range NOEs. This approach has been used to determine structures for a significant number of membrane proteins (Table 2, Method VII), with residual dipolar couplings (RDCs) and paramagnetic relaxation enhancement (PRE) data being the primary sources of long-range distant restraints.

5.3.1 Measurement of RDCs

Although rapid molecular reorientation rates in solution are responsible for the high resolution of liquid-state NMR spectra, this also gives rise to isotropic averaging that prevents the observation of orientation-dependent phenomena. However, it is possible to detect these orientation-dependent effects while retaining the advantages of solution phase spectroscopy by subjecting the protein of interest to a weakly aligning environment [311]. The resulting non-zero averaging allows data such as the RDC and residual chemical shift anisotropy (RCSA) to be acquired and used for structure determination.

From a practical viewpoint, the generation of weakly aligned membrane protein samples poses additional challenges relative to the case for water-soluble proteins. While the first RDC measurements on water-soluble proteins used liquid-crystalline phases of bicelles to induce weak alignment [312], the extension of this measurement to membrane protein samples awaited the development of detergent-resistant alignment media. The first, and one of the most widely used, type of detergent-compatible systems is the aqueous phase of mechanically strained polyacrylamide gels [313]. Asymmetry in gel pores can be straightforwardly created with the assistance of a commercially available gel-stretching apparatus [314], or simply by applying axial compression to gels with diameters smaller than that of the NMR tube interior [313, 315]. A specialized type of NMR tube has also been designed that makes it possible to measure both isotropic and aligned states in the same gel sample, a particularly important advance for accurate measurement of small RCSAs [316].

Variations in the degree of alignment can be obtained by using different gel densities, aspect ratios, and gel cross-linking ratios [313, 314]. However, at the gel concentrations required to maintain mechanical stability of the gel (>4% w/v), diffusion of large protein–detergent complexes into the pores can be hindered, reducing the amount of protein that can be introduced into this medium [68, 247]. Consequently, the first RDC measurements on membrane proteins were done with smaller protein–detergent complexes using higher gel concentrations [137, 160, 314, 317, 318]. More recently it was demonstrated that the ~100-kDa DAGK-DPC complex could be introduced into a 4% hydrated gel by allowing ~2 days for the system to come to equilibrium [38]. This method differed from typical protocols previously reported since a solution containing the NMR sample is usually used to rehydrate the dried gel, thereby reducing sample losses from dilution. However, this was not possible for the 4% gel, since the presence of detergent in the rehydrating solution did not allow the original gel dimensions to be restored [112]. This reflects the trade-off in mechanical stability that comes with the reduction in polyacrylamide concentrations required to make gels with larger pore sizes.

To address some of the difficulties in generating gel-aligned membrane protein samples, protocols have been developed to perform the gel polymerization in a solution already containing the protein sample, or to introduce an SDS-solubilized sample into a gel via electrophoresis [319]. In the case of copolymerization, a protocol would need to be carefully tailored to the specific features of each protein-detergent system being studied to minimize the potential for undesired covalent modifications during the relatively non-specific polymerization reaction [112]. However, a more general alternative to these neutral gels has also been developed using charged polyacrylamide-based copolymers [320, 321]. Rehydration of charged copolymers is facilitated by a strong electroosmotic effect, providing an increase in mechanical stability over neutral polyacrylamide gels. Consequently, alignment can be reproducibly achieved in the presence of detergents at gel concentrations as low as 2–3%. The larger pore sizes of these lower concentration gels, along with the higher mechanical stability, make it possible for larger protein–detergent complexes to be introduced into the gel by rehydration. This has allowed RDC measurements to be obtained for the 4-TM helix DsbB [39] and the 8-strand OmpA protein [320] both in complex with DPC micelles. In addition, alignment has been demonstrated using copolymers with either positive or negative charge, as well as mixtures of positively and negatively charged copolymers [320, 321]. This permits the electrostatic properties of the alignment medium to be tailored to the charge properties of the sample, and different alignment frame orientations to be generated for the measurement of more than one set of RDCs.

One aligning medium that has been extensively used for soluble proteins is filamentous bacteriophage [322], which was recently been shown to be compatible with modest concentrations (~100 mM) of phosphocholine-based membrane-mimetics so long as the pH of the solution exceeds ~6 [323]. These long negatively charged filamentous structures were also the inspiration for the design of a novel detergent-resistant aligning medium formed by DNA nanotubes [324]. These are self-assembled from concentrated solutions of a 7.3-kb "scaffold" strand mixed with a 170 base "staple" strand, and have been used to induce alignment in the $\zeta\zeta$ TM dimer [324], and the mitochondrial uncoupling protein 2 [41], both using detergent concentrations in excess of 150 mM. A potentially more convenient alternative has also been developed that generates a similar type of macromolecular structure from the potassium salt of the dinucleotide 2′-deoxyguanylyl-(3′,5′)-2′-deoxyguanosine [325]. These dinucleotides form G-tetrads that stack into columns with dimensions similar to those of bacteriophage, and have been used for the measurement of RDCs in the LMPG micelle-associated cytoplasmic domain of the influenza B proton channel [326] and the αIIbβ3 TM helix heterodimer in isotropic bicelles [184].

With the exception of the strained polyacrylamide gel, many of the detergent-resistant media described possess significant charge that can give rise to unfavorable interactions with the alignment media, potentially broadening spectra beyond detection [327]. An alternative that avoids this problem exploits the significant anisotropy of the magnetic susceptibility of lanthanide ions such as Yb^{3+} and Dy^{3+}. This leads to weak magnetic field-induced alignment that allows RDCs of lanthanide-bound species to be measured [328]. For proteins that do not have a high-affinity metal-binding site, it is possible to introduce a metal chelator via covalent modification of a unique cysteine residue [327, 329, 330] or by engineering a metal-binding EF-hand as was done for the HIV Vpu protein [331] and a modified form of OmpA [332]. However, to maximize the information content from these experiments, high affinity tags with a single susceptibility tensor that is minimally affected by dynamics should be used. For this purpose, sulfhydryl-reactive EDTA based analogs have been designed that, unlike the non-stereospecifically metal-binding precursor, create a single stereoisomer of the metal-bound complex [333].

5.3.2 Applications of RDCs in Structure Determination

The magnitude of the RDC depends on the degree of alignment induced in the sample, the inter-nuclear distance separating the two atoms involved, the orientation of the internuclear bond vector with respect to the external magnetic field, and the local dynamics of the atoms involved [311]. When measured for a set of covalently linked backbone atoms with similar dynamic properties, RDCs can be used to determine the relative orientation of the corresponding peptide planes. A large number of NMR experiments are now available to measure these types of RDCs (reviewed in [334]) and inclusion of at least one type of RDC is often performed in standard protein structure determinations (reviewed in [335, 336]), including those for membrane proteins [39, 152, 162, 166, 326, 337, 338]. However, RDCs alone are not enough to determine a unique protein fold when starting from an extended structure, since there is 16-fold degeneracy in the number of peptide plane orientations consistent with the measured couplings [339]. Consequently, RDCs cannot distinguish between structures of different quality at early points in a structure calculation when differences between calculated and actual structures are large (i.e., rmsd > 10 Å) [340, 341]. RDCs are instead used to refine NOE-based structures via a scalable energy function that gradually increases the magnitude of the RDC energy penalty as the NOEs drive the fold closer to its final state [342]. This allows the structure to be oriented into its alignment frame before the couplings refine the structure, an approach that can lead to an increase in the quality and precision of an NOE-based structure [342, 343]. However, these benefits are usually only realized when multiple RDCs per peptide plane, or RDCs from more than one type of alignment medium, are available [335, 339].

When structures of secondary structure elements or globular domains are known, it is possible to use RDCs to calculate the relative orientation of these structural units. First demonstrated for multi-domain proteins [344, 345], a similar strategy has since been used to determine average protein structures for small membrane proteins in the absence of other long-range restraints, using TM helices as the rigid, or tightly restrained bodies [137, 318, 346]. However, for each structural unit there will be four equivalent orientations that can be defined with respect to a right-handed alignment frame (assuming that the alignment tensor is not axially symmetric) [347]. Hence it is necessary to eliminate this degeneracy through the measurement of two or more sets of RDCs under different aligning conditions [348, 349] or by the recognition of structural restraints that can rule out extraneous solutions [318, 347, 350, 351].

A related application for RDCs has also been described based on the sequence-dependent pattern of RDCs along a helical structure, called a "dipolar wave" by the Opella group [317, 352, 353]. The magnitude and periodicity of the dipolar wave depends on the orientation of the helix, and can allow irregularities in helix structure to be identified. Most commonly, dipolar waves have been used to help determine the location of helices in a protein sequence, allowing these structural elements to be more rigidly restrained over the course of a structure calculation [57, 159, 323, 354]. This is particularly useful for larger helical membrane proteins, since α-helices are not well defined by the NOEs available in sparsely protonated samples [262].

Aside from helices, RDCs have a more general capacity to recognize related structures. This approach has been used to identify homologous structures in the protein databank [355], which could then be used as a starting point for RDC-based refinement to the final fold [356]. Having relatively few representatives in the PDB, a more useful application for membrane proteins is the identification of structurally homologous protein fragments from the structure database [357] or the structure modeling program Rosetta [358] to generate an initial model for refinement. This approach works well for smaller proteins having around four measured couplings per peptide plane, although ambiguities can arise with multiple structures being equally compatible with the RDC data for some fragments. Hence for larger proteins, it is necessary to supplement RDC data with distance restraints provided by NOEs [358] or paramagnetic relaxation enhancement (PRE) data [41, 351] to resolve these ambiguities.

The power of this fragment-replacement approach was recently demonstrated for the 6-TM helix UCP2 protein, for which an average of 2.2 RDCs per residue could be acquired (Fig. 6d) [41]. These RDCs were used to determine local and secondary structures using a PDB-based fragment database. This resulted in 15 continuous segments of UCP2 with local backbone structures that could be highly-restrained during structure calculation. RDC input at this stage helped to determine relative segment orientations, and PREs provided information on their relative positions. Although no structural information on side chains are obtained in this approach, the global folds that can be accessed without the requirement for NOE assignment have the exciting potential to increase the number of membrane proteins structures that can be characterized by solution NMR.

5.3.3 Paramagnetic Relaxation Enhancements

As shown in the UCP2 study, RDCs on their own are usually not sufficient to determine a unique protein conformation, and long-range distance restraints are often required to resolve structural ambiguities. PREs have frequently been used to provide this type of distance information for membrane proteins both in the absence and presence of RDC data, particularly when long-range NOEs are not available (listed in Table 2, Method VII). PREs arise from dipolar interactions involving the large dipole of an unpaired electron that leads to efficient loss of polarization and coherence in surrounding nuclear spins. There is a relatively straightforward relationship between the paramagnetically enhanced transverse relaxation rate of a proton and the inverse sixth power of its distance from the unpaired electron (r^{-6}), allowing this measurement to be converted into distance restraints for structure determination [359, 360]. This is usually achieved by measurement of paramagnetic spin-induced changes in backbone amide ^{1}H peak intensities [360], although greater accuracy can be achieved by either full [361] or two-point [362] R_2 relaxation measurements [363]. The range of distances that are accessed by the PRE measurement depends on the type of paramagnetic center introduced into the protein, but typically provides distance information on spins as far away as 25 Å or more.

There are several different types of spin labels that can be introduced into the target protein, usually via covalent modification of a unique cysteine residue and/or inclusion of a metal-chelating peptide in the protein sequence (reviewed in [364, 365]). Many of the metal-chelating tags used to induce alignment for measurement of RDCs (Sect. 5.3.1) can also be used to bind to paramagnetic metals with slow electronic relaxation properties (e.g., Mn^{2+} and Cu^{2+}) that give rise to a significant PRE. However, for membrane proteins, the vast majority of PRE-based restraints have used the sulfhydryl-reactive nitroxide spin label, 1-oxyl-2,2,5,5-tetramethyl-3-pyrroline-3-methyl-methanethiosulfonate (MTSL) [38, 39, 41, 67, 88, 136, 354, 366].

To obtain sufficient numbers of PRE-based restraints, it is usually necessary to introduce spin-labels at a diverse range of sites. For example, nine different MTSL-labeled DsbB samples (approximately 1 MTSL-label for every 20 residues), were required to generate an average of 6.5 PREs per residue [39]. Choosing where to place these spin-labels requires identification of sites that can provide informative PRE-based restraints without disrupting the structure. This makes it necessary to have some knowledge of the structure being studied, with secondary structure determination from backbone assignments being the first step in this process. Once labeled, the NMR spectrum is used to confirm that the probe does not alter structure, since only local shift changes should be observed in this case.

For any type of paramagnetic center used, accurate translation of PREs into distance restraints requires that the entire population be spin-labeled, since the extent of broadening will be underestimated if the labeling reaction is incomplete. In addition, the diamagnetic reference spectrum must not contain any residual paramagnetic contributions. These can be significant when water-soluble reducing agents such as ascorbic acid are used to reduce MTSL, particularly since the membrane-mimetic environment may shield the radical center. This was observed in the case of OmpA, with EPR spectra suggesting that only ~90% of the paramagnetic center was reduced, in spite of attempts to achieve higher levels of completion with a wide range of reducing agents [367]. To avoid this problem it is possible to prepare a separate sample labeled with a diamagnetic analog of MTSL that contains an acetyl group in place of the oxygen radical, allowing a purely diamagnetic reference spectrum to be acquired [367]. Similarly, the reference spectrum for PRE measurements with metal chelating tags is usually obtained with a diamagnetic ion (e.g., Ca^{2+}, Mg^{2+}) being bound in place of the paramagnetic metal [327, 329, 330].

If care is taken to ensure that samples are completely labeled and that an appropriate diamagnetic reference is used, then it is possible to generate accurate distance restraints from these PRE measurements. For example, residues showing paramagnetic:diamagnetic peak intensity ratios (I_{para}/I_{dia}) of 15–85% in MTSL-labeled OmpA had PRE-based distance measurements that were within 2 Å of the corresponding distances determined from the crystal structure [367]. This allows relatively tight restraint bounds to be applied during structure calculations, with ±2 Å for the 15–24 Å distance range, and an upper limit of 15 Å being applied for residues showing intensity ratios under 15%. It is also possible to impose lower distance limits of 20–22 Å for residues showing small PREs (>85%), although these did not improve the quality of the OmpA structure [367]. This was the approach used in the structure determination of the 4-TM helix DsbB, which along with RDCs and a handful of long-range NOEs, generated a highly precise ensemble (pairwise backbone rmsd of 0.80 Å) [39]. A comparable level of precision was also achieved for Rv1761, which used slightly looser bounds of ±3 Å, since the narrower bounds gave rise to distance violations [159]. More conservative bounds (e.g. ±4 Å [67, 351, 360] or greater [41]) or NOE-like assignments into distance bins (i.e., strong, medium weak PREs) have also been used [88, 368], although this does reduce the impact of the PRE on the structure [367].

An important consideration when converting PREs into distances is the influence of spin-label and backbone dynamics. Specifically, the r^{-6} dependence causes the measured PRE for a particular ^1H atom to be dominated by conformers having shorter distances to the spin label, even if this state is infrequently sampled [363]. Consequently, PREs involving flexible regions of a protein cannot be accurately converted into a single average distance. Since dynamic regions of the protein can be identified with standard backbone relaxation measurements, PREs involving these residues should be discarded or only used with very loose restraint bounds.

Less straightforward is the treatment of spin labels that are bound to the protein via a flexible linker, where exchange between various rotameric states can occur. To account for the different contributions to the PRE that can arise from this type of exchange, an ensemble approach was developed that refines the structures against PREs in place of PRE-derived distances [363], improving the accuracy of structures calculated for Mn^{2+}-EDTA labeled DNA in complex with SRY. This type of ensemble-based refinement has also been suggested to be necessary for more accurate use of PRE data from the MTLS spin label, particularly since some high-resolution crystal structures of MTSL-modified lysozyme show the presence of more than one spin-label conformer [369, 370]. However, the good agreement that was found between calculated and PRE-measured distances for the relatively rigid barrel region of MTSL-labeled OmpA implies that the presence of multiple conformations may be less significant for this particular spin label [367]. EPR line shapes of MTSL-modified lysozyme also show that the motion of this paramagnetic center is restricted [371]. Hence it appears that carefully measured PREs with MTSL in non-dynamic segments can in most cases be validly converted into long-range distance restraints with tighter bounds (i.e., ±2 Å).

Overall, PREs have significantly increased the number of larger membrane protein structures that have been determined by solution NMR in a relatively short period, becoming an established source of structural information. In addition, water-soluble paramagnetic relaxation agents can be used to reveal aqueous-exposed segments [136, 162, 184, 354, 366, 372–374], spin-labeled lipids and detergents can identify lipid-embedded regions [136, 372, 375–381], and intermolecular interactions can be probed with exquisite sensitivity using spin-labeled proteins [243, 382]. This versatility is one of the factors ensuring that paramagnetic effects in NMR will continue to make important contributions to our understanding of membrane protein structure.

6 Concluding Remarks

Although membrane proteins continue to present challenges for solution NMR, innovations in sample development, data acquisition, and structure determination strategies have allowed structural insights to be obtained from some of the most demanding systems tackled to date (e.g., GPCRs and large β-barrel channels). In addition to the advances outlined here, these achievements build upon contributions from the handful of veteran laboratories that have worked to apply solution NMR to membrane proteins, even before the introduction of more modern methods for study of large proteins (e.g., [140, 383–386]). Yet in spite of this long history, the field is still considered to be quite young, with only a limited number of groups fully exploiting the potential of solution NMR for membrane protein structure determination. Nonetheless, it is encouraging to note the progress that has been made since the first NMR structure was determined for an integral membrane protein comprised of more than one TM segment back in 1997 (i.e., the glycophorin A homodimer [140]). As of the end of 2010 the protein structure database holds approximately 24 integral membrane protein structures (counting only those with more than one membrane-spanning segment), putting the field at par with the state of membrane protein crystallography in the late 1990s. However, the rate of new structure accumulation closely follows the exponential rise seen for crystal

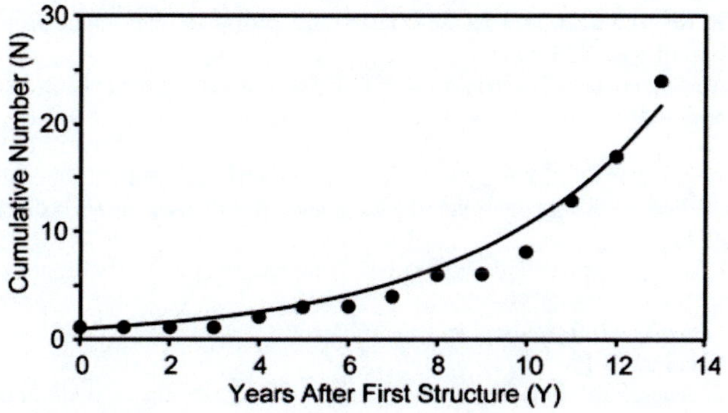

Fig. 7 Progress in integral membrane protein structure determination by solution NMR (structures with more than one membrane-spanning segment only). The cumulative number of integral membrane protein structures determined by solution NMR is shown for each year, starting with the year the first high-resolution structure of a TM-helix dimer was revealed in 1997 [140]. As had been shown for X-ray crystal structures of membrane proteins [399], this accumulation can be fit to the equation $N = \exp(aY)$, where N is the cumulative total of structures for each year following publication of the first structure (Y). The scaling parameter $a = 0.236$ for the best-fit curve (shown as the *solid line*) is almost the same as that previously determined for membrane protein X-ray structures ($a = 0.242$), showing similar rates of progress for the two structure determination techniques. Year 0 for X-ray crystal structures was recognized as 1985 [399], more than 10 years earlier than that for NMR (1997). A useful reference used in the compilation this data was provided by the on-line catalogue of membrane protein NMR structures (http://www.drorlist.com/nmr/MPNMR.html)

structures of membrane proteins (Fig. 7) [387], with the main disadvantage being its late start relative to X-ray crystallography. Assuming that solution NMR of membrane proteins continues at this pace, we can expect to see the 100 structure mark surpassed in approximately 6 years. The ability of long-range restraints to be acquired without the need for side chain assignment will be particularly important in meeting these projections, as will the implementation of new sensitivity-enhancing technologies by an increasing number of labs. This analysis shows that, while solution NMR studies of larger membrane proteins may not yet be routine, reaping the rewards of these endeavors will become increasingly feasible for a larger range of membrane proteins than ever before.

Acknowledgments This work was supported by the Natural Sciences and Engineering Research Council (NSERC).

References

1. Kim H, Melén K, Osterberg M, von Heijne G (2006) A global topology map of the *Saccharomyces cerevisiae* membrane proteome. Proc Natl Acad Sci USA 103:11142–11147
2. Almen MS, Nordstrom KJV, Fredriksson R, Schioth HB (2009) Mapping the human membrane proteome: a majority of the human membrane proteins can be classified according to function and evolutionary origin. BMC Biol 7:50

3. Sanders CR, Myers JK (2004) Disease-related misassembly of membrane proteins. Annu Rev Biophys Biomol Struct 33:25–51
4. Sanguinetti MC, Tristani-Firouzi M (2006) hERG potassium channels and cardiac arrhythmia. Nature 440:463–469
5. Rowe SM, Miller S, Sorscher EJ (2005) Cystic fibrosis. N Engl J Med 352:1992–2001
6. Suh YH, Checler F (2002) Amyloid precursor protein, presenilins, and alpha-synuclein: molecular pathogenesis and pharmacological applications in Alzheimer's disease. Pharmacol Rev 54:469–525
7. Gurrath M (2001) Peptide-binding G protein-coupled receptors: new opportunities for drug design. Curr Med Chem 8:1605–1648
8. Landry Y, Gies JP (2008) Drugs and the molecular targets: an updated overview. Fundam Clin Pharmacol 22:1–18
9. Yee AA, Savchenko A, Ignachenko A, Lukin J, Xu X, Skarina T, Evdokimova E, Liu CS, Semesi A, Guido V, Edwards AM, Arrowsmith CH (2005) NMR and X-ray crystallography, complementary tools in structural proteomics of small proteins. J Am Chem Soc 127:16512–16517
10. Yee AA, Gutmanas A, Arrowsmith CH (2006) Solution NMR in structural genomics. Curr Opin Struct Biol 16:611–617
11. Snyder DA, Chen Y, Denissova NG, Acton T, Aramini JM, Ciano M, Karlin R, Liu J, Manor P, Rajan PA, Rossi P, Swapna GV, Xiao R, Rost B, Hunt J, Montelione GT (2005) Comparisons of NMR spectral quality and success in crystallization demonstrate that NMR and X-ray crystallography are complementary methods for small protein structure determination. J Am Chem Soc 127:16505–16511
12. Baldwin AJ, Kay LE (2009) NMR spectroscopy brings invisible protein states into focus. Nat Chem Biol 5:808–814
13. Henzler-Wildman K, Kern D (2007) Dynamic personalities of proteins. Nature 450:964–972
14. Mittermaier AK, Kay LE (2006) Review - new tools provide new insights in NMR studies of protein dynamics. Science 312:224–228
15. Chill JH, Naider F (2011) A solution NMR view of protein dynamics in the biological membrane. Curr Opin Struct Biol 21:627-633
16. Renault M, Cukkemane A, Baldus M (2010) Solid-state NMR spectroscopy on complex biomolecules. Angew Chem Int Ed 49:8346–8357
17. Opella SJ, Marassi FM (2004) Structure determination of membrane proteins by NMR spectroscopy. Chem Rev 104:3587–3606
18. McDermott A (2009) Structure and dynamics of membrane proteins by magic angle spinning solid-state NMR. Ann Rev Biophys 38:385–403
19. Hong M (2007) Structure, topology, and dynamics of membrane peptides and proteins from solid-state NMR spectroscopy. J Phys Chem B 111:10340–10351
20. Gardner KH, Kay LE (1998) The use of ^{2}H, ^{13}C, ^{15}N multidimensional NMR to study the structure and dynamics of proteins. Annu Rev Biophys Biomol Struct 27:357–406
21. Baneyx F (1999) Recombinant protein expression in *Escherichia coli*. Curr Opin Biotechnol 10:411–425
22. Jana S, Deb JK (2005) Strategies for efficient production of heterologous proteins in *Escherichia coli*. Appl Microbiol Biotechnol 67:289–298
23. Sorensen HP, Mortensen KK (2005) Advanced genetic strategies for recombinant protein expression in *Escherichia coli*. J Biotechnol 115:113–128
24. Vaiphei ST, Tang Y, Montelione GT, Inouye M (2011) The use of the condensed single protein production system for isotope-labeled outer membrane proteins OmpA and OmpX in *E. coli*. Mol Biotechnol 47:205–210
25. Schneider WM, Inouye M, Montelione GT, Roth MJ (2009) Independently inducible system of gene expression for condensed single protein production (cSPP) suitable for high efficiency isotope enrichment. J Struct Funct Genomics 10:219–225
26. Suzuki M, Zhang J, Liu M, Woychik NA, Inouye M (2005) Single protein production in living cells facilitated by an mRNA interferase. Mol Cell 18:253–261

27. Mao L, Tang Y, Vaiphei ST, Shimazu T, Kim SG, Mani R, Fakhoury E, White E, Montelione GT, Inouye M (2009) Production of membrane proteins for NMR studies using the condensed single protein (cSPP) production system. J Struct Funct Genomics 10:281–289
28. Schneider WM, Tang Y, Vaiphei ST, Mao L, Maglaqui M, Inouye M, Roth MJ, Montelione GT (2010) Efficient condensed-phase production of perdeuterated soluble and membrane proteins. J Struct Funct Genomics 11:143–154
29. Sarramegna V, Talmont F, Demange P, Milon A (2003) Heterologous expression of G-protein-coupled receptors: comparison of expression systems from the standpoint of large-scale production and purification. Cell Mol Life Sci 60:1529–1546
30. Yokoyama S (2003) Protein expression systems for structural genomics and proteomics. Curr Opin Chem Biol 7:39–43
31. Fan Y, Shi L, Ladizhansky V, Brown LS (2011) Uniform isotope labeling of a eukaryotic seven-transmembrane helical protein in yeast enables high-resolution solid-state NMR studies in the lipid environment. J Biomol NMR 49:151–161
32. Egorova-Zachernyuk TA, Bosman GJ, DeGrip WJ (2011) Uniform stable-isotope labeling in mammalian cells: formulation of a cost-effective culture medium. Appl Microbiol Biotechnol 89:397–406
33. Takahashi H, Shimada I (2010) Production of isotopically labeled heterologous proteins in non-*E. coli* prokaryotic and eukaryotic cells. J Biomol NMR 46:3–10
34. Wang DN, Safferling M, Lemieux MJ, Griffith H, Chen Y, Li XD (2003) Practical aspects of overexpressing bacterial secondary membrane transporters for structural studies. Biochim Biophys Acta 1610:23–36
35. Duquesne K, Sturgis JN (2010) Membrane protein solubilization. Methods Mol Biol 601:205–217
36. Mancia F, Love J (2010) High-throughput expression and purification of membrane proteins. J Struct Biol 172:85–93
37. Columbus L, Lipfert J, Klock H, Millett I, Doniach S, Lesley SA (2006) Expression, purification, and characterization of *Thermotoga maritima* membrane proteins for structure determination. Protein Sci 15:961–975
38. Van Horn WD, Kim HJ, Ellis CD, Hadziselimovic A, Sulistijo ES, Karra MD, Tian C, Sönnichsen FD, Sanders CR (2009) Solution nuclear magnetic resonance structure of membrane-integral diacylglycerol kinase. Science 324:1726–1729
39. Zhou Y, Cierpicki T, Jimenez RH, Lukasik SM, Ellena JF, Cafiso DS, Kadokura H, Beckwith J, Bushweller JH (2008) NMR solution structure of the integral membrane enzyme DsbB: functional insights into DsbB-catalyzed disulfide bond formation. Mol Cell 31:896–908
40. Gautier A, Mott HR, Bostock MJ, Kirkpatrick JP, Nietlispach D (2010) Structure determination of the seven-helix transmembrane receptor sensory rhodopsin II by solution NMR spectroscopy. Nat Struct Mol Biol 17:768–775
41. Berardi MJ, Shih WM, Harrison SC, Chou JJ (2011) Mitochondrial uncoupling protein 2 structure determined by NMR molecular fragment searching. Nature 476:109–113
42. Laage R, Langosch D (2001) Strategies for prokaryotic expression of eukaryotic membrane proteins. Traffic 2:99–104
43. Wagner S, Bader ML, Drew D, de Gier JW (2006) Rationalizing membrane protein overexpression. Trends Biotechnol 24:364–371
44. von Meyenburg K, Jørgensen BB, Michelsen O, Sørensen L, McCarthy JE (1985) Proton conduction by subunit a of the membrane-bound ATP synthase of *Escherichia coli* revealed after induced overproduction. EMBO J 4:2357–2363
45. Wagner S, Baars L, Ytterberg AJ, Klussmeier AJ, Wagner CS, Nord O, Nygren P, van Wijk K, de Gier JW (2007) Consequences of membrane protein overexpression in *Escherichia coli*. Mol Cell Proteomics 6:1527–1550
46. Klepsch MM, Persson JO, de Gier JW (2011) Consequences of the overexpression of a eukaryotic membrane protein, the human KDEL receptor, in *Escherichia coli*. J Mol Biol 407:532–542

47. Drew D, Fröderberg L, Baars L, de Gier JW (2003) Assembly and overexpression of membrane proteins in *Escherichia coli*. Biochim Biophys Acta 1610:3–10
48. Miroux B, Walker JE (1996) Over-production of proteins in *Escherichia coli*: mutant hosts that allow synthesis of some membrane proteins and globular proteins at high levels. J Mol Biol 260:289–298
49. Steinfels E, Orelle C, Dalmas O, Penin F, Miroux B, Di Pietro A, Jault JM (2002) Highly efficient over-production in *E. coli* of YVcC, a multidrug-like ATP-binding cassette transporter from *Bacillus subtilis*. Biochim Biophys Acta 1565:1–5
50. Arechaga I, Miroux B, Karrasch S, Huijbregts R, de Kruijff B, Runswick MJ, Walker JE (2000) Characterisation of new intracellular membranes in *Escherichia coli* accompanying large scale over-production of the b subunit of F1Fo ATP synthase. FEBS J 482:215–219
51. Kiefer H (2003) In vitro folding of alpha-helical membrane proteins. Biochim Biophys Acta 1610:57–62
52. Junge F, Schneider B, Reckel S, Schwarz D, Dötsch V, Bernhard F (2008) Large-scale production of functional membrane proteins. Cell Mol Life Sci 65:1729–1755
53. Grisshammer R, Tate CG (1995) Overexpression of integral membrane proteins for structural studies. Q Rev Biophys 28:315–422
54. Bannwarth M, Schulz GE (2003) The expression of outer membrane proteins for crystallization. Biochim Biophys Acta 1610:37–45
55. Palmieri L, Runswick MJ, Fiermonte G, Walker JE, Palmieri F (2000) Yeast mitochondrial carriers: bacterial expression, biochemical identification and metabolic significance. J Bioenerg Biomembr 32:67–77
56. Kiefer H, Krieger J, Olszewski JD, Von Heijne G, Prestwich GD, Breer H (1996) Expression of an olfactory receptor in *Escherichia coli*: purification, reconstitution, and ligand binding. Biochemistry 35:16077–16084
57. Park SH, Mrse AA, Nevzorov AA, Mesleh MF, Oblatt-Montal M, Montal M, Opella SJ (2003) Three-dimensional structure of the channel-forming trans-membrane domain of virus protein "u" (Vpu) from HIV-1. J Mol Biol 333:409–424
58. Miozzari GF, Yanofsky C (1978) Translation of the leader region of the *Escherichia coli* tryptophan operon. J Bacteriol 133:1457–1466
59. Staley JP, Kim PS (1994) Formation of a native-like subdomain in a partially folded intermediate of bovine pancreatic trypsin inhibitor. Protein Sci 3:1822–1832
60. Ma C, Marassi FM, Jones DH, Straus SK, Bour S, Strebel K, Schubert U, Oblatt-Montal M, Montal M, Opella SJ (2002) Expression, purification, and activities of full-length and truncated versions of the integral membrane protein Vpu from HIV-1. Protein Sci 11:546–557
61. Crowell KJ, Franzin CM, Koltay A, Lee S, Lucchese AM, Snyder BC, Marassi FM (2003) Expression and characterization of the FXYD ion transport regulators for NMR structural studies in lipid micelles and lipid bilayers. Biochim Biophys Acta 1645:15–21
62. Xie XQ, Zhao J, Zheng H (2004) Expression, purification, and isotope labeling of cannabinoid CB2 receptor fragment, CB2(180–233). Protein Expr Purif 38:61–68
63. Zheng H, Zhao J, Wang S, Lin CM, Chen T, Jones DH, Ma C, Opella S, Xie XQ (2005) Biosynthesis and purification of a hydrophobic peptide from transmembrane domains of G-protein-coupled CB2 receptor. J Pept Res 65:450–458
64. Neumoin A, Cohen LS, Arshava B, Tantry S, Becker JM, Zerbe O, Naider F (2009) Structure of a double transmembrane fragment of a G-protein-coupled receptor in micelles. Biophys J 96:3187–3196
65. Call ME, Wucherpfennig KW, Chou JJ (2010) The structural basis for intramembrane assembly of an activating immunoreceptor complex. Nat Immunol 11:1023–1029
66. Arevalo E, Estephan R, Madeo J, Arshava B, Dumont M, Becker JM, Naider F (2003) Biosynthesis and biophysical analysis of domains of a yeast G protein-coupled receptor. Biopolymers 71:516–531
67. Kang C, Tian C, Sönnichsen FD, Smith JA, Meiler J, George AL Jr, Vanoye CG, Kim HJ, Sanders CR (2008) Structure of KCNE1 and implications for how it modulates the KCNQ1 potassium channel. Biochemistry 47:7999–8006

68. Hwang PM, Choy WY, Lo EI, Chen L, Forman-Kay JD, Raetz CR, Privé GG, Bishop RE, Kay LE (2002) Solution structure and dynamics of the outer membrane enzyme PagP by NMR. Proc Natl Acad Sci USA 99:13560–13565
69. Arora A, Abildgaard F, Bushweller JH, Tamm LK (2001) Structure of outer membrane protein A transmembrane domain by NMR spectroscopy. Nat Struct Biol 8:334–338
70. Renault M, Saurel O, Czaplicki J, Demange P, Gervais V, Löhr F, Réat V, Piotto M, Milon A (2009) Solution state NMR structure and dynamics of KpOmpA, a 210 residue transmembrane domain possessing a high potential for immunological applications. J Mol Biol 385:117–130
71. Fernández C, Hilty C, Wider G, Güntert P, Wüthrich K (2004) NMR structure of the integral membrane protein OmpX. J Mol Biol 336:1211–1221
72. Liang B, Tamm LK (2007) Structure of outer membrane protein G by solution NMR spectroscopy. Proc Natl Acad Sci USA 104:16140–16145
73. Hiller S, Garces RG, Malia TJ, Orekhov VY, Colombini M, Wagner G (2008) Solution structure of the integral human membrane protein VDAC-1 in detergent micelles. Science 321:1206–1210
74. Park SH, Prytulla S, De Angelis AA, Brown JM, Kiefer H, Opella SJ (2006) High-resolution NMR spectroscopy of a GPCR in aligned bicelles. J Am Chem Soc 128:7402–7403
75. Schmidt P, Berger C, Scheidt HA, Berndt S, Bunge A, Beck-Sickinger AG, Huster D (2010) A reconstitution protocol for the in vitro folded human G protein-coupled Y-2 receptor into lipid environment. Biophys Chem 150:29–36
76. Braiman MS, Stern LJ, Chao BH, Khorana HG (1987) Structure-function studies on bacteriorhodopsin. IV. Purification and renaturation of bacterio-opsin polypeptide expressed in *Escherichia coli*. J Biol Chem 262:9271–9276
77. Jekabsons MB, Echtay KS, Arechaga I, Brand MD (2003) Molecular properties of purified human uncoupling protein 2 refolded from bacterial inclusion bodies. J Bioenerg Biomembr 35:409–418
78. Pebay-Peyroula E, Dahout-Gonzalez C, Kahn R, Trézéguet V, Lauquin GJ, Brandolin G (2003) Structure of mitochondrial ADP/ATP carrier in complex with carboxyatractyloside. Nature 426:39–44
79. Kim DM, Choi CY (1996) A semicontinuous prokaryotic coupled transcription/translation system using a dialysis membrane. Biotechnol Prog 12:645–649
80. Knapp KG, Swartz JR (2004) Cell-free production of active *E. coli* thioredoxin reductase and glutathione reductase. FEBS Lett 559:66–70
81. Tyler RC, Aceti DJ, Bingman CA, Cornilescu CC, Fox BG, Frederick RO, Jeon WB, Lee MS, Newman CS, Peterson FC, Phillips GN, Shahan MN, Singh S, Song J, Sreenath HK, Tyler EM, Ulrich EL, Vinarov DA, Vojtik FC, Volkman BF, Wrobel RL, Zhao Q, Markley JL (2005) Comparison of cell-based and cell-free protocols for producing target proteins from the *Arabidopsis thaliana* genome for structural studies. Proteins 59:633–643
82. Vinarov DA, Lytle BL, Peterson FC, Tyler EM, Volkman BF, Markley JL (2004) Cell-free protein production and labeling protocol for NMR-based structural proteomics. Nat Methods 1:149–153
83. Spirin AS (2004) High-throughput cell-free systems for synthesis of functionally active proteins. Trends Biotechnol 22:538–545
84. Baranov VI, Spirin AS (1993) Gene-expression in cell-free system on preparative-scale. Methods Enzymol 217:123–142
85. Koglin A, Klarnmt C, Trbovic N, Schwarz D, Schneider B, Schäfer B, Löhr F, Bernhard F, Dötsch V (2006) Combination of cell-free expression and NMR spectroscopy as a new approach for structural investigation of membrane proteins. Magn Reson Chem 44:S17–S23
86. Kigawa T, Yabuki T, Yoshida Y, Tsutsui M, Ito Y, Shibata T, Yokoyama S (1999) Cell-free production and stable-isotope labeling of milligram quantities of proteins. FEBS Lett 442:15–19
87. Kigawa T, Muto Y, Yokoyama S (1995) Cell-free synthesis and amino acid-selective stable-isotope labeling of proteins for NMR analysis. J Biomol NMR 6:129–134

88. Maslennikov I, Klammt C, Hwang E, Kefala G, Okamura M, Esquivies L, Mörs K, Glaubitz C, Kwiatkowski W, Jeon YH, Choe S (2010) Membrane domain structures of three classes of histidine kinase receptors by cell-free expression and rapid NMR analysis. Proc Natl Acad Sci USA 107:10902–10907
89. Abdine A, Verhoeven MA, Park KH, Ghazi A, Guittet E, Berrier C, Van Heijenoort C, Warschawski DE (2010) Structural study of the membrane protein MscL using cell-free expression and solid-state NMR. J Magn Reson 204:155–159
90. Sobhanifar S, Reckel S, Junge F, Schwarz D, Kai L, Karbyshev M, Löehr F, Bernhard F, Dötsch V (2010) Cell-free expression and stable isotope labelling strategies for membrane proteins. J Biomol NMR 46:33–43
91. Schwarz D, Junge F, Durst F, Frölich N, Schneider B, Reckel S, Sobhanifar S, Dötsch V, Bernhard F (2007) Preparative scale expression of membrane proteins in *Escherichia coli*-based continuous exchange cell-free systems. Nat Protoc 2:2945–2957
92. Staunton D, Schlinkert R, Zanetti G, Colebrook SA, Campbell ID (2006) Cell-free expression and selective isotope labelling in protein NMR. Magn Reson Chem 44:S2–S9
93. Bakke CK, Jungbauer LM, Cavagnero S (2006) In vitro expression and characterization of native apomyoglobin under low molecular crowding conditions. Protein Expr Purif 45:381–392
94. Ozawa K, Jergic S, Crowther JA, Thompson PR, Wijffels G, Otting G, Dixon NA (2005) Cell-free protein synthesis in an autoinduction system for NMR studies of protein-protein interactions. J Biomol NMR 32:235–241
95. Ozawa K, Headlam MJ, Schaeffer PM, Henderson BR, Dixon NE, Otting G (2004) Optimization of an *Escherichia coli* system for cell-free synthesis of selectively ^{15}N-labelled proteins for rapid analysis by NMR spectroscopy. Eur J Biochem 271:4084–4093
96. Klammt C, Löhr F, Schäfer B, Haase W, Dötsch V, Ruterjans H, Glaubitz C, Bernhard F (2004) High level cell-free expression and specific labeling of integral membrane proteins. Eur J Biochem 271:568–580
97. Torizawa T, Shimizu M, Taoka M, Miyano H, Kainosho M (2004) Efficient production of isotopically labeled proteins by cell-free synthesis: a practical protocol. J Biomol NMR 30:311–325
98. Ishihara G, Goto M, Saeki M, Ito K, Hori T, Kigawa T, Shirouzu M, Yokoyama S (2005) Expression of G protein coupled receptors in a cell-free translational system using detergents and thioredoxin-fusion vectors. Protein Expr Purif 41:27–37
99. Schwarz D, Klammt C, Koglin A, Löhr F, Schneider B, Dötsch V, Bernhard F (2007) Preparative scale cell-free expression systems: new tools for the large scale preparation of integral membrane proteins for functional and structural studies. Methods 41:355–369
100. Berrier C, Park KH, Abes S, Bibonne A, Betton JM, Ghazi A (2004) Cell-free synthesis of a functional ion channel in the absence of a membrane and in the presence of detergent. Biochemistry 43:12585–12591
101. Klammt C, Schwarz D, Fendler K, Haase W, Dötsch V, Bernhard F (2005) Evaluation of detergents for the soluble expression of α-helical and β-barrel-type integral membrane proteins by a preparative scale individual cell-free expression system. FEBS J 272:6024–6038
102. Trbovic N, Klammt C, Koglin A, Löhr F, Bernhard F, Dötsch V (2005) Efficient strategy for the rapid backbone assignment of membrane proteins. J Am Chem Soc 127: 13504–13505
103. Elbaz Y, Steiner-Mordoch S, Danieli T, Schuldiner S (2004) In vitro synthesis of fully functional EmrE, a multidrug transporter, and study of its oligomeric state. Proc Natl Acad Sci USA 101:1519–1524
104. McIntosh LP, Dahlquist FW (1990) Biosynthetic incorporation of ^{15}N and ^{13}C for assignment and interpretation of nuclear magnetic resonance spectra of proteins. Q Rev Biophys 23:1–38
105. Le Master DM (1994) Isotope labeling in solution protein assignment and structural-analysis. Prog Nucl Magn Reson Spectrosc 26:371–419

106. Kainosho M, Tsuji T (1982) Assignment of the 3 methionyl carbonyl carbon resonances in *Streptomyces* subtilisin inhibitor by a ^{13}C and ^{15}N double-labeling technique - a new strategy for structural studies of proteins in solution. Biochemistry 21:6273–6279
107. Parker MJ, Aulton-Jones M, Hounslow AM, Craven CJ (2004) A combinatorial selective labeling method for the assignment of backbone amide NMR resonances. J Am Chem Soc 126:5020–5021
108. Shi J, Pelton JG, Cho HS, Wemmer DE (2004) Protein signal assignments using specific labeling and cell-free synthesis. J Biomol NMR 28:235–247
109. Hefke F, Bagaria A, Reckel S, Ullrich SJ, Dötsch V, Glaubitz C, Güntert P (2011) Optimization of amino acid type-specific ^{13}C and ^{15}N labeling for the backbone assignment of membrane proteins by solution- and solid-state NMR with the UPLABEL algorithm. J Biomol NMR 49:75–84
110. Fernández C, Wider G (2003) TROSY in NMR studies of the structure and function of large biological macromolecules. Curr Opin Struct Biol 13:570–580
111. Sanders CR, Sönnichsen F (2006) Solution NMR of membrane proteins: practice and challenges. Magn Reson Chem 44:S24–S40
112. Kim HJ, Howell SC, Van Horn WD, Jeon YH, Sanders CR (2009) Recent advances in the application of solution NMR spectroscopy to multi-span integral membrane proteins. Prog Nucl Magn Reson Spectrosc 55:335–360
113. Linke D (2009) Detergents: An overview. Methods Enzymol 463:603–617
114. Otzen DE (2002) Protein unfolding in detergents: effect of micelle structure, ionic strength, pH, and temperature. Biophys J 83:2219–2230
115. Privé GG (2007) Detergents for the stabilization and crystallization of membrane proteins. Methods 41:388–397
116. Henry GD, Sykes BD (1992) Assignment of amide ^{1}H and ^{15}N NMR resonances in detergent-solubilized M13 coat protein: a model for the coat protein dimer. Biochemistry 31:5284–5297
117. Schiksnis RA, Bogusky MJ, Tsang P, Opella SJ (1987) Structrue and dynamics of the Pf1 filamentous bacteriophage coat protein in micelles. Biochemistry 26:1373–1381
118. Otzen DE (2011) Protein-surfactant interactions: a tale of many states. Biochim Biophys Acta 1814:562–591
119. Popot JL, Engelman DM (2000) Helical membrane protein folding, stability, and evolution. Annu Rev Biochem 69:881–922
120. White SH, Ladokhin A, Jayasinghe S, Hristova K (2001) How membranes shape protein structure. J Biol Chem 276:32395–32398
121. Stanley AM, Fleming KG (2008) Process of folding proteins into membranes: challenges and progress. Arch Biochem Biophys 469:46–66
122. Damberg P, Jarvet J, Graslund A (2001) Micellar systems as solvents in peptide and protein structure determination. Methods Enzymol 339:271–285
123. Bordag N, Keller S (2010) α-Helical transmembrane peptides: a "divide and conquer" approach to membrane proteins. Chem Phys Lipids 163:1–26
124. Haney EF, Hunter HN, Matsuzaki K, Vogel HJ (2009) Solution NMR studies of amphibian antimicrobial peptides: linking structure to function? Biochim Biophys Acta 1788:1639–1655
125. Haney EF, Vogel HJ (2009) NMR of antimicrobial peptides. In: Webb G (ed) Annu Rep NMR Spectrosc, vol 65, pp 1–51
126. Löw C, Weininger U, Lee H, Schweimer K, Neundorf I, Beck-Sickinger AG, Pastor RW, Balbach J (2008) Structure and dynamics of helix-0 of the N-BAR domain in lipid micelles and bilayers. Biophys J 95:4315–4323
127. Bourbigot S, Dodd E, Horwood C, Cumby N, Fardy L, Welch WH, Ramjan Z, Sharma S, Waring AJ, Yeaman MR, Booth V (2009) Antimicrobial peptide RP-1 structure and interactions with anionic versus zwitterionic micelles. Biopolymers 91:1–13
128. Tulumello DV, Deber CM (2009) SDS micelles as a membrane-mimetic environment for transmembrane segments. Biochemistry 48:12096–12103

129. Lawrie CM, Sulistijo ES, MacKenzie KR (2010) Intermonomer hydrogen bonds enhance GxxxG-driven dimerization of the BNIP3 transmembrane domain: roles for sequence context in helix-helix association in membranes. J Mol Biol 396:924–936
130. Wehbi H, Gasmi-SeabrooK G, Choi MY, Deber CM (2008) Positional dependence of non-native polar mutations on folding of CFTR helical hairpins. Biochim Biophys Acta 1778: 79–87
131. Rath A, Glibowicka M, Nadeau VG, Chen G, Deber CM (2009) Detergent binding explains anomalous SDS-PAGE migration of membrane proteins. Proc Natl Acad Sci USA 106: 1760–1765
132. Langosch D, Brosig B, Kolmar H, Fritz HJ (1996) Dimerisation of the glycophorin A transmembrane segment in membranes probed with the ToxR transcription activator. J Mol Biol 263:525–530
133. Li R, Gorelik R, Nanda V, Law PD, Lear JD, DeGrado WF, Bennett JS (2004) Dimerization of the transmembrane domain of integrin αIIb subunit in cell membranes. J Biol Chem 279: 26666–26673
134. Melnyk RA, Kim S, Curran AR, Engelman DM, Bowie JU, Deber CM (2004) The affinity of GXXXG motifs in transmembrane helix-helix interactions is modulated by long-range communication. J Biol Chem 279:16591–16597
135. Partridge AW, Melnyk RA, Yang D, Bowie JU, Deber CM (2003) A transmembrane segment mimic derived from *Escherichia coli* diacylglycerol kinase inhibits protein activity. J Biol Chem 278:22056–22060
136. Sobhanifar S, Schneider B, Löhr F, Gottstein D, Ikeya T, Mlynarczyk K, Pulawski W, Ghoshdastider U, Kolinski M, Filipek S, Güntert P, Bernhard F, Dötsch V (2010) Structural investigation of the C-terminal catalytic fragment of presenilin 1. Proc Natl Acad Sci USA 107:9644–9649
137. Howell SC, Mesleh MF, Opella SJ (2005) NMR structure determination of a membrane protein with two transmembrane helices in micelles: MerF of the bacterial mercury detoxification system. Biochemistry 44:5196–5206
138. Chill JH, Louis JM, Miller C, Bax A (2006) NMR study of the tetrameric KcsA potassium channel in detergent micelles. Protein Sci 15:684–698
139. Kallick DA, Tessmer MR, Watts CR, Li CY (1995) The use of dodecylphosphocholine micelles in solution NMR. J Magn Reson B 109:60–65
140. MacKenzie KR, Prestegard JH, Engelman DM (1997) A transmembrane helix dimer: structure and implications. Science 276:131–133
141. Traaseth NJ, Verardi R, Torgersen KD, Karim CB, Thomas DD, Veglia G (2007) Spectroscopic validation of the pentameric structure of phospholamban. Proc Natl Acad Sci USA 104:14676–14681
142. Lauterwein J, Bösch C, Brown LR, Wüthrich K (1979) Physicochemical studies of the protein-lipid interactions in melittin-containing micelles. Biochim Biophys Acta 556: 244–264
143. Hauser H (2000) Short-chain phospholipids as detergents. Biochim Biophys Acta 1508: 164–181
144. Tausk RJ, van Esch J, Karmiggelt J, Voordouw G, Overbeek JT (1974) Physical chemical studies of short-chain lecithin homologues. II. Micellar weights of dihexanoyl- and diheptanoyllecithin. Biophys Chem 1974:184–203
145. Gautier A, Kirkpatrick JP, Nietlispach D (2008) Solution-state NMR spectroscopy of a seven-helix transmembrane protein receptor: backbone assignment, secondary structure, and dynamics. Angew Chem Int Ed 47:7297–7300
146. Vinogradova O, Sönnichsen F, Sanders CR (1998) On choosing a detergent for solution NMR studies of membrane proteins. J Biomol NMR 11:381–386
147. Rossi P, Swapna GT, Huang YJ, Aramini JM, Anklin C, Conover K, Hamilton K, Xiao R, Acton TB, Ertekin A, Everett JK, Montelione GT (2010) A microscale protein NMR sample screening pipeline. J Biomol NMR 46:11–22

148. Koehler J, Sulistijo ES, Sakakura M, Kim FJ, Ellis CD, Sanders CR (2010) Lysophospholipid micelles sustain the stability and catalytic activity of diacylglycerol kinase in the absence of lipids. Biochemistry 49:7089–7099
149. Krueger-Koplin RD, Sorgen PL, Krueger-Koplin ST, Rivera-Torres IO, Cahill SM, Hicks DB, Grinius L, Krulwich TA, Girvin ME (2004) An evaluation of detergents for NMR structural studies of membrane proteins. J Biomol NMR 28:43–57
150. Tian C, Vanoye CG, Kang C, Welch RC, Kim HJ, George AL, Sanders CR (2007) Preparation, functional characterization, and NMR studies of human KCNE1, a voltage-gated potassium channel accessory subunit associated with deafness and long QT syndrome. Biochemistry 46:11459–11472
151. Zhang Q, Horst R, Geralt M, Ma X, Hong WX, Finn MG, Stevens RC, Wüthrich K (2008) Microscale NMR screening of new detergents for membrane protein structural biology. J Am Chem Soc 130:7357–7363
152. Call ME, Schnell JR, Xu C, Lutz RA, Chou JJ, Wucherpfennig KW (2006) The structure of the ζζ transmembrane dimer reveals features essential for its assembly with the T cell receptor. Cell 127:355–368
153. Shenkarev ZO, Paramonov AS, Lyukmanova EN, Shingarova LN, Yakimov SA, Dubinnyi MA, Chupin VV, Kirpichnikov MP, Blommers MJ, Arseniev AS (2010) NMR structural and dynamical investigation of the isolated voltage-sensing domain of the potassium channel KvAP: implications for voltage gating. J Am Chem Soc 132:5630–5637
154. Zou C, Naider F, Zerbe O (2008) Biosynthesis and NMR-studies of a double transmembrane domain from the Y4 receptor, a human GPCR. J Biomol NMR 42:257–269
155. Columbus L, Lipfert J, Jambunathan K, Fox DA, Sim AY, Doniach S, Lesley SA (2009) Mixing and matching detergents for membrane protein NMR structure determination. J Am Chem Soc 131:7320–7326
156. Lipfert J, Columbus L, Chu VB, Lesley SA, Doniach S (2007) Size and shape of detergent micelles determined by small-angle X-ray scattering. J Phys Chem B 111:12427–12438
157. Ahn VE, Lo EI, Engel CK, Chen L, Hwang PM, Kay L, Bishop RE, Privé GG (2004) A hydrocarbon ruler measures palmitate in the enzymatic acylation of endotoxin. EMBO J 23:2931–2941
158. Stouffer AL, Acharya R, Salom D, Levine AS, Di Costanzo L, Soto CS, Tereshko V, Nanda V, Stayrook S, DeGrado WF (2008) Structural basis for the function and inhibition of an influenza virus proton channel. Nature 451:596–599
159. Page RC, Lee S, Moore JD, Opella SJ, Cross TA (2009) Backbone structure of a small helical integral membrane protein: a unique structural characterization. Protein Sci 18:134–146
160. Chou JJ, Kaufman JD, Stahl SJ, Wingfield PT, Bax A (2002) Micelle-induced curvature in a water-insoluble HIV-1 Env peptide revealed by NMR dipolar coupling measurement in stretched polyacrylamide gel. J Am Chem Soc 124:2450–2451
161. Lee SY, Lee A, Chen J, MacKinnon R (2005) Structure of the KvAP voltage-dependent K + channel and its dependence on the lipid membrane. Proc Natl Acad Sci USA 102: 15441–15446
162. Lau TL, Partridge AW, Ginsberg MH, Ulmer TS (2008) Structure of the integrin beta3 transmembrane segment in phospholipid bicelles and detergent micelles. Biochemistry 47: 4008–4016
163. Lundquist A, Wessman P, Rennie AR, Edwards K (2008) Melittin-lipid interaction: a comparative study using liposomes, micelles and bilayer disks. Biochim Biophys Acta 1778:2210–2216
164. Vos WL, Nazarov PV, Koehorst RB, Spruijt RB, Hemminga MA (2009) From 'I' to 'L' and back again: the odyssey of membrane-bound M13 protein. Trends Biochem Sci 34:249–255
165. Sherratt AR, Braganza MV, Nguyen E, Ducat T, Goto NK (2009) Insights into the effect of detergents on the full-length rhomboid protease from Pseudomonas aeruginosa and its cytosolic domain. Biochim Biophys Acta 1788:2444–2453

166. Schnell JR, Chou JJ (2008) Structure and mechanism of the M2 proton channel of influenza A virus. Nature 451:591–612
167. Cady SD, Schmidt-Rohr K, Wang J, Soto CS, Degrado WF, Hong M (2010) Structure of the amantadine binding site of influenza M2 proton channels in lipid bilayers. Nature 463:689–692
168. Cross TA, Sharma M, Yi M, Zhou HX (2011) Influence of solubilizing environments on membrane protein structures. Trends Biochem Sci 36:117–125
169. Miller C (2008) Ion channels: coughing up flu's proton channels. Nature 451:532–533
170. Pielak RM, Schnell JR, Chou JJ (2011) Mechanism of drug inhibition and drug resistance of influenza A M2 channel. Proc Natl Acad Sci USA 106:7379–7384
171. Cady SD, Wang J, Wu Y, DeGrado WF, Hong M (2011) Specific binding of adamantane drugs and direction of their polar amines in the pore of the influenza M2 transmembrane domain in lipid bilayers and dodecylphosphocholine micelles determined by NMR spectroscopy. J Am Chem Soc 133:4274–4284
172. Poget SF, Girvin ME (2007) Solution NMR of membrane proteins in bilayer mimics: small is beautiful, but sometimes bigger is better. Biochim Biophys Acta 1768:3098–3106
173. Prosser RS, Evanics F, Kitevski JL, Al-Abdul-Wahid MS (2006) Current applications of bicelles in NMR studies of membrane-associated amphiphiles and proteins. Biochemistry 45:8453–8465
174. Marcotte I, Auger M (2005) Bicelles as model membranes for solid- and solution-state NMR studies of membrane peptides and proteins. Concepts Magn Reson Part A 24A:17–37
175. Vold RR, Prosser RS, Deese AJ (1997) Isotropic solutions of phospholipid bicelles: a new membrane mimetic for high-resolution NMR studies of polypeptides. J Biomol NMR 9: 329–335
176. Sanders CR, Schwonek JP (1992) Characterization of magnetically orientable bilayers in mixtures of dihexanoylphosphatidylcholine and dimyristoylphosphatidylcholine by solid-state NMR. Biochemistry 31:8898–8905
177. Andersson A, Mäler L (2006) Size and shape of fast-tumbling bicelles as determined by translational diffusion. Langmuir 22:2447–2449
178. Luchette PA, Vetman TN, Prosser RS, Hancock RE, Nieh MP, Glinka CJ, Krueger S, Katsaras J (2001) Morphology of fast-tumbling bicelles: a small angle neutron scattering and NMR study. Biochim Biophys Acta 1513:83–94
179. Glover KJ, Whiles JA, Wu G, Yu N, Deems R, Struppe JO, Stark RE, Komives EA, Vold RR (2001) Structural evaluation of phospholipid bicelles for solution-state studies of membrane-associated biomolecules. Biophys J 81:2163–2171
180. Chou JJ, Baber JL, Bax A (2004) Characterization of phospholipid mixed micelles by translational diffusion. J Biomol NMR 29:299–308
181. Papadopoulos E, Oglecka K, Mäler L, Jarvet J, Wright PE, Dyson HJ, Gräslund A (2006) NMR solution structure of the peptide fragment 1–30, derived from unprocessed mouse Doppel protein, in DHPC micelles. Biochemistry 45:159–166
182. Biverståhl H, Andersson A, Gräslund A, Mäler L (2004) NMR solution structure and membrane interaction of the N-terminal sequence (1–30) of the bovine prion protein. Biochemistry 43:14940–14947
183. Poget SF, Cahill SM, Girvin ME (2007) Isotropic bicelles stabilize the functional form of a small multidrug-resistance pump for NMR structural studies. J Am Chem Soc 129: 2432–2433
184. Lau TL, Kim C, Ginsberg MH, Ulmer TS (2009) The structure of the integrin alphaIIbbeta3 transmembrane complex explains integrin transmembrane signalling. EMBO J 28: 1351–1361
185. Li R, Babu CR, Lear JD, Wand AJ, Bennett JS, DeGrado WF (2001) Oligomerization of the integrin alphaIIbbeta3: roles of the transmembrane and cytoplasmic domains. Proc Natl Acad Sci USA 98:12462–12467
186. De Angelis AA, Opella SJ (2007) Bicelle samples for solid-state NMR of membrane proteins. Nat Protoc 2:2332–2338

187. Sanders CR, Landis GC (1995) Reconstitution of membrane-proteins into lipid-rich bilayered mixed micelles for NMR-studies. Biochemistry 34:4030–4040
188. Bocharov EV, Pustovalova YE, Pavlov KV, Volynsky PE, Goncharuk MV, Ermolyuk YS, Karpunin DV, Schulga AA, Kirpichnikov MP, Efremov RG, Maslennikov IV, Arseniev AS (2007) Unique dimeric structure of BNip3 transmembrane domain suggests membrane permeabilization as a cell death trigger. J Biol Chem 282:16256–16266
189. Bocharov EV, Mayzel ML, Volynsky PE, Goncharuk MV, Ermolyuk YS, Schulga AA, Artemenko EO, Efremov RG, Arseniev AS (2008) Spatial structure and pH-dependent conformational diversity of dimeric transmembrane domain of the receptor tyrosine kinase EphA1. J Biol Chem 283:29385–29395
190. Kang C, Vanoye CG, Welch RC, Van Horn WD, Sanders CR (2010) Functional delivery of a membrane protein into oocyte membranes using bicelles. Biochemistry 49:653–655
191. Park SH, Casagrande F, Das BB, Albrecht L, Chu M, Opella SJ (2011) Local and global dynamics of the G protein-coupled receptor CXCR1. Biochemistry 50:2371–2380
192. Ottiger M, Bax A (1999) Bicelle-based liquid crystals for NMR-measurement of dipolar couplings at acidic and basic pH values. J Biomol NMR 13:187–191
193. Bertelsen K, Vad B, Nielsen EH, Hansen SK, Skrydstrup T, Otzen DE, Vosegaard T, Nielsen NC (2011) Long-term-stable ether-lipid vs conventional ester-lipid bicelles in oriented solid-state NMR: altered structural information in studies of antimicrobial peptides. J Phys Chem B 115:1767–1774
194. Struppe J, Whiles JA, Vold RR (2000) Acidic phospholipid bicelles: a versatile model membrane system. Biophys J 78:281–289
195. Losonczi JA, Prestegard JH (1998) Improved dilute bicelle solutions for high-resolution NMR of biological macromolecules. J Biomol NMR 12:447–451
196. Anglister J, Grzesiek S, Ren H, Klee CB, Bax A (1993) Isotope-edited multidimensional NMR of calcineurin-B in the presence of the non-deuterated detergent CHAPS. J Biomol NMR 3:121–126
197. Wu H, Su K, Guan X, Sublette ME, Stark RE (2010) Assessing the size, stability, and utility of isotropically tumbling bicelle systems for structural biology. Biochim Biophys Acta 1798:482–488
198. Czerski L, Sanders CR (2000) Functionality of a membrane protein in bicelles. Anal Biochem 284:327–333
199. Denisov IG, McLean MA, Shaw AW, Grinkova YV, Sligar SG (2005) Thermotropic phase transition in soluble nanoscale lipid bilayers. J Phys Chem B 109:15580–15588
200. Bayburt TH, Sligar SG (2010) Membrane protein assembly into nanodiscs. FEBS Lett 584:1721–1727
201. Civjan NR, Bayburt TH, Schuler MA, Sligar SG (2003) Direct solubilization of heterologously expressed membrane proteins by incorporation into nanoscale lipid bilayers. Biotechniques 35:556–560, 562–563
202. Bayburt TH, Sligar SG (2010) Self-assembly of single integral membrane proteins into soluble nanoscale phospholipid bilayers. Protein Sci 12:2476–2481
203. Kijac AZ, Li Y, Sligar SG, Rienstra CM (2007) Magic-angle spinning solid-state NMR spectroscopy of nanodisc-embedded human CYP3A4. Biochemistry 46:13696–13703
204. Lyukmanova EN, Shenkarev ZO, Paramonov AS, Sobol AG, Ovchinnikova TV, Chupin VV, Kirpichnikov MP, Blommers MJ, Arseniev AS (2008) Lipid-protein nanoscale bilayers: a versatile medium for NMR investigations of membrane proteins and membrane-active peptides. J Am Chem Soc 130:2140–2141
205. Glück JM, Wittlich M, Feuerstein S, Hoffmann S, Willbold D, Koenig BW (2009) Integral membrane proteins in nanodiscs can be studied by solution NMR spectroscopy. J Am Chem Soc 131:12060–12061
206. Kobashigawa Y, Harada K, Yoshida N, Ogura K, Inagaki F (2011) Phosphoinositide-incorporated lipid-protein nanodiscs: a tool for studying protein-lipid interactions. Anal Biochem 410:77–83

207. Shenkarev ZO, Lyukmanova EN, Solozhenkin OI, Gagnidze IE, Nekrasova OV, Chupin VV, Tagaev AA, Yakimenko ZA, Ovchinnikova TV, Kirpichnikov MP, Arseniev AS (2009) Lipid-protein nanodiscs: possible application in high-resolution NMR investigations of membrane proteins and membrane-active peptides. Biochemistry (Moscow) 74:756–765
208. Ritchie TK, Grinkova YV, Bayburt TH, Denisov IG, Zolnerciks JK, Atkins WM, Sligar SG (2009) Reconstitution of membrane proteins in phospholipid bilayer nanodiscs. Methods Enzymol 464:211–231
209. Shenkarev ZO, Lyukmanova EN, Paramonov AS, Shingarova LN, Chupin VV, Kirpichnikov MP, Blommers MJ, Arseniev AS (2010) Lipid-protein nanodiscs as reference medium in detergent screening for high-resolution NMR studies of integral membrane proteins. J Am Chem Soc 132:5628–5629
210. Yoshiura C, Kofuku Y, Ueda T, Mase Y, Yokogawa M, Osawa M, Terashima Y, Matsushima K, Shimada I (2010) NMR analyses of the interaction between CCR5 and its ligand using functional reconstitution of CCR5 in lipid bilayers. J Am Chem Soc 132:6768–6777
211. Tribet C, Audebert R, Popot JL (1996) Amphipols: polymers that keep membrane proteins soluble in aqueous solutions. Proc Natl Acad Sci USA 93:15047–15050
212. Sanders CR, Hoffmann A, Grayn DN, Keyes MH, Ellis CD (2004) French swimwear for membrane proteins. Chembiochem 5:423–426
213. Popot JL, Berry EA, Charvolin D, Creuzenet C, Ebel C, Engelman DM, Flötenmeyer M, Giusti F, Gohon Y, Herve P, Hong Q, Lakey JH, Leonard K, Shuman HA, Timmins P, Warschawski DE, Zito F, Zoonens M, Pucci B, Tribet C (2003) Amphipols: polymeric surfactants for membrane biology research. Cell Mol Life Sci 60:1559–1574
214. Popot JL (2010) Amphipols, nanodiscs, and fluorinated surfactants: three nonconventional approaches to studying membrane proteins in aqueous solutions. Annu Rev Biochem 79:737–775
215. Tribet C, Diab C, Dahmane T, Zoonens M, Popot JL, Winnik FM (2009) Thermodynamic characterization of the exchange of detergents and amphipols at the surfaces of integral membrane proteins. Langmuir 25:12623–12634
216. Pocanschi CL, Dahmane T, Gohon Y, Rappaport F, Apell HJ, Kleinschmidt JH, Popot JL (2006) Amphipathic polymers: tools to fold integral membrane proteins to their active form. Biochemistry 45:13954–13961
217. Gorzelle BM, Hoffman AK, Keyes MH, Gray DN, Ray DG, Sanders CR (2002) Amphipols can support the activity of a membrane enzyme. J Am Chem Soc 124:11594–11595
218. Martinez KL, Gohon Y, Corringer PJ, Tribet C, Mérola F, Changeux JP, Popot JL (2002) Allosteric transitions of Torpedo acetylcholine receptor in lipids, detergent and amphipols: molecular interactions vs. physical constraints. FEBS Lett 528:251–256
219. Dahmane T, Damian M, Mary S, Popot JL, Banéres JL (2009) Amphipol-assisted in vitro folding of G protein-coupled receptors. Biochemistry 48:6516–6521
220. Catoire LJ, Damian M, Giusti F, Martin A, van Heijenoort C, Popot JL, Guittet E, Banéres JL (2010) Structure of a GPCR ligand in its receptor-bound state: leukotriene B4 adopts a highly constrained conformation when associated to human BLT2. J Am Chem Soc 132:9049–9057
221. Zoonens M, Catoire LJ, Giusti F, Popot JL (2005) NMR study of a membrane protein in detergent-free aqueous solution. Proc Natl Acad Sci USA 102:8893–8898
222. Wand AJ, Ehrhardt M, Flynn P (1998) High-resolution NMR of encapsulated proteins dissolved in low-viscosity fluids. Proc Natl Acad Sci USA 95:15299–15302
223. Babu CR, Flynn PF, Wand AJ (2003) Preparation, characterization, and NMR spectroscopy of encapsulated proteins dissolved in low viscosity fluids. J Biomol NMR 25:313–323
224. Babu CR, Flynn PF, Wand AJ (2001) Validation of protein structure from preparations of encapsulated proteins dissolved in low viscosity fluids. J Am Chem Soc 123:2691–2692
225. Kielec JM, Valentine KG, Wand AJ (2010) A method for solution NMR structural studies of large integral membrane proteins: reverse micelle encapsulation. Biochim Biophys Acta 1798:150–160

226. Peterson RW, Lefebvre BG, Wand AJ (2005) High-resolution NMR studies of encapsulated proteins in liquid ethane. J Am Chem Soc 127:10176–10177
227. Van Horn WD, Ogilvie ME, Flynn PF (2008) Use of reverse micelles in membrane protein structural biology. J Biomol NMR 40:203–211
228. Kielec JM, Valentine KG, Babu CR, Wand AJ (2009) Reverse micelles in integral membrane protein structural biology by solution NMR spectroscopy. Structure 17:345–351
229. Binks BP, Chatenay D, Nicot C, Urbach W, Waks M (1989) Structural parameters of the myelin transmembrane proteolipid in reverse micelles. Biophys J 55:949–955
230. Flynn PF, Mattiello DL, Hill HDW, Wand AJ (2000) Optimal use of cryogenic probe technology in NMR studies of proteins. J Am Chem Soc 122:4823–4824
231. Krüger J, Fischer WB (2010) Structural implications of mutations assessed by molecular dynamics: Vpu(1–32) from HIV-1. Eur Biophys J 39:1069–1077
232. Stanczak P, Horst R, Serrano P, Wüthrich K (2009) NMR characterization of membrane protein-detergent micelle solutions by use of microcoil equipment. J Am Chem Soc 131:18450–18456
233. Lee D, Hilty C, Wider G, Wüthrich K (2006) Effective rotational correlation times of proteins from NMR relaxation interference. J Magn Reson 178:72–76
234. Maslennikov I, Kefala G, Johnson C, Riek R, Choe S, Kwiatkowski W (2007) NMR spectroscopic and analytical ultracentrifuge analysis of membrane protein detergent complexes. BMC Struct Biol 7:74
235. Kovacs H, Moskau D, Spraul M (2005) Cryogenically cooled probes - a leap in NMR technology. Prog Nucl Magn Reson Spectrosc 46:131–155
236. McDonnell PA, Shon K, Kim Y, Opella SJ (1993) fd coat protein-structure in membrane environments. J Mol Biol 233:447–463
237. Olson DL, Peck TL, Webb AG, Magin RL, Sweedler JV (1995) High-resolution microcoil ^{1}H-NMR for mass-limited, nanoliter-volume samples. Science 270:1967–1970
238. Peck TL, Magin RL, Laterbur PC (1995) Design and analysis of microcoils for NMR spectroscopy. J Magn Reson B 108:114–124
239. Webb AG, Grant SC (1996) Signal-to-noise and magnetic susceptibility trade-offs in solenoidal microcoils for NMR. J Magn Reson B 113:83–87
240. Li Y, Logan TT, Edison AS, Webb A (2003) Design of small volume HX and triple-resonance probes for improved limits of detection in protein NMR experiments. J Magn Reson 164:128–135
241. Peti W, Norcross J, Eldridge G, O'Neil-Johnson M (2004) Biomolecular NMR using a microcoil NMR probe - new technique for the chemical shift assignment of aromatic side chains in proteins. J Am Chem Soc 126:5873–5878
242. Aramini JM, Rossi P, Anklin C, Xiao R, Montelione GT (2007) Microgram-scale protein structure determination by NMR. Nat Methods 4:491–493
243. Wu Y, Shih SC, Goto NK (2007) Probing the structure of the Ff bacteriophage major coat protein transmembrane helix dimer by solution NMR. Biochim Biophys Acta 1768: 3206–3215
244. King G, Oates J, Patel D, van den Berg HA, Dixon AM (2011) Towards a structural understanding of the smallest known oncoprotein: investigation of the bovine papillomavirus E5 protein using solution-state NMR. Biochim Biophys Acta 1808:1493–1501
245. Fleming KG (2002) Standardizing the free energy change of transmembrane helix–helix interactions. J Mol Biol 323:563–571
246. Strop P, Brunger AT (2005) Refractive index-based determination of detergent concentration and its application to the study of membrane proteins. Protein Sci 14:2207–2211
247. Oxenoid K, Kim HJ, Jacob J, Sönnichsen FD, Sanders CR (2004) NMR assignments for a helical 40 kDa membrane protein. J Am Chem Soc 126:5048–5049
248. Goto NK, Kay LE (2000) New developments in isotope labeling strategies for protein solution NMR spectroscopy. Curr Opin Struct Biol 10:585–592

249. Lian LY, Middleton DA (2001) Labelling approaches for protein structural studies by solution-state and solid-state NMR. Prog Nucl Magn Reson Spectrosc 39:171–190
250. Löhr F, Katsemi V, Hartleib J, Günther U, Rüterjans H (2003) A strategy to obtain backbone resonance assignments of deuterated proteins in the presence of incomplete amide ^{2}H/^{1}H back-exchange. J Biomol NMR 25:291–311
251. Gardner KH, Rosen MK, Kay LE (1997) Global folds of highly deuterated, methyl-protonated proteins by multidimensional NMR. Biochemistry 36:1389–1401
252. Venters RA, Metzler WJ, Spicer LD, Mueller L, Farmer BT (1995) Use of ^{1}HN NOES to determine protein global folds in perdeuterated proteins. J Am Chem Soc 117:9592–9593
253. Janin J, Miller S, Chothia C (1988) Surface, subunit interfaces and interior of oligomeric proteins. J Mol Biol 204:155–164
254. Metzler W, Wittekind MJ, Goldfarb V, Mueller L, Farmer BT (1996) Incorporation of ^{1}H/^{13}C/^{15}N-{Ile, Leu, Val} into a perdeuterated, ^{15}N-labeled protein: potential in structure determination of large proteins by NMR. J Am Chem Soc 118:6800–6801
255. Zheng D, Huang YJ, Moseley HNB, Xiao R, Aramini J, Swapna GVT, Montelione GT (2003) Automated protein fold determination using a minimal NMR constraint strategy. Protein Sci 12:1232–1246
256. Kay LE (2011) Solution NMR spectroscopy of supra-molecular systems, why bother? A methyl-TROSY view. J Magn Reson 210:159–170
257. Ruschak AM, Kay LE (2010) Methyl groups as probes of supra-molecular structure, dynamics and function. J Biomol NMR 46:75–87
258. Tugarinov V, Kay LE (2005) Methyl groups as probes of structure and dynamics in NMR studies of high-molecular-weight proteins. Chembiochem 6:1567–1577
259. Goto NK, Gardner KH, Mueller GA, Willis RC, Kay LE (1999) A robust and cost-effective method for the production of Val, Leu, Ile (delta 1) methyl-protonated ^{15}N, ^{13}C, ^{2}H-labeled proteins. J Biomol NMR 13:369–374
260. Gardner KH, Kay LE (1997) Production and incorporation of ^{15}N, ^{13}C, ^{2}H (^{1}H-δ 1 methyl) isoleucine into proteins for multidimensional NMR studies. J Am Chem Soc 119:7599–7600
261. Hilty C, Fernández C, Wider G, Wüthrich K (2002) Side chain NMR assignments in the membrane protein OmpX reconstituted in DHPC micelles. J Biomol NMR 23:289–301
262. Shih SC, Stoica I, Goto NK (2008) Investigation of the utility of selective methyl protonation for determination of membrane protein structures. J Biomol NMR 42:49–58
263. Ayala I, Sounier R, Usé N, Gans P, Boisbouvier J (2009) An efficient protocol for the complete incorporation of methyl-protonated alanine in perdeuterated protein. J Biomol NMR 43:111–119
264. Godoy-Ruiz R, Guo C, Tugarinov V (2010) Alanine methyl groups as NMR probes of molecular structure and dynamics in high-molecular-weight proteins. J Am Chem Soc 132:18340–18350
265. Isaacson RL, Simpson PJ, Liu M, Cota E, Zhang X, Freemont P, Matthews S (2007) A new labeling method for methyl transverse relaxation-optimized spectroscopy NMR spectra of alanine residues. J Am Chem Soc 129:15428–15429
266. Gelis I, Bonvin AM, Keramisanou D, Koukaki M, Gouridis G, Karamanou S, Economou A, Kalodimos CG (2007) Structural basis for signal-sequence recognition by the translocase motor SecA as determined by NMR. Cell 131:756–769
267. Fischer M, Kloiber K, Häusler J, Ledolter K, Konrat R, Schmid W (2007) Synthesis of a ^{13}C-methyl-group-labeled methionine precursor as a useful tool for simplifying protein structural analysis by NMR spectroscopy. Chembiochem 8:610–612
268. Ruschak AM, Velyvis A, Kay LE (2010) A simple strategy for ^{13}C, ^{1}H labeling at the Ile-gamma 2 methyl position in highly deuterated proteins. J Biomol NMR 48:129–135
269. Takeda M, Ikeya T, Güntert P, Kainosho M (2007) Automated structure determination of proteins with the SAIL-FLYA NMR method. Nat Protoc 2:2896–2902
270. Kainosho M, Torizawa T, Iwashita Y, Terauchi T, Mei Ono A, Guntert P (2006) Optimal isotope labelling for NMR protein structure determinations. Nature 440:52–57

271. Religa TL, Sprangers R, Kay LE (2010) Dynamic regulation of archaeal proteasome gate opening as studied by TROSY NMR. Science 328:98–102
272. Ruschak AM, Religa TL, Breuer S, Witt S, Kay LE (2010) The proteasome antechamber maintains substrates in an unfolded state. Nature 467:868-872
273. Imai S, Osawa M, Takeuchi K, Shimada I (2010) Structural basis underlying the dual gate properties of KcsA. Proc Natl Acad Sci USA 107:6216–6221
274. Bodner CR, Dobson CM, Bax A (2009) Multiple tight phospholipid-binding modes of alpha-synuclein revealed by solution NMR spectroscopy. J Mol Biol 390:775–790
275. Bokoch MP, Zou Y, Rasmussen SG, Liu CW, Nygaard R, Rosenbaum DM, Fung JJ, Choi HJ, Thian FS, Kobilka TS, Puglisi JD, Weis WI, Pardo L, Prosser RS, Mueller L, Kobilka BK (2010) Ligand-specific regulation of the extracellular surface of a G-protein-coupled receptor. Nature 463:108–112
276. Pervushin K, Riek R, Wider G, Wüthrich K (1997) Attenuated T2 relaxation by mutual cancellation of dipole-dipole coupling and chemical shift anisotropy indicates an avenue to NMR structures of very large biological macromolecules in solution. Proc Natl Acad Sci USA 94:12366–12371
277. Pervushin K, Riek R, Wider G, Wüthrich K (1998) Transverse relaxation-optimized spectroscopy (TROSY) for NMR studies of aromatic spin systems in ^{13}C-labeled proteins. J Am Chem Soc 120:6394–6400
278. Yang D, Kay LE (1999) TROSY triple-resonance four-dimensional NMR spectroscopy of a 46 ns tumbling protein. J Am Chem Soc 121:2571–2575
279. Yang D, Kay LE (1999) Improved ^{1}HN-detected triple resonance TROSY-based experiments. J Biomol NMR 13:3–10
280. Permi P, Annila A (2004) Coherence transfer in proteins. Prog Nucl Magn Reson Spectrosc 44:97–137
281. Riek R, Wider G, Pervushin K, Wüthrich K (1999) Polarization transfer by cross-correlated relaxation in solution NMR with very large molecules. Proc Natl Acad Sci USA 96:4918–4923
282. Riek R, Fiaux J, Bertelsen EB, Horwich AL, Wüthrich K (2002) Solution NMR techniques for large molecular and supramolecular structures. J Am Chem Soc 124:12144–12153
283. Pervushin K, Vögeli B, Eletsky A (2002) Longitudinal ^{1}H relaxation optimization in TROSY NMR spectroscopy. J Am Chem Soc 124:12898–12902
284. Tugarinov V, Hwang PM, Kay LE (2004) Nuclear magnetic resonance spectroscopy of high-molecular-weight proteins. Annu Rev Biochem 73:107–146
285. Tian C, Karra MD, Ellis CD, Jacob J, Oxenoid K, Sönnichsen F, Sanders CR (2005) Membrane protein preparation for TROSY NMR screening. Methods Enzymol 394:321–334
286. Page RC, Moore JD, Nguyen HB, Sharma M, Chase R, Gao FP, Mobley CK, Sanders CR, Ma L, Sönnichsen FD, Lee S, Howell SC, Opella SJ, Cross TA (2006) Comprehensive evaluation of solution nuclear magnetic resonance spectroscopy sample preparation for helical integral membrane proteins. J Struct Funct Genomics 7:51–64
287. Bocharov EV, Mineev KS, Volynsky PE, Ermolyuk YS, Tkach EN, Sobol AG, Chupin VV, Kirpichnikov MP, Efremov RG, Arseniev AS (2008) Spatial structure of the dimeric transmembrane domain of the growth factor receptor ErbB2 presumably corresponding to the receptor active state. J Biol Chem 283:6950–6956
288. Tugarinov V, Hwang PM, Ollerenshaw JE, Kay LE (2003) Cross-correlated relaxation enhanced ^{1}H-^{13}C NMR spectroscopy of methyl groups in very high molecular weight proteins and protein complexes. J Am Chem Soc 125:10420–10428
289. Ollerenshaw JE, Tugarinov V, Kay LE (2003) Methyl TROSY: explanation and experimental verification. Magn Reson Chem 41:843–852
290. Tugarinov V, Kay LE (2004) An isotope labeling strategy for methyl TROSY spectroscopy. J Biomol NMR 28:165–172
291. Tugarinov V, Ollerenshaw JE, Kay LE (2005) Probing side-chain dynamics in high molecular weight proteins by deuterium NMR spin relaxation: an application to an 82-kDa enzyme. J Am Chem Soc 127:8214–8225

292. Velyvis A, Yang YR, Schachman HK, Kay LE (2007) A solution NMR study showing that active site ligands and nucleotides directly perturb the allosteric equilibrium in aspartate transcarbamoylase. Proc Natl Acad Sci USA 104:8815–8820
293. Tugarinov V, Kay LE (2003) Ile, Leu, and Val methyl assignments of the 723-residue malate synthase G using a new labeling strategy and novel NMR methods. J Am Chem Soc 125:13868–13878
294. Tugarinov V, Kay LE (2003) Side chain assignments of Ile delta 1 methyl groups in high molecular weight proteins: an application to a 46 ns tumbling molecule. J Am Chem Soc 125:5701–5706
295. Sprangers R, Kay LE (2007) Quantitative dynamics and binding studies of the 20S proteasome by NMR. Nature 445:618–622
296. Xu Y, Liu M, Simpson PJ, Isaacson R, Cota E, Marchant J, Yang D, Zhang X, Freemont P, Matthews S (2009) Automated assignment in selectively methyl-labeled proteins. J Am Chem Soc 131:9480–9481
297. Bayrhuber M, Meins T, Habeck M, Becker S, Giller K, Villinger S, Vonrhein C, Griesinger C, Zweckstetter M, Zeth K (2008) Structure of the human voltage-dependent anion channel. Proc Natl Acad Sci USA 105:15370–15375
298. Yu L, Sun C, Song D, Shen J, Xu N, Gunasekera A, Hajduk PJ, Olejniczak ET (2005) Nuclear magnetic resonance structural studies of a potassium channel-charybdotoxin complex. Biochemistry 44:15834–15841
299. Cornilescu G, Delaglio F, Bax A (1999) Protein backbone angle restraints from searching a database for chemical shift and sequence homology. J Biomol NMR 13:289–302
300. Rovnyak D, Frueh DP, Sastry M, Sun ZY, Stern AS, Hoch JC, Wagner G (2004) Accelerated acquisition of high resolution triple-resonance spectra using non-uniform sampling and maximum entropy reconstruction. J Magn Reson 170:15–21
301. Barna JCJ, Laue ED (1987) Conventional and exponential sampling for 2D NMR experiments with application to a 2D NMR-spectrum of a protein. J Magn Reson 75:384–389
302. Gryk MR, Vyas J, Maciejewski MW (2010) Biomolecular NMR data analysis. Prog Nucl Magn Reson Spectrosc 56:329–345
303. Sibisi S, Skilling J, Brereton R, Laue E, Staunton J (1984) Maximum-entropy signal processing in practical NMR spectroscopy. Nature 311:446–447
304. Donoho DL, Johnstone IM, Stern AS, Hoch JC (1990) Does the maximum-entropy method improve sensitivity. Proc Natl Acad Sci USA 87:5066–5068
305. Felli IC, Brutscher B (2009) Recent advances in solution NMR: fast methods and heteronuclear direct detection. Chemphyschem 10:1356–1368
306. Tugarinov V, Kay L, Ibraghimov I, Orekhov V (2005) High-resolution four-dimensional ^{1}H-^{13}C NOE spectroscopy using methyl-TROSY, sparse data acquisition, and multidimensional decomposition. J Am Chem Soc 127:2767–2775
307. Orekhov VY, Ibraghimov I, Billeter M (2001) MUNIN: a new approach to multi-dimensional NMR spectra interpretation. J Biomol NMR 20:49–60
308. Orekhov VY, Ibraghimov I, Billeter M (2003) Optimizing resolution in multidimensional NMR by three-way decomposition. J Biomol NMR 27:165–173
309. Luan T, Jaravine V, Yee A, Arrowsmith CH, Orekhov VY (2005) Optimization of resolution and sensitivity of 4D NOESY using multi-dimensional decomposition. J Biomol NMR 33:1–14
310. Hiller S, Ibraghimov I, Wagner G, Orekhov VY (2009) Coupled decomposition of four-dimensional NOESY spectra. J Am Chem Soc 131:12970–12978
311. Tjandra N, Bax A (1997) Direct measurement of distances and angles in biomolecules by NMR in a dilute liquid crystalline medium. Science 278:1111–1114
312. Bax A, Tjandra N (1997) High-resolution heteronuclear NMR of human ubiquitin in an aqueous liquid crystalline medium. J Biomol NMR 10:289–292
313. Tycko R, Blanco FJ, Ishii Y (2000) Alignment of biopolymers in strained gels: a new way to create detectable dipole-dipole couplings in high-resolution biomolecular NMR. J Am Chem Soc 122:9340–9341

314. Chou JJ, Gaemers S, Howder B, Louis JM, Bax A (2001) A simple apparatus for generating stretched polyacrylamide gels, yielding uniform alignment of proteins and detergent micelles. J Biomol NMR 21:377–382
315. Sass HJ, Musco G, Stahl SJ, Wingfield PT, Grzesiek S (2000) Solution NMR of proteins within polyacrylamide gels: diffusional properties and residual alignment by mechanical stress or embedding of oriented purple membranes. J Biomol NMR 18:303–309
316. Liu Y, Prestegard JH (2010) A device for the measurement of residual chemical shift anisotropy and residual dipolar coupling in soluble and membrane-associated proteins. J Biomol NMR 47:249–258
317. Mesleh MF, Veglia G, DeSilva TM, Marassi FM, Opella SJ (2002) Dipolar waves as NMR maps of protein structure. J Am Chem Soc 124:4206–4207
318. Lee S, Mesleh MF, Opella SJ (2003) Structure and dynamics of a membrane protein in micelles from three solution NMR experiments. J Biomol NMR 26:327–334
319. Jones DH, Opella SJ (2004) Weak alignment of membrane proteins in stressed polyacrylamide gels. J Magn Reson 171:258–269
320. Cierpicki T, Bushweller JH (2004) Charged gels as orienting media for measurement of residual dipolar couplings in soluble and integral membrane proteins. J Am Chem Soc 126:16259–16266
321. Meier S, Häussinger D, Grzesiek S (2002) Charged acrylamide copolymer gels as media for weak alignment. J Biomol NMR 24:351–356
322. Hansen MR, Mueller L, Pardi A (1998) Tunable alignment of macromolecules by filamentous phage yields dipolar coupling interactions. Nat Struct Biol 5:1065–1074
323. Park SH, Son WS, Mukhopadhyay R, Valafar H, Opella SJ (2009) Phage-induced alignment of membrane proteins enables the measurement and structural analysis of residual dipolar couplings with dipolar waves and lambda-maps. J Am Chem Soc 131:14140–14141
324. Douglas SM, Chou JJ, Shih WM (2007) DNA-nanotube-induced alignment of membrane proteins for NMR structure determination. Proc Natl Acad Sci USA 104:6644–6648
325. Lorieau J, Yao L, Bax A (2008) Liquid crystalline phase of G-tetrad DNA for NMR study of detergent-solubilized proteins. J Am Chem Soc 130:7536–7537
326. Wang J, Pielak RM, McClintock MA, Chou JJ (2009) Solution structure and functional analysis of the influenza B proton channel. Nat Struct Mol Biol 16:1267–1271
327. Kamen DE, Cahill SM, Girvin ME (2007) Multiple alignment of membrane proteins for measuring residual dipolar couplings using lanthanide ions bound to a small metal chelator. J Am Chem Soc 129:1846–1847
328. Contreras M, Ubach J, Millet O, Rizo J, Pons M (1999) Measurement of one bond dipolar couplings through lanthanide-induced orientation of a calcium-binding protein. J Am Chem Soc 121:8947–8948
329. Pintacuda G, Moshref A, Leonchiks A, Sharipo A, Otting G (2004) Site-specific labelling with a metal chelator for protein-structure refinement. J Biomol NMR 29:351–361
330. Ikegami T, Verdier L, Sakhaii P, Grimme S, Pescatore B, Saxena K, Fiebig K, Griesinger C (2004) Novel techniques for weak alignment of proteins in solution using chemical tags coordinating lanthanide ions. J Biomol NMR 29:339–349
331. Ma C, Opella SJ (2000) Lanthanide ions bind specifically to an added "EF-hand" and orient a membrane protein in micelles for solution NMR spectroscopy. J Magn Reson 146: 381–384
332. Johansson MU, Alioth S, Hu K, Walser R, Koebnik R, Pervushin K (2007) A minimal transmembrane beta-barrel platform protein studied by nuclear magnetic resonance. Biochemistry 46:1128–1140
333. Haberz P, Rodriguez-Castaneda F, Junker J, Becker S, Leonov A, Griesinger C (2006) Two new chiral EDTA-based metal chelates for weak alignment of proteins in solution. Org Lett 8:1275–1278
334. Prestegard JH, Bougault CM, Kishore AI (2004) Residual dipolar couplings in structure determination of biomolecules. Chem Rev 104:3519–3540

335. Bax A, Grishaev A (2005) Weak alignment NMR: a hawk-eyed view of biomolecular structure. Curr Opin Struct Biol 15:563–570
336. Blackledge M (2005) Recent progress in the study of biomolecular structure and dynamics in solution from residual dipolar couplings. Prog Nucl Magn Reson Spectrosc 46:23–61
337. Cierpicki T, Liang B, Tamm LK, Bushweller JH (2006) Increasing the accuracy of solution NMR structures of membrane proteins by application of residual dipolar couplings. High-resolution structure of outer membrane protein A. J Am Chem Soc 128:6947–6951
338. Oxenoid K, Chou JJ (2005) The structure of phospholamban pentamer reveals a channel-like architecture in membranes. Proc Natl Acad Sci USA 102:10870–10875
339. Hus JC, Salmon L, Bouvignies G, Lotze J, Blackledge M, Brüschweiler R (2008) 16-Fold degeneracy of peptide plane orientations from residual dipolar couplings: analytical treatment and implications for protein structure determination. J Am Chem Soc 130:15927–15937
340. Meiler J, Blomberg N, Nilges M, Griesinger C (2000) A new approach for applying residual dipolar couplings as restraints in structure elucidation. J Biomol NMR 16:245–252
341. Skrynnikov NR, Kay LE (2000) Assessment of molecular structure using frame-independent orientational restraints derived from residual dipolar couplings. J Biomol NMR 18:239–252
342. Tjandra N, Omichinski JG, Gronenborn AM, Clore GM, Bax A (1997) Use of dipolar ^{1}H-^{15}N and ^{1}H-^{13}C couplings in the structure determination of magnetically oriented macromolecules in solution. Nat Struct Biol 4:732–738
343. Clore GM, Gronenborn AM, Tjandra N (1998) Direct structure refinement against residual dipolar couplings in the presence of rhombicity of unknown magnitude. J Magn Reson 131:159–162
344. Losonczi JA, Andrec M, Fischer MW, Prestegard JH (1999) Order matrix analysis of residual dipolar couplings using singular value decomposition. J Magn Reson 138:334–342
345. Fischer MW, Losonczi JA, Weaver JL, Prestegard JH (1999) Domain orientation and dynamics in multidomain proteins from residual dipolar couplings. Biochemistry 38:9013–9022
346. Chill JH, Louis JM, Delaglio F, Bax A (2007) Local and global structure of the monomeric subunit of the potassium channel KcsA probed by NMR. Biochim Biophys Acta 1768: 3260–3270
347. Brüschweiler R, Liao X, Wright PE (1995) Long-range motional restrictions in a multidomain zinc-finger protein from anisotropic tumbling. Science 268:886–889
348. Al-Hashimi HM, Valafar H, Terrell M, Zartler ER, Eidsness MK, Prestegard JH (2000) Variation of molecular alignment as a means of resolving orientational ambiguities in protein structures from dipolar couplings. J Magn Reson 143:402–406
349. Ramirez BE, Bax A (1998) Modulation of the alignment tensor of macromolecules dissolved in a dilute liquid crystalline medium. J Am Chem Soc 120:9106–9107
350. Skrynnikov NR, Goto NK, Yang D, Choy WY, Tolman JR, Mueller GA, Kay LE (2000) Orienting domains in proteins using dipolar couplings measured by liquid-state NMR: differences in solution and crystal forms of maltodextrin binding protein loaded with beta-cyclodextrin. J Mol Biol 295:1265–1273
351. Shi L, Traaseth NJ, Verardi R, Gustavsson M, Gao J, Veglia G (2011) Paramagnetic-based NMR restraints lift residual dipolar coupling degeneracy in multidomain detergent-solubilized membrane proteins. J Am Chem Soc 133:2232–2241
352. Mesleh MF, Lee S, Veglia G, Thiriot DS, Marassi FM, Opella SJ (2003) Dipolar waves map the structure and topology of helices in membrane proteins. J Am Chem Soc 125:8928–8935
353. Mesleh MF, Opella SJ (2003) Dipolar waves as NMR maps of helices in proteins. J Magn Reson 163:288–299
354. Franzin CM, Yu J, Thai K, Choi J, Marassi FM (2005) Correlation of gene and protein structures in the FXYD family proteins. J Mol Biol 354:743–750
355. Annila A, Aitio H, Thulin E, Drakenberg T (1999) Recognition of protein folds via dipolar couplings. J Biomol NMR 14:223–230

356. Chou JJ, Li S, Bax A (2000) Study of conformational rearrangement and refinement of structural homology models by the use of heteronuclear dipolar couplings. J Biomol NMR 18:217–227
357. Delaglio F, Kontaxis G, Bax A (2000) Protein structure determination using molecular fragment replacement and NMR dipolar couplings. J Am Chem Soc 122:2142–2143
358. Raman S, Lange OF, Rossi P, Tyka M, Wang X, Aramini J, Liu G, Ramelot TA, Eletsky A, Szyperski T, Kennedy MA, Prestegard J, Montelione GT, Baker D (2010) NMR structure determination for larger proteins using backbone-only data. Science 327:1014–1018
359. Solomon I, Bloembergen N (1956) Nuclear magnetic interactions in the HF molecule. J Chem Phys 25:261–266
360. Battiste JL, Wagner G (2000) Utilization of site-directed spin labeling and high-resolution heteronuclear nuclear magnetic resonance for global fold determination of large proteins with limited nuclear overhauser effect data. Biochemistry 39:5355–5365
361. Donaldson LW, Skrynnikov NR, Choy WY, Muhandiram DR, Sarkar B, Forman-Kay JD, Kay LE (2001) Structural characterization of proteins with an attached ATCUN motif by paramagnetic relaxation enhancement NMR spectroscopy. J Am Chem Soc 123:9843–9847
362. Iwahara J, Tang CD, Clore GM (2007) Practical aspects of ^{1}H transverse paramagnetic relaxation enhancement measurements on macromolecules. J Magn Reson 184:185–195
363. Iwahara J, Schwieters CD, Clore GM (2004) Ensemble approach for NMR structure refinement against H-1 paramagnetic relaxation enhancement data arising from a flexible paramagnetic group attached to a macromolecule. J Am Chem Soc 126:5879–5896
364. Keizers PM, Ubbink M (2011) Paramagnetic tagging for protein structure and dynamics analysis. Prog Nucl Magn Reson Spectrosc 58:88–96
365. Su XC, Otting G (2010) Paramagnetic labelling of proteins and oligonucleotides for NMR. J Biomol NMR 46:101–112
366. Teriete P, Franzin CM, Choi J, Marassi FM (2007) Structure of the Na, K-ATPase regulatory protein FXYD1 in micelles. Biochemistry 46:6774–6783
367. Liang B, Bushweller JH, Tamm LK (2006) Site-directed parallel spin-labeling and paramagnetic relaxation enhancement in structure determination of membrane proteins by solution NMR spectroscopy. J Am Chem Soc 128:4389–4397
368. Roosild TT, Greenwald J, Vega M, Castronovo S, Riek R, Choe S (2005) NMR structure of Mistic, a membrane-integrating protein for membrane protein expression. Science 307: 1317–1321
369. Guo Z, Cascio D, Hideg K, Kálái T, Hubbell WL (2007) Structural determinants of nitroxide motion in spin-labeled proteins: tertiary contact and solvent-inaccessible sites in helix G of T4 lysozyme. Prot Sci 16:1069–1086
370. Langen R, Oh KJ, Cascio D, Hubbell WL (2000) Crystal structures of spin labeled T4 lysozyme mutants: implications for the interpretation of EPR spectra in terms of structure. Biochemistry 39:8396–8405
371. Mchaourab HA, Lietzow MA, Hideg KJ, Hubbell WL (1996) Motion of spin-labeled side chains in T4 lysozyme, correlation with protein structure and dynamics. Biochemistry 35: 7692–7704
372. Lindberg M, Gräslund A (2001) The position of the cell penetrating peptide penetratin in SDS micelles determined by NMR. FEBS Lett 497:39–44
373. Tzeng J, Lee BL, Sykes BD, Fliegel L (2010) Structural and functional analysis of transmembrane segment VI of the NHE1 isoform of the Na+/H + exchanger. J Biol Chem 285:36656–36665
374. Cook GA, Opella SJ (2011) Secondary structure, dynamics, and architecture of the p7 membrane protein from hepatitis C virus by NMR spectroscopy. Biochim Biophys Acta 1808:1448–1453
375. Papavoine CH, Aelen JM, Konings RN, Hilbers CW, Van de Ven FJ (1995) NMR studies of the major coat protein of the bacteriophage M13. Structural information of gVIIIp in dodecylphosphocoline micelles. Eur J Biochem 232:490–500

376. Williams KA, Farrow NA, Deber CM, Kay LE (1996) Structure and dynamics of bacteriophage IKe major coat protein in LMPG micelles by solution NMR. Biochemistry 35:5145–5157
377. Papavoine CH, Konings RN, Hilbers CW, Van de Ven FJ (1994) Location of M13 coat protein in sodium dodecyl-sulfate micelles as determined by NMR. Biochemistry 33: 12990–12997
378. Zamoon J, Nitu F, Karim C, Thomas DD, Veglia G (2005) Mapping the interaction surface of a membrane protein: unveiling the conformational switch of phospholamban in calcium pump regulation. Proc Natl Acad Sci USA 102:4747–4752
379. Brown LR, Braun W, Kuman A, Wüthrich K (1982) High-resolution nuclear magnetic resonance studies of the conformation and orientation of melittin bound to a lipid-water interface. Biophys J 37:319–328
380. Brown LR, Bosch C, Wüthrich K (1981) Location and orientation relative to the micelle surface for glucagon in mixed micelles with dodecylphosphocholine: EPR and NMR studies. Biochim Biophys Acta 642:296–312
381. Neumoin A, Arshava B, Becker J, Zerbe O, Naider F (2007) NMR studies in dodecylphosphocholine of a fragment containing the seventh transmembrane helix of a G-protein-coupled receptor from *Saccharomyces cerevisiae*. Biophys J 93:467–482
382. Clore GM, Tang C, Iwahara J (2007) Elucidating transient macromolecular interactions using paramagnetic relaxation enhancement. Curr Opin Struct Biol 17:603–616
383. Shon KJ, Kim Y, Colnago LA, Opella SJ (1991) NMR studies of the structure and dynamics of membrane-bound bacteriophage Pfl coat protein. Science 252:1303–1304
384. Henry GD, Sykes BD (1994) Methods to study membrane protein structure in solution. Methods Enzymol 239:515–535
385. Vinogradova O, Badola P, Czerski L, Sönnichsen FD, Sanders CR (1997) *Escherichia coli* diacylglycerol kinase: a case study in the application of solution NMR methods to an integral membrane protein. Biophys J 72:2688–2701
386. Girvin ME, Fillingame RH (1995) Determination of local protein structure by spin-label difference 2D NMR: the region neighboring Asp61 of subunit-c of the F1F0 ATP synthase. Biochemistry 34:1635–1645
387. White SH (2009) Biophysical dissection of membrane proteins. Nature 459:344–346
388. Lin TL, Chen SH, Gabriel NE, Roberts MF (1987) Small-angle neutron scattering techniques applied to the study of polydisperse rodlike diheptanoylphosphatidylcholine micelles. J Phys Chem 91:406–413
389. le Maire M, Champeil P, Moller JV (2000) Interaction of membrane proteins and lipids with solubilizing detergents. Biochim Biophys Acta 1508:86–111
390. Stafford RE, Fanni T, Dennis EA (1989) Interfacial properties and critical micelle concentration of lysophospholipids. Biochemistry 1989:5113–5120
391. MacKenzie KR, Prestegard JH, Engelman DM (1996) Leucine side-chain rotamers in a glycophorin A transmembrane peptide as revealed by three-bond carbon-carbon couplings and ^{13}C chemical shifts. J Biomol NMR 7:256–260
392. Bocharov EV, Mayzel ML, Volynsky PE, Mineev KS, Tkach EN, Ermolyuk YS, Schulga AA, Efremov RG, Arseniev AS (2010) Left-handed dimer of EphA2 transmembrane domain: helix packing diversity among receptor tyrosine kinases. Biophys J 98:881–889
393. Mineev KS, Khabibullina NF, Lyukmanova EN, Dolgikh DA, Kirpichnikov MP, Arseniev AS (2011) Spatial structure and dimer-monomer equilibrium of the ErbB3 transmembrane domain in DPC micelles. Biochim Biophys Acta 1808:2081–2088
394. Buck-Koehntopa BA, Mascionia A, Buffya JJ, Veglia G (2005) Structure, dynamics, and membrane topology of stannin: a mediator of neuronal cell apoptosis induced by trimethyltin chloride. J Mol Biol 354:652–665
395. Franzin CM, Teriete P, Marassi FM (2007) Structural similarity of a membrane protein in micelles and membranes. J Am Chem Soc 129:8078–8079

396. Wittlich M, Thiagarajan P, Koenig BW, Hartmann R, Willbold D (2010) NMR structure of the transmembrane and cytoplasmic domains of human CD4 in micelles. Biochim Biophys Acta 1798:122–127
397. Zamoon J, Mascioni A, Thomas DD, Veglia G (2003) NMR solution structure and topological orientation of monomeric phospholamban in dodecylphosphocholine micelles. Biophys J 85:2589–2598
398. Koradi R, Billeter M, Wüthrich K (1996) MOLMOL: a program for display and analysis of macromolecular structures. J Mol Graph 14:51–55
399. White SH (2004) The progress of membrane protein structure determination. Protein Sci 13:1948–1949

Top Curr Chem (2012) 326: 187–214
DOI: 10.1007/128_2011_287

Published online: 9 December 2011

Protein Structure Determination by Solid-State NMR

Xin Zhao

Abstract Membrane proteins are a large, diverse group of proteins, representing about 20–30% of the proteomes of most organisms, serving a multitude of cellular functions and more than 40% of drug targets. Knowledge of a membrane protein structure enables us insight into its function and dynamics, and can be used for further rational drug design. Owing to the intrinsic hydrophobicity, flexibility, and instability of membrane proteins, solid-state NMR may offer an unique opportunity to study membrane protein structure, ligand binding, and activation at atomic resolution in the native membrane environment on a wide ranging time scale. Over the past several years, solid-state NMR has made tremendous progress, showing its capability of determining membrane protein structure, ligand binding, and protein dynamic conformation on a variety of time scales at atomic resolution. In this chapter we will mainly discuss some recent achievements on membrane protein structure determination, ligand conformation and binding, structure changes upon activation, and structure of insoluble fibrous proteins investigated by using magic-angle spinning solid-state NMR from the structural biology point of view. Protein dynamics, sensitivity enhancement, and the possibility of chemical shift-based structure determination in solid-state NMR are also briefly touched upon.

Keywords Membrane proteins · MAS solid-state NMR · Structure determination · ligand conformation · structural changes and dynamics

---

X. Zhao (✉)
Shanghai Key Lab of Magnetic Resonance, East China Normal University, 200062 Shanghai, People's Republic of China
e-mail: xzhao@phy.ecnu.edu.cn

Contents

1 Introduction ... 188
2 Basic Experimental Techniques Used in Solid-State NMR ... 189
2.1 Magic-Angle Spinning ... 189
2.2 High Power Heteronuclear Decoupling ... 190
2.3 Cross-Polarization ... 190
2.4 Recoupling in MAS Solid-State NMR ... 190
3 Protein Structure Determination by MAS Solid-State NMR ... 191
3.1 Labeling Strategy and Sample Environments ... 191
3.2 Three-Dimensional Structure Determination ... 194
3.3 Ligand Conformation and Binding ... 196
3.4 Structural Changes upon Activation ... 199
3.5 Protein Dynamics ... 201
3.6 Sensitivity Enhancement ... 203
3.7 Structure Determination Based on Chemical Shifts ... 204
4 Perspective ... 204
References ... 205

1 Introduction

Membrane proteins are a large, diverse group of proteins, representing about 20–30% of the proteomes of most organisms, serving a multitude of cellular functions, and more than 40% of drug targets [1]. For example, membrane proteins often serve as receptors or provide channels for polar or charged molecules to pass through the cell membrane and maintain a variety of biological processes [2]. Knowledge of a membrane protein structure enables us insight into its function and dynamics, and can be used for further rational drug design. Therefore it is always desirable to have an accurate picture of protein structure in the highest resolution possible. However, owing to their intrinsic hydrophobicity, flexibility, and instability, many fewer structures have been solved by X-ray crystallography for membrane proteins compared to soluble proteins [3, 4]. Moreover, the conformational change of the transmembrane helices upon activation increase the difficulty of capturing the activation state of a membrane protein to a higher resolution by X-ray crystallography [5, 6]. In contrast, solid-state NMR (SSNMR) is a suitable technique to study molecular structure and interactions at atomic level in a variety of sample forms; it can be used to determine a membrane protein structure and probe its conformational dynamics in the native membrane environments.

Over the past several years, SSNMR has made tremendous progress, showing its capability of determining membrane protein structure, ligand binding, and protein dynamic conformation on a variety of time scales at atomic resolution. Many membrane proteins have been investigated by magic-angle spinning (MAS) SSNMR, factors investigated including the following: activation, inhibition, and dynamics of the potassium channel KcsA-Kv1.3 [7–11]; structure, ligand conformation, activation, and dynamics of the G protein coupled receptor – rhodopsin [12–25]; protonation switch mechanism of the human H1 receptor[26]; the influenza M2 proton channel structure, function and ligand binding [27, 28]; the human prion protein [29]; the structural conversion of neurotoxic amyloid beta oligomers to fibrils [30]; and the structure and dynamics of the retinylidene proteins from bacteria, including bacteriorhodopsin, sensory rhodopsin, halorhodopsin, and proteorhodopsin [31–41], and the structure of the HET-s(218–289) fibril [42, 43]. Some other membrane proteins, such as membrane-embedded enzymes [44–46], histidine kinases [47], ABC transporters [48], and bacterial outer membrane proteins [49, 50], have also been investigated through multidimensional correlation experiments in SSNMR.

In this chapter we will briefly review some of the recent progress in studying membrane proteins by using magic-angle spinning solid-state NMR from biological structure point of view; for a complete overview of the achievements in this field, please refer to the following excellent reviews [19, 51–67].

2 Basic Experimental Techniques Used in Solid-State NMR

Unlike solution-state NMR, the resolution and sensitivity of SSNMR are affected heavily by orientation dependent anisotropic spin interactions such as chemical shift anisotropy, homonuclear and heteronuclear dipolar couplings, quadrupole coupling, etc. These interactions generally cannot be averaged out by the molecular tumbling motions in solids, presenting a very broad line shape with poor sensitivity of SSNMR spectra. Combining MAS with cross-polarization (CP), high power proton decoupling, recoupling, and isotopic labeling can achieve high resolution and signal-to-noise ratio in SSNMR for protein structure determination.

2.1 Magic-Angle Spinning

MAS is an essential technique in SSNMR for obtaining a high resolution spectrum [68–70]. The basic idea is to spin the sample container (rotor) about an axis, which subtends an angle of 54.74°, the magic-angle, with respect to the static field B_0. The spatial rotation of the sample introduces time-dependence to anisotropic spin interactions, such as chemical shift anisotropy, homonuclear dipole–dipole couplings, and heteronuclear dipole–dipole couplings, which are averaged out more efficiently as the sample spinning frequency increases. Due to the periodic time-dependence of the CSA and dipolar spin interactions, the broad static line shape breaks up into a center band at the isotropic position and a set of spinning sidebands separated by the spinning frequency. As the spinning frequency increases, the time averaging is more effective, which leads to a decrease in the sideband intensities and an increase in the center band intensity. The advantage of MAS is that both resolution and sensitivity are greatly increased.

The disadvantage is that the spectrum loses all the anisotropic dependent molecular geometry information.

2.2 *High Power Heteronuclear Decoupling*

Hydrogen is a very common element and has a large γ value and almost 100% natural abundance. Its coupling to rare spins with low γ complicates the observation of the rare spins, such as ^{13}C and ^{15}N. It is usually necessary to remove heteronuclear dipolar couplings between protons and the observed rare spins by using strong rf irradiation on proton spins, i.e., so-called high power proton decoupling. Continuous-wave (CW) decoupling is a standard scheme for decoupling of heteronuclear dipolar spin interactions. The rf field induces a fast rotation of the proton spin states, averaging out their interaction with the rare spins [71, 72]. For fast MAS, more sophisticated pulse sequences, such as TPPM, C12, R24, XiX, and Spinal64, will achieve a better decoupling efficiency [73–77].

2.3 *Cross-Polarization*

The sensitivity of the rare spin species (*S*-spins) with low magnetogyric ratio, such as ^{13}C and ^{15}N, can be enhanced by transferring magnetization from abundant spin species (*I*-spins) with high magnetogyric ratio, such as ^{1}H. Cross-polarization (CP) is the most widely used method in solids to transfer polarization between unlike spin species through heteronuclear dipolar couplings [78, 79]. Magnetization transfer is achieved when the strengths of the two fields match the Hartmann-Hahn (HH) condition $|\gamma_I B_{rf}^I| = |\gamma_S B_{rf}^S|$, where γ_I and γ_S are the magnetogyric ratios of the *I*-spins and *S*-spins, respectively [80]. The enhancement of the *S*-spin magnetization is roughly proportional to the ratio of the two magnetogyric ratios $|\gamma_I/\gamma_S|$. Ramping or adiabatic-passing one of the *rf* fields can improve the reliability and reproducibility of HH-CP, especially in fast MAS experiments [81–84].

2.4 *Recoupling in MAS Solid-State NMR*

Fast MAS leads to high resolution and sensitivity, the basic requirement for sequential assignment of protein structures in SSNMR. However, all anisotropic spin interactions which can be used to extract molecular geometry information are averaged out by MAS. In order to get back the anisotropic spin interactions in the presence of MAS, recoupling techniques were developed to retrieve selectively the anisotropic spin interactions for rotating solids [189–190]. Generally, there are two approaches to reintroduce anisotropic dipolar interactions, either mechanically, where the recoupling is achieved through rotational resonance, or by rf pulse driven methods, where the recoupling is achieved by applying rf pulse trains. By applying recoupling sequences, certain anisotropic interaction terms can be retrieved in the presence of MAS. Magic-angle spinning NMR in combination with recoupling has become a widely used experimental technique for obtaining molecular structural information in non-crystalline or disordered materials. MAS makes it possible to obtain SSNMR spectra with good resolution and sensitivity, while the recoupling techniques cause a selective restoration of informative anisotropic nuclear spin interactions, which are normally suppressed by magic angle rotation. A wide range of recoupling methods has been developed for homonuclear and heteronuclear spin systems [85–88]. With a combination of recoupling sequences and multidimensional experiment schemes, membrane protein structure and dynamic conformation upon activation can be revealed through the short and long range semi-quantitative distance restraints from multidimensional homonuclear/heteronuclear correlation experiments, such as PDSD/DARR, PAR, TEDOR, PAIN-CP, and CHHC/NHHC type experiments [18, 89–92, 127, 191–192], etc.

3 Protein Structure Determination by MAS Solid-State NMR

3.1 Labeling Strategy and Sample Environments

Isotopic labeling plays a very important role in molecular structure determination in SSNMR. It not only enhances the spectral sensitivity and improves the spectral resolution, but also helps with the resonance assignments of NMR spectra and tackles the specific problem with the structure and dynamics of proteins through the designed labeling scheme.

There are three main approaches to isotopically label proteins: specific, selective, and uniform labeling. All are used extensively in SSNMR studies of protein structure and ligand conformation, depending on the questions that the designed experiment needs to be answered.

Specific labeling normally refers to incorporate a non-uniformly ^{13}C, ^{15}N labeled amino acid into a polypeptide/protein or a ligand at a certain position. It requires solid-phase peptide synthesis or chemical synthesis. This approach has been used extensively to study amyloid peptides, membrane peptides, and GPCR ligand conformation [93–99].

Selective labeling, including forward and reverse labeling, refers to the biosynthetic incorporation of a single type or several types of labeled amino acid(s) – with the rest unlabeled – into the protein expression media. Selective labeling is widely used for the protein resonance assignments and structure calculations in SSNMR. For example, a selectively and extensively labeling scheme using [1, 3-^{13}C] and [2-^{13}C]-glycerol as the sole carbon source in the bacterial growth medium has been used to determine the α-spectrin SH3 domain, αB-crystalline, and outer membrane protein G (OmpG) [100, 101]. A selective labeling scheme has also been used in the HEK293 system to express labeled rhodopsin for structure and function study [12, 18, 22, 102–104]. The reverse labeling approach has been designed to study structure and dynamics of the sensory rhodopsin II and the potassium ion channel KcsA-Kv1.3 [105, 106]. This approach has the advantage of decreased numbers of overlapping resonances from the hydrophobic transmembrane helices, making the extensive assignments possible. By taking advantage of the glycolysis pathway, the TEASE ^{13}C selective labeling scheme has been suggested to probe the transmembrane segments of membrane proteins [107], and this reverse labeling approach may possibly be adaptable for other different applications for its flexibility.

A uniformly labeling scheme is the simplest and most cost-effective biosynthetic labeling method for protein SSNMR. Normally, uniformly ^{13}C-labeled glucose or glycerol, and ^{15}N-labeled ammonium chloride or ammonium sulfate are used as the labeled precursors in the growth medium. With a single sample, all the structural constraints can be obtained through a set of correlation experiments and the protein structure can be calculated thereby. This approach has been first demonstrated on microcrystalline proteins of known structure [108–111], and then successfully applied to membrane protein for structure determination [7, 38, 41, 105, 112–114].

Recently, predeuteration of proteins has been exploited to gain even higher resolution for U-^{13}C and ^{15}N labeled proteins in SSNMR by taking advantage of minimizing the ^{1}H-^{1}H dipolar couplings causing spectral broadening. Perdeuterated soluble microcrystalline proteins have been used to study protein dynamics and interactions [115–117], and structure determination [118]. This approach has also been attempted to express the fully labeled 7-transmembrane protein bacteriorhodopsin for structure investigation, as shown in Fig. 1 [119].

However, with increase of deuteration level, protein expression level may drop and some strains may even be difficult to grow on D_2O [120]. Increasing deuteration level may also influence the resonance frequencies and CP efficiency. Therefore, a good balance needs to be considered from both the NMR and biological sides.

Choosing an appropriate sample environment is not only critical for generating a high resolution NMR spectrum, but also critical for the protein to have correct folding and activity. Obtaining homogeneous sample preparation leads to improved linewidths and therefore spectral resolution, while heterogeneous samples can result in artifacts such as unexpected peaks and peak doubling [120]. Furthermore, it is not enough to show that a protein construct is functional to validate a structure unless the functional assay is performed in the same environment as that used for the structural characterization [121]. For small proteins, nano-/micro-crystalline or nano-disk samples have been proven a good choice for yielding high quality spectra in solids [122–124]. For membrane proteins, samples can be prepared in either detergent micelles, bicelles, or lipid bilayers. Given that membrane proteins function within a bilayer environment, it is more biologically applicable to be able to carry out structural investigations in lipids [120]. Structural data obtained in an appropriate lipid bilayer environment can serve as benchmarks for validating structures determined in other mimetic environments [121]. Figure 2 shows the

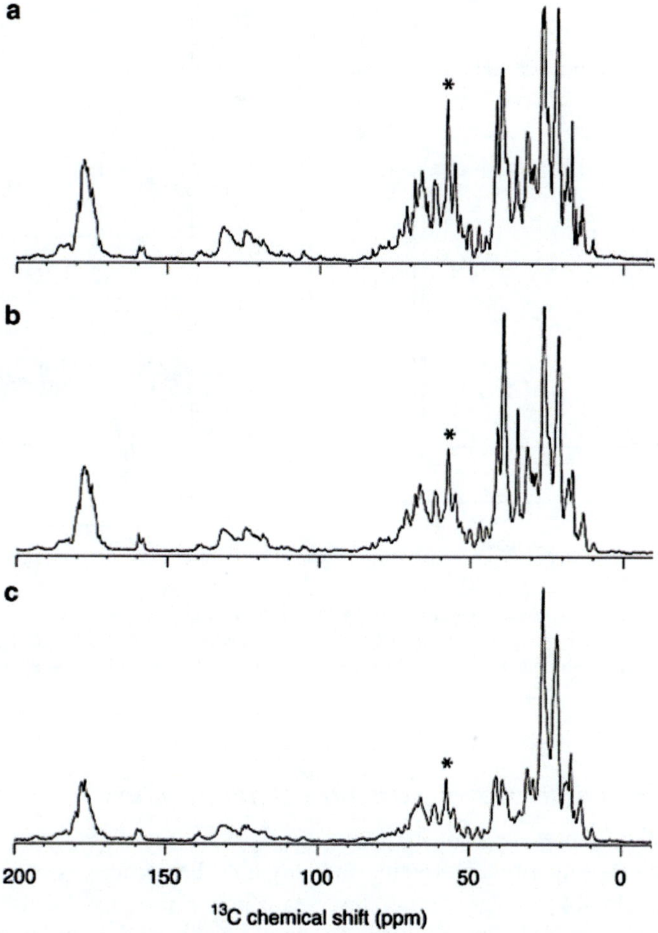

Fig. 1 1D ^{13}C NMR spectra (800 MHz for ^{1}H) of CN-bR (**a**), HDCN-bR (**b**), and DCN-bR (**c**), fully hydrated in sodium citrate buffer and collected under very similar conditions. Adapted from [119] with permission from Elsevier B.V

^{13}C-DQ/SQ correlation experiment on the fully hydrated [$^{13}C_6$, ^{15}N]-Leu-BR sample using a Bruker AV-III 600 MHz wide-bore spectrometer with an MAS rate of 8 kHz (Fig. 2a). The sample activity is carefully assessed through detection of the M state and the proton pumping cycle at 412nm and 456nm by optical dynamic spectroscopy, respectively. It is clear that the resolved narrow spectral linewidths are attributed to the predominant distribution of the Leu residues on the helical segments with similar local environments, and the up-field shifted peak is attributed to the Leu residue located at the loop region. This small shift clear indicates a homogenous sample condition. The expressed protein is fully functioning as confirmed by capture of the M state signal and the proton pumping cycle signal shown in Fig 2b and c.

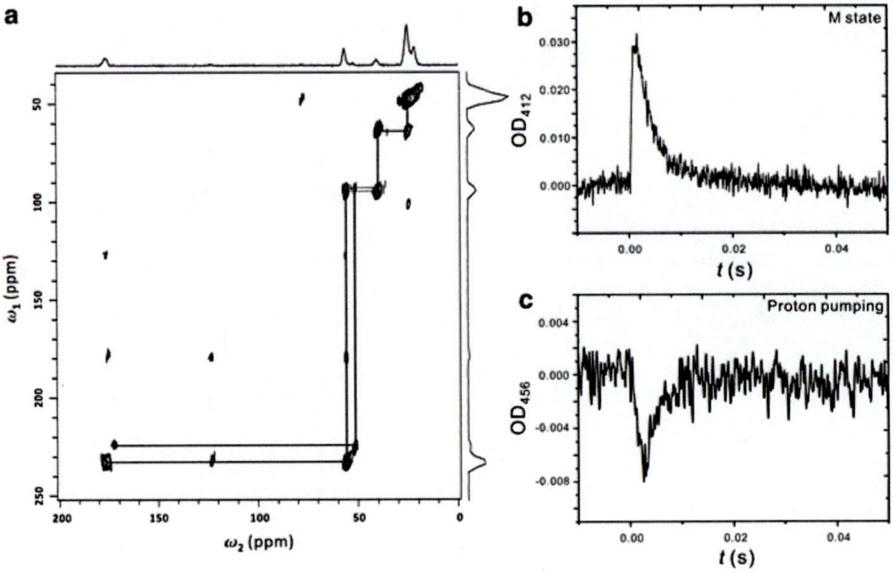

Fig. 2 2D ^{13}C DQ/SQ NMR correlation of the fully hydrated [$^{13}C_6$, ^{15}N]-Leu-BR sample using a Bruker AV-III 600 MHz wide-bore spectrometer with an MAS rate of 8 kHz (**a**) and the protein function assay through detection of the M state signal and proton pumping cycle signal at 412nm (b) and 456nm (c) by optical dynamic spectroscopy, respectively

3.2 Three-Dimensional Structure Determination

Determination of membrane proteins and amyloid fiber structures is still a frontier of structural biology. Over the last few years, much progress has been made in various areas from NMR methodology to sample preparation in order to reduce the dipolar truncation, establish reliable short and long range distance constraints, and improve spectral linewidths and resolution for a full structure determination. The Baldus group has established a method to probe through-space (^{1}H, ^{1}H) contacts of protonated solid-phase systems in high spectral resolution [125]. The so-called CHHC/NHHC experiment was first demonstrated to solve a micro-crystalline Crh protein [126], and then successfully applied to the membrane bound potassium channel KcsA-Kv1.3 to study its structure activation and inhibition [7]. By combing the ^{1}H...^{1}H distance restraints, chemical shift changes, and molecular modeling, they have obtained the first membrane protein structure by SSNMR, as shown in Fig. 3, which provides a deep understanding of channel activation and inhibition [7].

Full or nearly full SSNMR spectral assignments have not been achieved on membrane proteins for quite some time due to the signal overlapping and fast longitudinal and transverse relaxation which causes line broadening and signal intensity loss in many correlation experiments, impeding the pace of SSNMR structure determination of large membrane proteins and protein-complexes. Recently, the Oschkinat group has proposed a novel specific labeling scheme by

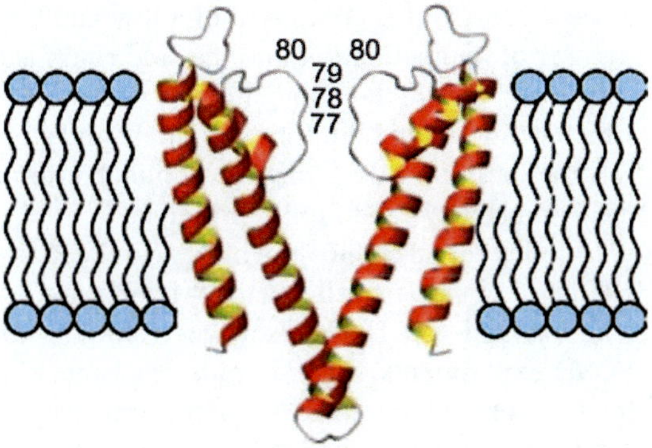

Fig. 3 Solid-state NMR structure of membrane bound potassium channel KcsA-Kv1.3. Adapted from [7] with permission from Nature Publishing Group

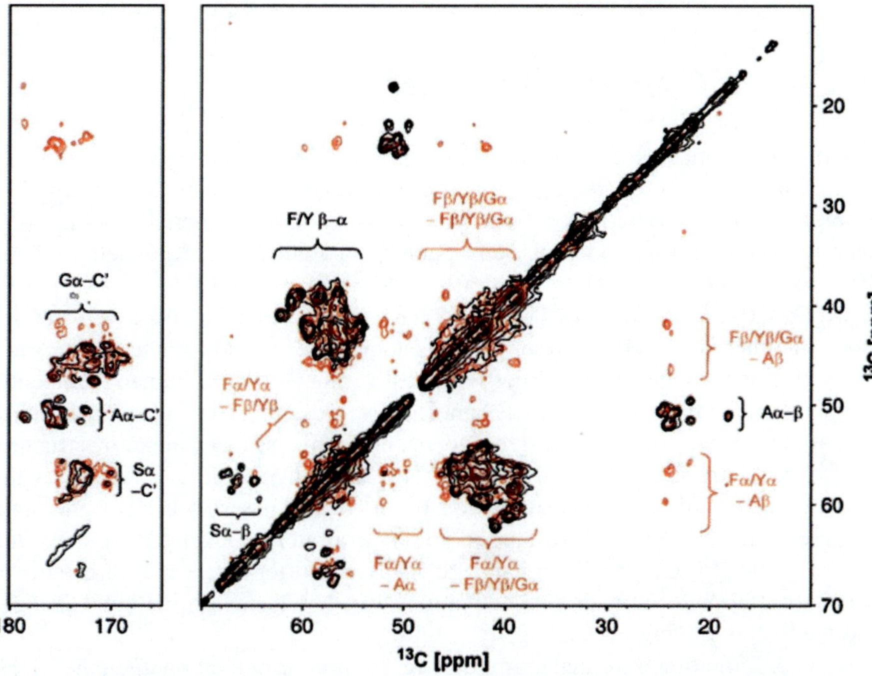

Fig. 4 ^{13}C–^{13}C proton-driven spin diffusion spectrum of OmpG-GAFY recorded at 900 MHz with 20 ms (*black*) and 700 ms (*red*) mixing time. Intra- (*black*) and inter-residual (*red*) cross-peak regions are indicated. Adapted from [49] with permission from the American Chemical Society

using [^{15}N, 2, 3-^{13}C]-labeled Phe and Tyr residues and fully labeled Gly and Ala residues to restore favorable cross-relaxation properties of the glycerol samples in order to obtain inter-residue cross-peaks [49]. They have expressed the 281-residue OmpG with this labeling strategy (referred to as OmpG-GAFY) and demonstrated its advantage by the ^{13}C-^{13}C proton-driven spin diffusion (PDSD) experiment [127, 191–192] as shown in Fig. 4. It is very clear that the Ca-Cb signals for Phe, Tyr, and Ala, and the Ca-C' cross-peaks for Gly and Ala are all well resolved and the signal intensities are increased significantly [49]. This labeling scheme has several advantages: (1) it is composed of a low number of small, isolated spin systems, (2) transfer of magnetization into the side chain is thus eliminated and spectral quality enhanced, and (3) the number of inter-residue cross-peaks is significantly increased which is important both for assignment and structure calculation [49]. This labeling scheme or similar ones may be applicable to other large membrane proteins, e.g., the 7-transmembrane family proteins.

Another great achievement by SSNMR is the structure determination of the HET-s (218–289) fibril by the Meier group [42], as shown in Fig. 5. Total 90 ^{13}C–^{13}C and 44 ^{1}H–^{1}H distance restraints obtained by the CHHC, NHHC and PDSD experiments, and 74 angle restraints obtained by TALOS [193], were used for structure calculations. The extraordinarily high order in the HET-s prion fibrils can be explained by the well-organized structure obtained by SSNMR.

3.3 Ligand Conformation and Binding

The molecular mechanisms of membrane protein activation are at the center of interest in the study of cellular responses to biogenic stimulus and drugs. For example, GPCRs are activated by a wide range of stimuli, including hormones, neurotransmitters, ions, odorants, and photons of light [128]. Knowledge of the three-dimensional structures of several GPCRs, such as rhodopsin, β_1AR, β_2AR, A_{2a}R, CXCR4, D_3R, and H_1R [129–141] have been resolved by X-ray crystallography in either an inactive state or agonists/antagonists bound form at high resolution which open up new possibilities for investigating GPCRs of human therapeutic significance and for rational drug design. However, despite the availability of those crystal structures, a comprehensive understanding of the mechanism of structure activation is still a challenge due to the lack of high resolution structure at the activated state. For example, activation of rhodopsin has still been challenging because of the lack of high resolution structure at atomic level for the activated Meta II state [6]. On the other hand, SSNMR can offer direct measurements at atomic resolution to study protein activation caused by the conformational changes and the ligand binding interactions [65, 120].

The conformation of retinal chromophore and protein activation upon the 11-*cis* to all-*trans* isomerization of the retinal have been studied extensively by MAS SSNMR through the chemical shift measurements, distance measurements, correlation experiments, and ^{2}H NMR [12–18, 20, 22, 23, 25, 95, 103, 104, 142–144]. Complete assignment of the retinal carbons in ground state and partially in the Batho-, Meta I, and Meta II intermediates have been achieved [14, 17, 145–151]. These valuable data allow us insight into the conformational changes of the retinal protonated Schiff base (PSB) complex and the related transmembrane helices through the transition from the ground state to the Batho, Meta I, and Meta II intermediates in the binding pocket, as shown in Fig. 6.

Large down-field or up-field changes have been observed at the C16, C17 of the β-ionone ring, and on the retinal polyene chain around C9–C10–C11–C12 and

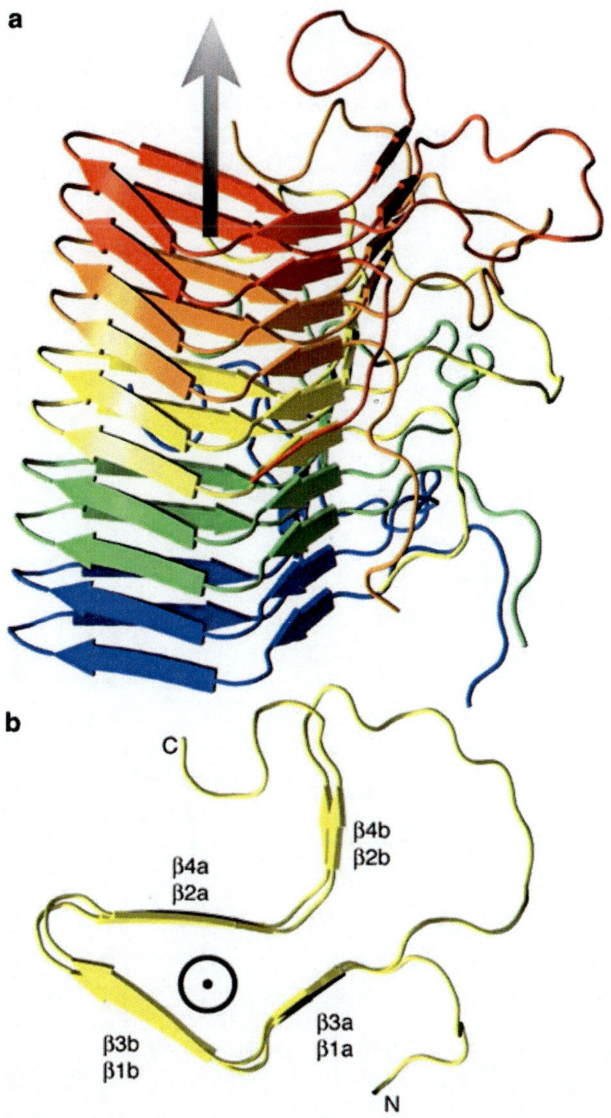

Fig. 5 Solid-state NMR structure of the HET-s (218–289) fibril. (A) Side view of the five central molecules of the lowestenergy structure of the HET-s(218–289) heptamer calculated from the NMR restraints. (B) Top view of the central molecule from (A). beta3 and beta4 lie on top of beta1 and beta2, respectively. Adapted from [42] with permission from the American Association for the Advancement of Science

C13–C14–C15 regions (Fig. 6). The large change at C13, C9, and C17 may be attributed to the break of the Glu113 salt bridge, van der Waals contact, and local hydrophobic environmental changes due to the displacements of the helices upon activation. In addition to the studies on the retinal in rhodopsin, the structure of

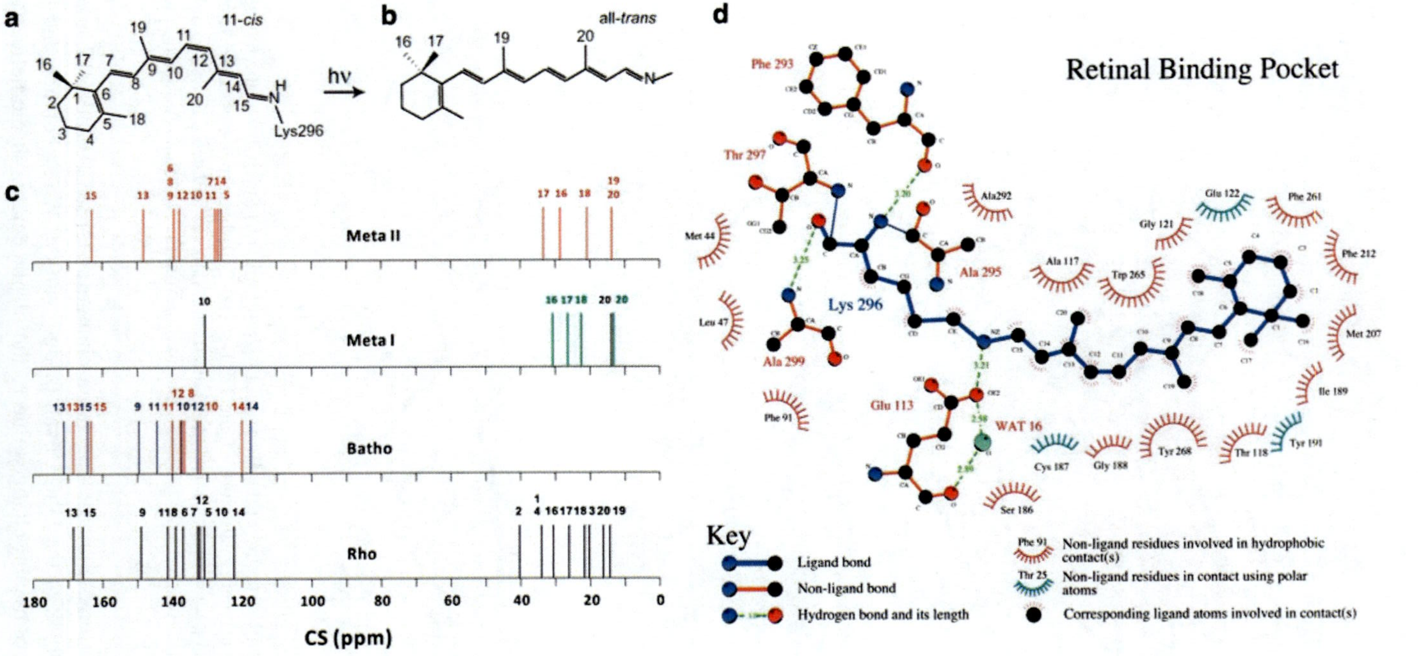

Fig. 6 Comparison of retinal protonated Schiff base chemical shifts in rhodopsin different photocycle states. The chemical structure of retinal chromophore in 11-*cis* (**a**) and all-*trans* configuration (**b**). Chemical shifts of the retinal atoms in ground state, and Batho-, Meta I, and Meta II intermediate states (**c**) and a schematic drawing of the retinal binding pocket containing all residues within 4 Å to the retinal and Lys296 (**d**). The chemical shifts are adapted from the following references: [146–148] (*black*); [14] (*blue*); [149–151] (*green*); and [17, 145] (*red*). (**d**) is adapted from [188] with permission from the Elsevier B.V

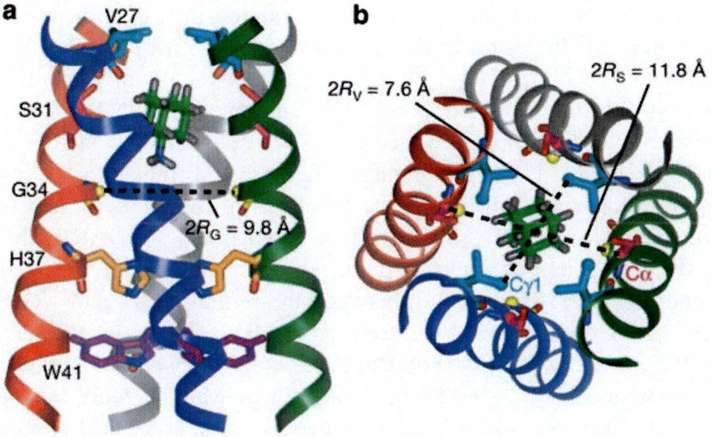

Fig. 7 The solid-state NMR structure of Amt-bound M2 in lipid bilayers. (A) Side view showing several key residues and Amt in the high-affinity luminal site. The time-averaged Amt orientation is parallel to the channel axis. (B) Top view showing the Ser 31 and Val 27 pore radii. Adapted from [27] with permission from Nature Publishing Group

neurotensin bound or unbound to the NTS1 receptor has also been investigated by SSNMR, revealing a β-strand conformation upon binding [94]. The conformation of the bradykinin (BK) peptide bound to the human bradykinin B2 receptor in DDM, on the other hand, has recently been proposed to have a double S-shape structure [152]. In the case of the human histamine H1 receptor, changes in the protonation state of the ligand histamine binding to the receptor, SSNMR experiments have revealed that the ligand can bind in a different cationic form and a protonation switch might be part of the activation mechanism [26]. Another outstanding example of these types of work is the study of influenza M2 proton channel structure, function, and ligand binding conducted by the group of Hong [27, 28]; through the extensive heteronuclear distance measurements and orientation measurements, they have successfully proposed a structure model of the amantadine binding to M2 in phospholipid bilayers, as shown in Fig. 7. This study has clearly demonstrated the ability of SSNMR to elucidate drug-membrane protein interactions at atomic resolution and this is useful when conducting novel drug design for human therapeutics.

3.4 Structural Changes upon Activation

Upon binding of agonists, which typically occurs in proximity to the extracellular opening of the helical bundle, GPCRs undergo a series of structural changes that cascade from the extracellular to the intracellular part of the receptor and ultimately lead to G protein activation [153]. The multiple structural "switches" in rhodopsin that trigger the conformational changes are involved in activation and formation of the G protein binding site. In the last few years, the Smith group has focused on SSNMR studies of structural changes on the extracellular side of the receptor caused by retinal isomerization. By combined SSNMR chemical shift measurements and 2D-dipolar assisted rotational resonance (DARR) [154] NMR measurements with selective labeling schemes and mutagenesis, they have published several papers addressing the functions of the displacement of EL2 on rhodopsin activation. Figure 8 shows the two-dimensional ^{13}C DARR NMR spectra of retinal-EL2 interactions – close contact between the retinal ^{13}C14 and ^{13}C15 carbons and ^{13}Cβ-Ser186 (Fig. 8a), between the retinal ^{13}C12 and ^{13}C20 carbons and ^{13}C1-Cys187 (Fig. 8b), and between the retinal ^{13}C12 and ^{13}C20 carbons and ^{13}Cα-Gly188 in rhodopsin (Fig. 8c). But the contacts between the retinal ^{13}C9 and ^{13}C12 carbons and U-$^{13}C_6$-Ile189 in rhodopsin or Meta II were not observable (Fig. 8d) [18]. All the results indicate that the formation of Meta II is accompanied by the displacement of EL2 away from the retinal binding site and that there is a rearrangement in the hydrogen-bonding networks connecting EL2 with the extracellular ends of transmembrane helices H4, H5, and H6. This displacement is coupled to the rotation of TM5 and breaking of the ionic lock connecting TM3 and TM6 [18]. These comprehensive results may lead to further investigation of the molecular mechanism of the cavity formation between H3, H5, and H6 for G protein binding [155].

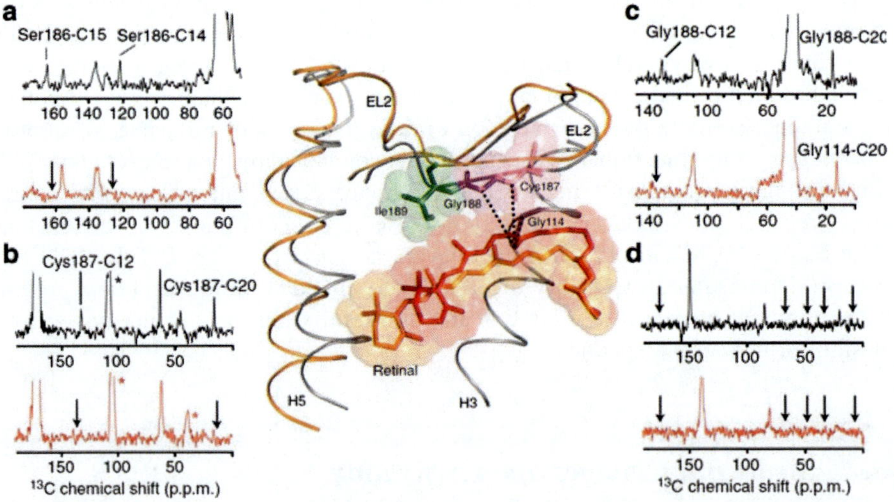

Fig. 8 Two-dimensional ^{13}C DARR NMR spectra of retinal-EL2 interactions. Rows from the two-dimensional ^{13}C DARR NMR spectra of rhodopsin (*black*) and Meta II (*red*) are shown. The rhodopsin crystal structure (*gray*) with the Meta II model (*orange*) obtained from molecular dynamic simulations are shown in the middle of the figure. Adapted from [18] with permission from Nature Publishing Group

3.5 Protein Dynamics

Transmission of signals between cells, within cells, and from the extracellular environment to the cellular interior is essential to life, and the dynamic properties of the signaling proteins are crucial to their functions [156]. Therefore, understanding the dynamic nature of a membrane protein within lipid bilayers is crucial to reveal its function mechanism at the molecular level.

SSNMR may be the best technique to study membrane protein dynamics in the native lipid environment [56, 157, 158], and site specific ^{2}H labeling at the methyl groups of Ala, Leu, and Val is most commonly used in SSNMR, and is an excellent probe of the dynamics of membrane proteins. The Glaubitz group has recently reported a dynamic picture of the green proteorhodopsin structure using through-space and through-bond correlation experiments in SSNMR [41]. They have used U-[^{13}C, ^{15}N]-PR and reversely labeled U-[^{13}C, ^{15}N]$_{\backslash WHYFI}$-PR samples to establish a clear correlation between hydration water and the mobile J-residues (mainly in flexible loops and tails), as indicated in Fig. 9. Hydration water plays an essential role for enhancing molecular dynamics of residues in tails and interhelical loops, but not in transmembrane domains or rigid, structured loop segments. The result is very important for understanding the dynamic proton pumping mechanism of proteorhodopsin. It also provides an approach to study the site-resolved effects of water and lipid bilayers on the dynamic properties of membrane proteins in general.

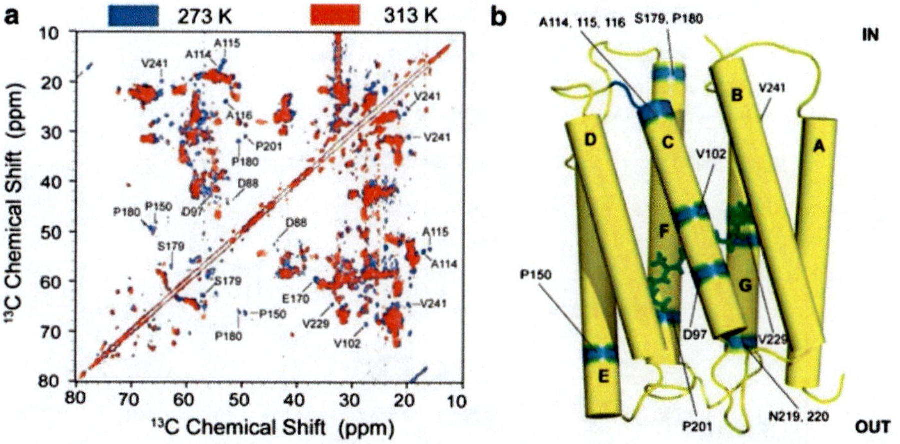

Fig. 9 ^{13}C-DARR spectra of the aliphatic regions of U-[^{13}C, ^{15}N]$_{\backslash WHYFI}$-PR at 273 (*blue*, gel phase) and 313 K (*red*, liquid crystal phase) (**a**) and the modified homology model of *green* PR (**b**). Helical residues influenced by changes in membrane elasticity (labeled in *blue*) are found in helices C, E, F, and G as well as in loops EC and EF. These residues disappear in the fluid membrane but are visible in the gel phase. This indicates that especially helices C and G but also E and F undergo thermal equilibrium fluctuations in the ground state of PR. Adapted from [41] with permission from the American Chemical Society

Contact and association of hydrophilic peptides and peripheral proteins with cellular membranes are commonly found through the sequence motifs which contain basic and aromatic amino acid residues. Those sequence motifs are not only critical for protein binding but also important for local disruption and penetration of membranes, for recruitment of lipids, and for membrane fusion [159–165]. However, the locations and insertion depths in membrane lipids are quite different as compared with those residues in for transmembrane proteins. It is not straightforward to apply those well established principles generalized from transmembrane proteins to peripheral membrane proteins. For example, we have recently investigated the insertion depth of Trp residue into POPC lipid bilayers in different peptide sequencing, as shown in Fig. 10. Clearly the penetration depth of the side chain of Try residue in AWA and VAMP2 peptides are deeper than the

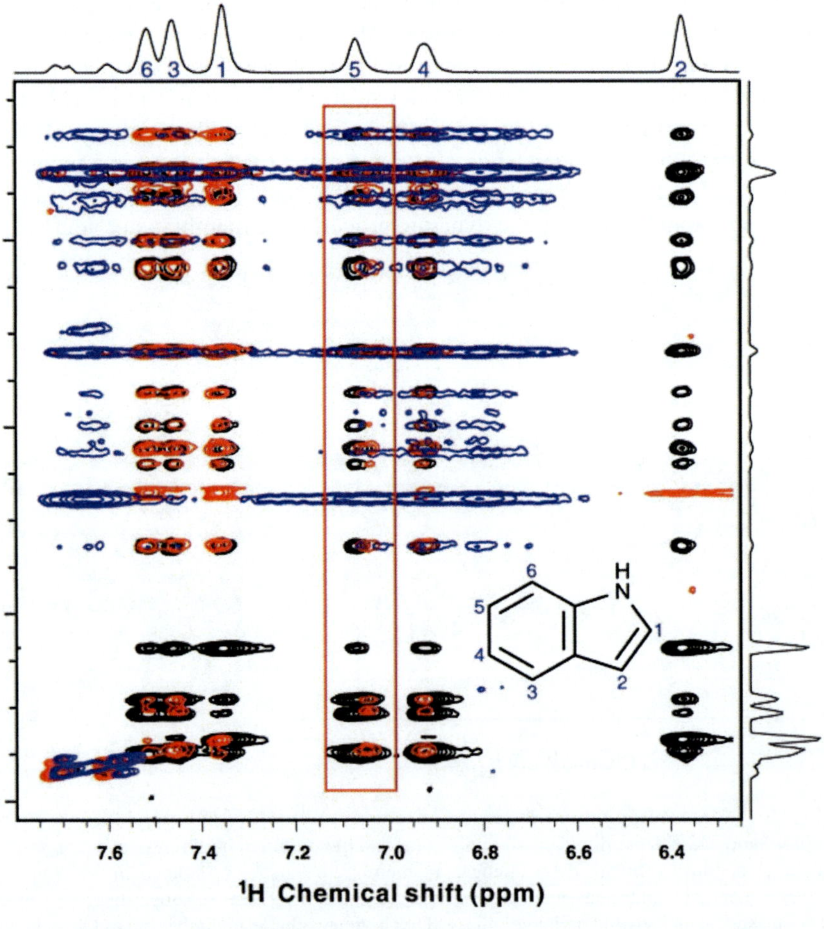

Fig. 10 ^{1}H MAS NOESY spectra of indole (*black*), AWA (*red*), and VAMP2 (*blue*). The indole structure and numbering are for aid of peak assignment

free indole ring in lipid bilayers, as indicated by an up-field shift of the C5 at the indole ring due to the even strong hydrophobic interaction by the ^{1}H MAS NOESY experiments. Currently, we are conducting further research on peptide sequencing, composition, and insertion depth with basic and aromatic amino acid residues in different membrane lipids.

^{1}H MAS NOESY experiments have been widely used in SSNMR to study peptide–lipid interactions because of the fast axially rotation and segmental motion of membrane lipids in the liquid crystalline phase which average out efficiently the ^{1}H-^{1}H dipolar couplings, resulting in a high resolution ^{1}H spectrum of membrane lipids under the slow MAS frequencies [166], which leads to the rapid applications of the NOESY-type [167] of solution NMR methods to study peptide-membrane interactions in MAS SSNMR [168–170].

The insertion depth of aromatic residues in membrane lipids depends not only on the peptide sequencing and charge but also on the lipid composition, state, hydration level, and peptide/lipid molar ratio, so the insertion presents a complicated dynamic mechanism.

3.6 *Sensitivity Enhancement*

SSNMR sensitivity can be enhanced by manipulating the Boltzmann factor by such means as increasing the field strength, decreasing the system temperature, and transferring proton polarization to a rare nucleus. However, the maximum polarization transferred in the CP experiment is determined by γ_I/γ_S and the overall enhancement is not big enough for the free use of multidimensional correlation experiments on low-concentrated membrane proteins, especially for identifying the dynamic switch of GPCRs between an inactive (R) state and an active (R*) conformation [171]. Dynamic nuclear polarization (DNP) [172, 173], chemically induced nuclear polarization (CIDNP), or photo-chemically induced nuclear polarization (photo-CIDNP) [174–176], and spin-exchange optical pumping (SEOP) [177, 178] have been exploited to deliver much higher spin polarizations. For example, by transferring the polarization of electron spins to nuclei, the MAS-DNP has been successfully applied to study the intermediate states in the photocycle of bacteriorhodopsin, a 7TM light-driven proton pump. The enhanced sensitivity of DNP permitted for the first time the characterization of the retinal conformation in the K, L, and M states [35, 179, 180]. For detailed description of the DNP experiment, please refer to the chapter of "Dynamic Nuclear Polarization: New Methodology and Applications" in this book.

The NMR signal-to-noise ratio is directly proportional to the square root of the number of transitions and the total experimental time is mainly determined by the repetition time between two successive transitions, that is about five times of the spin longitudinal relaxation time T_1. Therefore, NMR sensitivity enhancement can be achieved by shortening the spin longitudinal relaxation time T_1. Ishii and colleagues have demonstrated that ^{1}H T_1 values of the two model proteins, lysozyme and ubiquitin, in microcrystals can be reduced to 60 ms by Cu–EDTA doping without major degradation in the resolution of their ^{13}C CP-MAS spectra [181]. They have further presented the paramagnetic relaxation-assisted condensed data collection (PACC) experiment on β-amyloid fibrils and ubiquitin to obtain the two-dimensional ^{13}C-^{13}C and ^{13}C-^{15}N SSNMR spectra in 1–2 d by reducing the ^{1}H T_1 down to 50–100 ms with carefully adjusted paramagnetic doping [182]. This sensitivity enhancement through shortening data acquisition time will make those sophisticated and time-consuming multidimensional correlation experiments on macrobiomolecules, such as membrane proteins become realistic.

3.7 Structure Determination Based on Chemical Shifts

The chemical shift of a nucleus directly reflects the electron density around it and also reflects the influence of the local environment on it, including the nature of the neighboring atoms and hydrogen-bonding, etc. Therefore, chemical shifts are valuable parameters for protein structure determination. Very recently, three chemical shift-based methods, Cheshire [183], CS-Rosetta [184], and CS23D [185] have been reported to determine the native state structures of proteins in solution up to 130 residues and to a resolution of 2 Å or better. These approaches have also been successfully demonstrated to determine the native structures of GB1, GB3, ubiquitin, and SH3 using SSNMR chemical shift data to a relatively high resolution [186, 187]. These methods may have the potential to determine the properties of even bigger proteins, for example membrane proteins, to a higher resolution with further improvement of the methods.

4 Perspective

Magic-angle spinning solid-state NMR has made tremendous progress, showing its capability of determining membrane protein structure, ligand binding, and protein dynamic conformation on a variety of time scales at atomic resolution. Further developments may be threefold:

1. Sample preparation. New attempts might focus on developing new labeling schemes to reduce further the spectral overlap and dipolar truncation, expressing a sufficient amount of functioning membrane proteins using a cell-free expression system, and a new approach to produce high ordered NMR samples in membrane native environment.
2. New NMR methods which should be suitable for fully hydrated membrane samples with better measurement accuracy and less dipolar truncation effects.

3. Integration of all newly developed techniques, including cell-free expression, specific labeling, new pulse sequences, molecular dynamic simulations and DNP.

With all the integrated techniques, we should be in a good position to solve a membrane protein structure based purely on the SSNMR constraints.

Acknowledgments Financial support by the National Natural Science foundation (grant 30970657), Shanghai Pujiang Program (grant 09PJ1404300), and the start funds by the East China Normal University (grants 79003A29, 79301207, and 79301411) are gratefully acknowledged. We thank Professor Jiandong Ding from Fudan University for the help with optical dynamic measurements of BR samples. Also, we thank our laboratory members, especially Xiaoyan Ding, for great assistance with this project.

References

1. Carpenter EP, Beis K, Cameron AD et al (2008) Overcoming the challenges of membrane protein crystallography. Curr Opin Struct Biol 18:581–586
2. Lodish H, Berk A, Kaiser CA et al (2007) Molecular cell biology, 6th edn. W.H. Freeman, New York
3. White SH (2009) Biophysical dissection of membrane proteins. Nature 459:344–346
4. Bill RM, Henderson PJF, Iwata S et al (2011) Overcoming barriers to membrane protein structure determination. Nat Biotechnol 29:335–340
5. Okada T, Le Trong I, Fox BA et al (2000) X-ray diffraction analysis of three-dimensional crystals of bovine rhodopsin obtained from mixed micelles. J Struct Biol 130:73–80
6. Salom D, Lodowski DT, Stenkamp RE et al (2006) Crystal structure of a photoactivated deprotonated intermediate of rhodopsin. Proc Natl Acad Sci USA 103:16123–16128
7. Lange A, Giller K, Hornig S et al (2006) Toxin-induced conformational changes in a potassium channel revealed by solid-state NMR. Nature 440:959–962
8. Ader C, Schneider R, Hornig S et al (2008) A structural link between inactivation and block of a K+ channel. Nat Struct Mol Biol 15:605–612
9. Korukottu J, Schneider R, Vijayan V et al (2008) High-resolution 3D structure determination of kaliotoxin by solid-state NMR spectroscopy. PLoS One 3:e2359
10. Zachariae U, Schneider R, Velisetty P et al (2008) The molecular mechanism of toxin-induced conformational changes in a potassium channel: relation to C-type inactivation. Structure 16:747–754
11. Ader C, Pongs O, Becker S et al (2010) Protein dynamics detected in a membrane-embedded potassium channel using two-dimensional solid-state NMR spectroscopy. Biochim Biophys Acta 1798:286–290
12. Crocker E, Eilers M, Ahuja S et al (2006) Location of Trp265 in metarhodopsin II: implications for the activation mechanism of the visual receptor rhodopsin. J Mol Biol 357:163–172
13. Struts AV, Salgado GFJ, Tanaka K et al (2007) Structural analysis and dynamics of retinal chromophore in dark and metal states of rhodopsin from H-2 NMR of aligned membranes. J Mol Biol 372:50–66
14. Concistrè M, Gansmüller A, McLean N et al (2008) Double-quantum 13C nuclear magnetic resonance of bathorhodopsin, the first photointermediate in mammalian vision. J Am Chem Soc 130:10490–10491
15. Mahalingam M, Martinez-Mayorga K, Brown MF et al (2008) Two protonation switches control rhodopsin activation in membranes. Proc Natl Acad Sci USA 105:17795–17800

16. Ahuja S, Crocker E, Eilers M et al (2009) Location of the retinal chromophore in the activated state of rhodopsin. J Biol Chem 284:10190–10201
17. Ahuja S, Eilers M, Hirshfeld A et al (2009) 6-s-cis Conformation and polar binding pocket of the retinal chromophore in the photoactivated state of rhodopsin. J Am Chem Soc 131:15160–15169
18. Ahuja S, Hornak V, Yan ECY et al (2009) Helix movement is coupled to displacement of the second extracellular loop in rhodopsin activation. Nat Struct Mol Biol 16:168–175
19. Ahuja S, Smith SO (2009) Multiple switches in G protein-coupled receptor activation. Trends Pharmacol Sci 30:494–502
20. Concistrè M, Gansmüller A, McLean N et al (2009) Light penetration and photoisomerization in rhodopsin studied by numerical simulations and double-quantum solid-state NMR spectroscopy. J Am Chem Soc 131:6133–6140
21. Gansmuller A, Concistre M, McLean N et al (2009) Towards an interpretation of 13C chemical shifts in bathorhodopsin, a functional intermediate of a G-protein coupled receptor. Biochim Biophys Acta 1788:1350–1357
22. Goncalves JA, South K, Ahuja S et al (2010) Highly conserved tyrosine stabilizes the active state of rhodopsin. Proc Natl Acad Sci USA 107:19861–19866
23. Hornak V, Ahuja S, Eilers M et al (2010) Light activation of rhodopsin: insights from molecular dynamics simulations guided by solid-state NMR distance restraints. J Mol Biol 396:510–527
24. Bedford L, Lowe J, Dick LR et al (2011) Ubiquitin-like protein conjugation and the ubiquitin–proteasome system as drug targets. Nat Rev Drug Discov 10:29–46
25. Struts AV, Salgado GFJ, Martínez-Mayorga K et al (2011) Retinal dynamics underlie its switch from inverse agonist to agonist during rhodopsin activation. Nat Struct Mol Biol 18:392–394
26. Ratnala VRP, Kiihne SR, Buda F et al (2007) Solid-state NMR evidence for a protonation switch in the binding pocket of the H1 receptor upon binding of the agonist histamine. J Am Chem Soc 129:867–872
27. Cady SD, Schmidt-Rohr K, Wang J et al (2010) Structure of the amantadine binding site of influenza M2 proton channels in lipid bilayers. Nature 463:689–692
28. Hu F, Luo W, Hong M (2010) Mechanisms of proton conduction and gating in influenza M2 proton channels from solid-state NMR. Science 330:505–508
29. Helmus JJ, Surewicz K, Nadaud PS et al (2008) Molecular conformation and dynamics of the Y145Stop variant of human prion protein. Proc Natl Acad Sci USA 105:6284–6289
30. Ahmed M, Davis J, Aucoin D et al (2010) Structural conversion of neurotoxic amyloid-[beta] 1-42 oligomers to fibrils. Nat Struct Mol Biol 17:561–567
31. Kawamura I, Ikeda Y, Sudo Y et al (2007) Participation of the surface structure of Pharaonis phoborhodopsin, ppR and its A149S and A149V mutants, consisting of the C-terminal alpha-helix and E-F loop, in the complex-formation with the cognate transducer pHtrII, as revealed by site-directed C-13 solid-state NMR. Photochem Photobiol 83:339–345
32. Saito H, Naito A (2007) NMR studies on fully hydrated membrane proteins, with emphasis on bacteriorhodopsin as a typical and prototype membrane protein. Biochim Biophys Acta 1768:3145–3161
33. Pfleger N, Lorch M, Woerner AC et al (2008) Characterisation of Schiff base and chromophore in green proteorhodopsin by solid-state NMR. J Biomol NMR 40:15–21
34. Varga K, Aslimovska L, Watts A (2008) Advances towards resonance assignments for uniformly - C-13, N-15 enriched bacteriorhodopsin at 18.8 T in purple membranes. J Biomol NMR 41:1–4
35. Bajaj VS, Mak-Jurkauskas ML, Belenky M et al (2009) Functional and shunt states of bacteriorhodopsin resolved by 250 GHz dynamic nuclear polarization-enhanced solid-state NMR. Proc Natl Acad Sci USA 106:9244–9249

36. Hellmich UA, Pfleger N, Glaubitz C (2009) F-19-MAS NMR on proteorhodopsin: enhanced protocol for site-specific labeling for general application to membrane proteins. Photochem Photobiol 85:535–539
37. Shi L, Lake EMR, Ahmed MAM et al (2009) Solid-state NMR study of proteorhodopsin in the lipid environment: secondary structure and dynamics. Biochim Biophys Acta 1788:2563–2574
38. Shi LC, Ahmed MAM, Zhang WR et al (2009) Three-dimensional solid-state NMR study of a seven-helical integral membrane proton pump-structural insights. J Mol Biol 386:1078–1093
39. Etzkorn M, Seidel K, Li L et al (2010) Complex formation and light activation in membrane-embedded sensory rhodopsin II as seen by solid-state NMR spectroscopy. Structure 18:293–300
40. Shi L, Kawamura I, Jung K-H et al (2011) Conformation of a seven-helical transmembrane photosensor in the lipid environment. Angew Chem Int Ed 50:1302–1305
41. Yang J, Aslimovska L, Glaubitz C (2011) Molecular dynamics of proteorhodopsin in lipid bilayers by solid-state NMR. J Am Chem Soc 133:4874–4881
42. Wasmer C, Lange A, Van Melckebeke H et al (2008) Amyloid fibrils of the HET-s(218–289) prion form a beta solenoid with a triangular hydrophobic core. Science 319:1523–1526
43. Wasmer C, Benkemoun L, Sabate R et al (2009) Solid-state NMR spectroscopy reveals that E. coli inclusion bodies of HET-s(218–289) are amyloids. Angew Chem Int Ed 48:4858–4860
44. Zamoon J, Nitu F, Karim C et al (2005) Mapping the interaction surface of a membrane protein: unveiling the conformational switch of phospholamban in calcium pump regulation. Proc Natl Acad Sci USA 102:4747–4752
45. Li Y, Berthold DA, Frericks HL et al (2007) Partial 13C and 15N chemical-shift assignments of the disulfide-bond-forming enzyme DsbB by 3D magic-angle spinning NMR spectroscopy. ChemBioChem 8:434–442
46. Seidel K, Andronesi OC, Krebs J et al (2008) Structural characterization of Ca2+−ATPase-bound phospholamban in lipid bilayers by solid-state nuclear magnetic resonance (NMR) spectroscopy. Biochemistry (Mosc) 47:4369–4376
47. Etzkorn M, Kneuper H, Dunnwald P et al (2008) Plasticity of the PAS domain and a potential role for signal transduction in the histidine kinase DcuS. Nat Struct Mol Biol 15:1031–1039
48. Lange V, Becker-Baldus J, Kunert B et al (2010) A MAS NMR study of the bacterial ABC transporter ArtMP. ChemBioChem 11:547–555
49. Hiller M, Higman VA, Jehle S et al (2008) [2,3-C-13]-Labeling of aromatic residues-getting a head start in the magic-angle-spinning NMR assignment of membrane proteins. J Am Chem Soc 130:408–409
50. Schneider R, Etzkorn M, Giller K et al (2010) The native conformation of the human VDAC1 N terminus. Angew Chem Int Ed 49:1882–1885
51. Baldus M (2006) Molecular interactions investigated by multi-dimensional solid-state NMR. Curr Opin Struct Biol 16:618–623
52. Baldus M (2006) Solid-state NMR spectroscopy: molecular structure and organization at the atomic level. Angew Chem Int Ed Engl 45:1186–1188
53. Hong M (2006) Oligomeric structure, dynamics, and orientation of membrane proteins from solid-state NMR. Structure 14:1731–1740
54. Hong M (2006) Solid-state NMR studies of the structure, dynamics, and assembly of β-sheet membrane peptides and α-helical membrane proteins with antibiotic activities. Acc Chem Res 39:176–183
55. Baldus M (2007) ICMRBS founder's medal 2006: biological solid-state NMR, methods and applications. J Biomol NMR 39:73–86
56. Hong M (2007) Structure, topology, and dynamics of membrane peptides and proteins from solid-state NMR spectroscopy. J Phys Chem B 111:10340–10351
57. McDermott A, Polenova T (2007) Solid state NMR: new tools for insight into enzyme function. Curr Opin Struct Biol 17:617–622

58. Naito A, Kawamura I (2007) Solid-state NMR as a method to reveal structure and membrane-interaction of amyloidogenic proteins and peptides. Biochim Biophys Acta 1768:1900–1912
59. Ramamoorthy A (2007) NMR structural studies on membrane proteins. Biochim Biophys Acta 1768:2947–2948
60. Watts A (2007) Solid state NMR for studying membrane proteins. In: Pifat-Mrzljak G (ed) Supramolecular structure and function 9. Springer, Netherlands
61. Bockmann A (2008) 3D protein structures by solid-state NMR spectroscopy: Ready for high resolution. Angew Chem Int Ed 47:6110–6113
62. McDermott A (2009) Structure and dynamics of membrane proteins by magic angle spinning solid-state NMR. Annu Rev Biophys 38:385–403
63. Ramamoorthy A (2009) Beyond NMR spectra of antimicrobial peptides: dynamical images at atomic resolution and functional insights. Solid State Nucl Magn Reson 35:201–207
64. Brown MF, Salgado GFJ, Struts AV (2010) Retinal dynamics during light activation of rhodopsin revealed by solid-state NMR spectroscopy. Biochim Biophys Acta 1798:177–193
65. Goncalves JA, Ahuja S, Erfani S et al (2010) Structure and function of G protein-coupled receptors using NMR spectroscopy. Prog Nucl Magn Reson Spectrosc 57:159–180
66. Renault M, Cukkemane A, Baldus M (2010) Solid-state NMR spectroscopy on complex biomolecules. Angew Chem Int Ed 49:8346–8357
67. Smith SO (2010) Structure and activation of the visual pigment rhodopsin. Annu Rev Biophys 39:309–328
68. Andrew ER, Bradbury A, Eades RG (1958) Nuclear magnetic resonance spectra from a crystal rotated at high speed. Nature 182:1659
69. Andrew ER, Bradbury A, Eades RG (1959) Removal of dipolar broadening of nuclear magnetic resonance spectra of solids by specimen rotation. Nature 183:1802–1803
70. Lowe IJ (1959) Free induction decays of rotating solids. Phys Rev Lett 2:285–287
71. Sinning G, Mehring M, Pines A (1976) Dynamics of spin decoupling in carbon-13–proton NMR. Chem Phys Lett 43:382–386
72. Mehring M (1983) Principles of high resolution NMR in solids. Springer, Berlin
73. Bennett AE, Rienstra CM, Auger M et al (1995) Heteronuclear decoupling in rotating solids. J Chem Phys 103:6951–6958
74. Eden M, Levitt MH (1999) Pulse sequence symmetries in the nuclear magnetic resonance of spinning solids: application to heteronuclear decoupling. J Chem Phys 111:1511–1519
75. Carravetta M, Eden M, Zhao X et al (2000) Symmetry principles for the design of radiofrequency pulse sequences in the nuclear magnetic resonance of rotating solids. Chem Phys Lett 321:205–215
76. Fung BM, Khitrin AK, Ermolaev K (2000) An improved broadband decoupling sequence for liquid crystals and solids. J Magn Reson 142:97–101
77. Detken A, Hardy EH, Ernst M et al (2002) Simple and efficient decoupling in magic-angle spinning solid-state NMR. The XiX scheme. Chem Phys Lett 356:298–304
78. Pines A, Gibby MG, Waugh JS (1972) Proton-enhanced nuclear induction spectroscopy. A method for high resolution NMR of dilute spins in solids. J Chem Phys 56:1776–1777
79. Pines A, Gibby MG, Waugh JS (1973) Proton-enhanced NMR of dilute spins in solids. J Chem Phys 59:569–590
80. Hartmann SR, Hahn EL (1962) Nuclear double resonance in the rotating frame. Phys Rev 128:2042–2053
81. Peersen OB, Wu X, Kustanovich I et al (1993) Variable amplitude cross polarization MAS NMR. J Magn Reson 104:334–339
82. Hediger S, Meier BH, Kurur ND et al (1994) NMR cross polarization by adiabatic passage through the Hartmann–Hahn condition (APHH). Chem Phys Lett 223:283–288
83. Metz G, Wu X, Smith SO (1994) Ramped-amplitude cross polarization in magic angle spinning NMR. J Magn Reson A 110:219–227
84. Hediger S, Meier BH, Ernst RR (1995) Adiabatic passage Hartmann-Hahn cross polarization in NMR under magic angle sample spinning. Chem Phys Lett 240:449–456

85. Griffin RG (1998) Dipolar recoupling in MAS spectra of biological solids. Nat Struct Biol 5:508–512
86. Gullion T (1998) Introduction to rotational-echo, double-resonance NMR. Concepts Magn Reson 10:277–289
87. Dusold S, Sebald A (2000) Dipolar recoupling under magic-angle spinning conditions. Annu Rep NMR Spectrosc 41:185–264
88. Levitt M (2002) Symmetry-based pulse sequences in magic-angle spinning solid-state NMR. In: Encyclopedia of nuclear magnetic resonance, vol 9. Wiley, Chichester, UK, pp 165–196
89. Helmus JJ, Nadaud PS, Hofer N et al (2008) Determination of methyl [sup 13]C–[sup 15]N dipolar couplings in peptides and proteins by three-dimensional and four-dimensional magic-angle spinning solid-state NMR spectroscopy. J Chem Phys 128:052314
90. Loquet A, Bardiaux B, Gardiennet C et al (2008) 3D structure determination of the Crh protein from highly ambiguous solid-state NMR restraints. J Am Chem Soc 130:3579–3589
91. Loquet A, Laage S, Gardiennet C et al (2008) Methyl proton contacts obtained using heteronuclear through-bond transfers in solid-state NMR spectroscopy. J Am Chem Soc 130:10625–10632
92. Manolikas T, Herrmann T, Meier BH (2008) Protein structure determination from 13C spin-diffusion solid-state NMR spectroscopy. J Am Chem Soc 130:3959–3966
93. Petkova AT, Ishii Y, Balbach JJ et al (2002) A structural model for Alzheimer's β-amyloid fibrils based on experimental constraints from solid state NMR. Proc Natl Acad Sci USA 99:16742–16747
94. Luca S, White JF, Sohal AK et al (2003) The conformation of neurotensin bound to its G protein-coupled receptor. Proc Natl Acad Sci USA 100:10706–10711
95. Carravetta M, Zhao X, Johannessen OG et al (2004) Protein-induced bonding perturbation of the rhodopsin chromophore detected by double-quantum solid-state NMR. J Am Chem Soc 126:3948–3953
96. Petkova AT, Leapman RD, Guo ZH et al (2005) Self-propagating, molecular-level polymorphism in Alzheimer's β-amyloid fibrils. Science 307:262–265
97. Mani R, Cady SD, Tang M et al (2006) Membrane-dependent oligomeric structure and pore formation of a b-hairpin antimicrobial peptide in lipid bilayers from solid-state NMR. Proc Natl Acad Sci USA 103:16242–16247
98. Tang M, Waring AJ, Lehrer RL et al (2008) Effects of guanidinium-phosphate hydrogen bonding on the membrane-bound structure and activity of an arginine-rich membrane peptide from solid-state NMR spectroscopy. Angew Chem Int Ed 47:3202–3205
99. Song C, Tapaneeyakorn S, Murphy AC et al (2009) Enantioselective syntheses of α-Fmoc-Pbf-[2-13C]-l-arginine and Fmoc-[1,3-13C2]-l-proline and incorporation into the neurotensin receptor 1 lLigand, NT8−13. J Org Chem 74:8980–8987
100. Castellani F, van Rossum B, Diehl A et al (2002) Structure of a protein determined by solid-state magic-angle-spinning NMR spectroscopy. Nature 420:98–102
101. Higman V, Flinders J, Hiller M et al (2009) Assigning large proteins in the solid state: a MAS NMR resonance assignment strategy using selectively and extensively ^{13}C-labelled proteins. J Biomol NMR 44:245–260
102. Eilers M, Ying WW, Reeves PJ et al (2002) Magic angle spinning nuclear magnetic resonance of isotopically labeled rhodopsin. Methods Enzymol 343:212–222
103. Patel AB, Crocker E, Eilers M et al (2004) Coupling of retinal isomerization to the activation of rhodopsin. Proc Natl Acad Sci USA 101:10048–10053
104. Patel AB, Crocker E, Reeves PJ et al (2005) Changes in interhelical hydrogen bonding upon rhodopsin activation. J Mol Biol 347:803–812
105. Etzkorn M, Martell S, Andronesi OC et al (2007) Secondary structure, dynamics, and topology of a seven-helix receptor in native membranes, studied by solid-state NMR spectroscopy. Angew Chem Int Ed Engl 46:459–462

106. Schneider R, Ader C, Lange A et al (2008) Solid-state NMR spectroscopy applied to a chimeric potassium channel in lipid bilayers. J Am Chem Soc 130:7427–7435
107. Hong M, Jakes K (1999) Selective and extensive C-13 labeling of a membrane protein for solid-state NMR investigations. J Biomol NMR 14:71–74
108. Igumenova TI, McDermott AE, Zilm KW et al (2004) Assignments of carbon NMR resonances for microcrystalline ubiquitin. J Am Chem Soc 126:6720–6727
109. Franks WT, Zhou DH, Wylie BJ et al (2005) Magic-angle spinning solid-state NMR spectroscopy of the β1 immunoglobulin binding domain of protein G (GB1): ^{15}N and ^{13}C chemical shift assignments and conformational analysis. J Am Chem Soc 127:12291–12305
110. Heise H, Seidel K, Etzkorn M et al (2005) 3D NMR spectroscopy for resonance assignment and structure elucidation of proteins under MAS: novel pulse schemes and sensitivity considerations. J Magn Reson 173:64–74
111. Franks W, Kloepper K, Wylie B et al (2007) Four-dimensional heteronuclear correlation experiments for chemical shift assignment of solid proteins. J Biomol NMR 39:107–131
112. Huang L, McDermott AE (2008) Partial site-specific assignment of a uniformly C-13, N-15 enriched membrane protein, light-harvesting complex 1 (LH1), by solid state NMR. Biochim Biophys Acta 1777:1098–1108
113. Li Y, Berthold DA, Gennis RB et al (2008) Chemical shift assignment of the transmembrane helices of DsbB, a 20-kDa integral membrane enzyme, by 3D magic-angle spinning NMR spectroscopy. Protein Sci 17:199–204
114. Shi L, Traaseth NJ, Verardi R et al (2009) A refinement protocol to determine structure, topology, and depth of insertion of membrane proteins using hybrid solution and solid-state NMR restraints. J Biomol NMR 44:195–205
115. Morcombe CR, Gaponenko V, Byrd RA et al (2005) 13C CPMAS spectroscopy of deuterated proteins: CP dynamics, line shapes, and T1 relaxation. J Am Chem Soc 127:397–404
116. Lesage A, Emsley L, Penin F et al (2006) Investigation of dipolar-mediated water-protein interactions in microcrystalline Crh by solid-state NMR spectroscopy. J Am Chem Soc 128:8246–8255
117. Akbey Ü, Lange S, Trent Franks W et al (2010) Optimum levels of exchangeable protons in perdeuterated proteins for proton detection in MAS solid-state NMR spectroscopy. J Biomol NMR 46:67–73
118. Zhou Donghua H, Shea John J, Nieuwkoop Andrew J et al (2007) Solid-state protein-structure determination with proton-detected triple-resonance 3D magic-angle-spinning NMR spectroscopy. Angew Chem Int Ed 46:8380–8383
119. Varga K, Ashmovska L, Parrot I et al (2007) NMR crystallography: the effect of deuteration on high resolution C-13 solid state NMR spectra of a 7-TM protein. Biochim Biophys Acta 1768:3029–3035
120. Tapaneeyakorn S, Goddard AD, Oates J et al (2011) Solution- and solid-state NMR studies of GPCRs and their ligands. Biochim Biophys Acta 1808:1462–1475
121. Cross TA, Sharma M, Yi M et al (2011) Influence of solubilizing environments on membrane protein structures. Trends Biochem Sci 36:117–125
122. Pauli J, van Rossum B, Forster H et al (2000) Sample optimization and identification of signal patterns of amino acid side chains in 2D RFDR spectra of the alpha-spectrin SH3 domain. J Magn Reson 143:411–416
123. Igumenova TI, Wand AJ, McDermott AE (2004) Assignment of the backbone resonances for microcrystalline ubiquitin. J Am Chem Soc 126:5323–5331
124. Kijac AZ, Li Y, Sligar SG et al (2007) Magic-angle spinning solid-state NMR spectroscopy of nanodisc-embedded human CYP3A4. Biochemistry (Mosc) 46:13696–13703
125. Lange A, Luca S, Baldus M (2002) Structural constraints from proton-mediated rare-spin correlation spectroscopy in rotating solids. J Am Chem Soc 124:9704–9705
126. Lange A, Becker S, Seidel K et al (2005) A concept for rapid protein-structure determination by solid-state NMR spectroscopy. Angew Chem Int Ed 44:2089–2092

127. Suter D, Ernst RR (1985) Spin diffusion in resolved solid-state NMR spectra. Phys Rev B 32:5608–5627
128. Lundstrom K (2007) Structural genomics and drug discovery. J Cell Mol Med 11:224–238
129. Palczewski K, Kumasaka T, Hori T et al (2000) Crystal structure of rhodopsin: a G protein-coupled receptor. Science 289:739–745
130. Cherezov V, Rosenbaum DM, Hanson MA et al (2007) High-resolution crystal structure of an engineered human beta(2)-adrenergic G protein-coupled receptor. Science 318:1258–1265
131. Rosenbaum DM, Cherezov V, Hanson MA et al (2007) GPCR engineering yields high-resolution structural insights into beta2-adrenergic receptor function. Science 318:1266–1273
132. Murakami M, Kouyama T (2008) Crystal structure of squid rhodopsin. Nature 453:363–367
133. Park JH, Scheerer P, Hofmann KP et al (2008) Crystal structure of the ligand-free G-protein-coupled receptor opsin. Nature 454:183–187
134. Scheerer P, Park JH, Hildebrand PW et al (2008) Crystal structure of opsin in its G-protein-interacting conformation. Nature 455:497–502
135. Warne T, Serrano-Vega MJ, Baker JG et al (2008) Structure of a beta1-adrenergic G-protein-coupled receptor. Nature 454:486–492
136. Chien EYT, Liu W, Zhao Q et al (2010) Structure of the human dopamine D3 receptor in complex with a D2/D3 selective antagonist. Science 330:1091–1095
137. Wu B, Chien EYT, Mol CD et al (2010) Structures of the CXCR4 chemokine GPCR with small-molecule and cyclic peptide antagonists. Science 330:1066–1071
138. Rasmussen SGF, Choi H-J, Fung JJ et al (2011) Structure of a nanobody-stabilized active state of the β_2 adrenoceptor. Nature 469:175–180
139. Rosenbaum DM, Zhang C, Lyons JA et al (2011) Structure and function of an irreversible agonist-β_2 adrenoceptor complex. Nature 469:236–240
140. Warne T, Moukhametzianov R, Baker JG et al (2011) The structural basis for agonist and partial agonist action on a β1-adrenergic receptor. Nature 469:241–244
141. Xu F, Wu H, Katritch V et al (2011) Structure of an agonist-bound human A2A adenosine receptor. Science 332:322–327
142. Creemers AFL, Bovee-Geurts PHM, DeGrip WJ et al (2004) Solid-state NMR analysis of ligand-receptor interactions reveals an induced misfit in the binding site of isorhodopsin. Biochemistry (Mosc) 43:16011–16018
143. van Gammeren AJ, Hulsbergen FB, Hollander JG et al (2004) Biosynthetic site-specific C-13 labeling of the light-harvesting 2 protein complex: a model for solid state NMR structure determination of transmembrane proteins. J Biomol NMR 30:267–274
144. Higman V, Rösner H, Ugolini R et al (2009) Probing the urea dependence of residual structure in denatured human α-lactalbumin. J Biomol NMR 45:121–131
145. Smith SO, Courtin J, de Groot H et al (1991) 13C magic-angle spinning NMR studies of bathorhodopsin, the primary photoproduct of rhodopsin. Biochemistry (Mosc) 30:7409–7415
146. Verdegem PJE, Bovee-Geurts PHM, De Grip WJ et al (1999) Retinylidene ligand structure in bovine rhodopsin, metarhodopsin-I, and 10-methylrhodopsin from internuclear distance measurements using C-13-labeling and 1-D rotational resonance MAS NMR. Biochemistry (Mosc) 38:11316–11324
147. Verhoeven MA, Creemers AFL, Bovee-Geurts PHM et al (2001) Ultra-high-field MAS NMR assay of a multispin labeled ligand bound to its G-protein receptor target in the natural membrane environment: electronic structure of the retinylidene chromophore in rhodopsin. Biochemistry (Mosc) 40:3282–3288
148. Creemers AFL, Kiihne S, Bovee-Geurts PHM et al (2002) 1H and 13C MAS NMR evidence for pronounced ligand-protein interactions involving the ionone ring of the retinylidene chromophore in rhodopsin. Proc Natl Acad Sci USA 99:9101–9106
149. Spooner PJR, Sharples JM, Verhoeven MA et al (2002) Relative orientation between the β-ionone ring and the polyene chain for the chromophore of rhodopsin in native membranes. Biochemistry (Mosc) 41:7549–7555

150. Spooner PJR, Sharples JM, Goodall SC et al (2003) Conformational similarities in the β-ionone ring region of the rhodopsin chromophore in its ground state and after photoactivation to the metarhodopsin-I intermediate. Biochemistry (Mosc) 42:13371–13378
151. Spooner PJR, Sharples JM, Goodall SC et al (2004) The ring of the rhodopsin chromophore in a hydrophobic activation switch within the binding pocket. J Mol Biol 343:719–730
152. Lopez JJ, Shukla AK, Reinhart C et al (2008) The structure of the neuropeptide bradykinin bound to the human G-protein coupled receptor bradykinin B2 as determined by solid-state NMR spectroscopy. Angew Chem Int Ed 47:1668–1671
153. Tikhonova IG, Costanzi S (2009) Unraveling the structure and function of G protein-coupled receptors through NMR spectroscopy. Curr Pharm Des 15:4003–4016
154. Takegoshi K, Nakamura S, Terao T (2001) ^{13}C-^{1}H dipolar-assisted rotational resonance in magic-angle spinning NMR. Chem Phys Lett 344:631–637
155. Ambrosio M, Zürn A, Lohse MJ (2011) Sensing G protein-coupled receptor activation. Neuropharmacology 60:45–51
156. Smock RG, Gierasch LM (2009) Sending signals dynamically. Science 324:198–203
157. Saito H, Kawase Y, Kira A et al (2007) Surface and dynamic structures of bacteriorhodopsin in a 2D crystal, a distorted or disrupted lattice, as revealed by site-directed solid-state 13C NMR. Photochem Photobiol 83:253–262
158. Brown MF, Martinez-Mayorga K, Nakanishi K et al (2009) Retinal conformation and dynamics in activation of rhodopsin illuminated by solid-state H-2 NMR spectroscopy. Photochem Photobiol 85:442–453
159. Hurley JH, Misra S (2000) Signaling and subcellular targeting by membrane-binding domains. Annu Rev Biophys Biomol Struct 29:49–79
160. Tossi A, Sandri L, Giangaspero A (2000) Amphipathic, alpha-helical antimicrobial peptides. Biopolymers 55:4–30
161. Tamm LK, Han X, Li YL et al (2002) Structure and function of membrane fusion peptides. Biopolymers 66:249–260
162. Richard JP, Melikov K, Vives E et al (2003) Cell-penetrating peptides - a reevaluation of the mechanism of cellular uptake. J Biol Chem 278:585–590
163. Zimmerberg J, McLaughlin S (2004) Membrane curvature: how BAR domains bend bilayers. Curr Biol 14:R250–R252
164. Cho WH, Stahelin RV (2005) Membrane-protein interactions in cell signaling and membrane trafficking. Annu Rev Biophys Biomol Struct 34:119–151
165. Hong H, Park S, FloresJimenez RH et al (2007) Role of aromatic side chains in the folding and thermodynamic stability of integral membrane proteins. J Am Chem Soc 129:8320–8327
166. Oldfield E, Bowers JL, Forbes J (1987) High-resolution proton and carbon-13 NMR of membranes: why sonicate? Biochemistry 26:6919–6923
167. Jeener J, Meier BH, Bachmann P et al (1979) Investigation of exchange processes by two-dimensional NMR spectroscopy. J Chem Phys 71:4546–4553
168. Davis JH, Auger M, Hodges RS (1995) High resolution ^{1}H nuclear magnetic resonance of a transmembrane peptide. Biophys J 69:1917–1932
169. Huster D, Kuhn K, Kadereit D et al (2001) H-1 high-resolution magic angle spinning NMR spectroscopy for the investigation of a Ras lipopeptide in a lipid membrane. Angew Chem Int Ed Engl 40:1056–1058
170. Zhang W, Crocker E, McLaughlin S et al (2003) Binding of peptides with basic and aromatic residues to bilayer membranes: phenylalanine in the myristoylated alanine-rich C kinase substrate effector domain penetrates into the hydrophobic core of the bilayer. J Biol Chem 278:21459–21466
171. Sprang SR (2011) Cell signalling: binding the receptor at both ends. Nature 469:172–173
172. Abragam A, Goldman M (1978) Principles of dynamic nuclear polarisation. Rep Prog Phys 41:395
173. Bajaj VS, Farrar CT, Hornstein MK et al (2003) Dynamic nuclear polarization at 9 T using a novel 250 GHz gyrotron microwave source. J Magn Reson 160:85–90

174. Zysmilich MG, McDermott A (1994) Photochemically induced dynamic nuclear polarization in the solid-state 15N spectra of reaction centers from photosynthetic bacteria Rhodobacter sphaeroides R-26. J Am Chem Soc 116:8362–8363
175. Bargon J (2006) The inter-relationship between triplet energies and spin chemistry. Photochem Photobiol Sci 5:970–978
176. Daviso E, Diller A, Alia A et al (2008) Photo-CIDNP MAS NMR beyond the T-1 limit by fast cycles of polarization extinction and polarization generation. J Magn Reson 190:43–51
177. Raftery D (2006) Xenon NMR spectroscopy. In: Webb GA (ed) Annual reports on NMR spectroscopy. Vol 57, Acadamic Press, Salt Lake City, USA, pp. 205–270
178. Ruset IC, Ketel S, Hersman FW (2006) Optical pumping system design for large production of hyperpolarized. Phys Rev Lett 96:053002
179. Bajaj VS, Hornstein MK, Kreischer KE et al (2007) 250 GHz CW gyrotron oscillator for dynamic nuclear polarization in biological solid state NMR. J Magn Reson 189:251–279
180. Mak-Jurkauskas ML, Bajaj VS, Hornstein MK et al (2008) Energy transformations early in the bacteriorhodopsin photocycle revealed by DNP-enhanced solid-state NMR. Proc Natl Acad Sci USA 105:883–888
181. Wickramasinghe NP, Kotecha M, Samoson A et al (2007) Sensitivity enhancement in 13C solid-state NMR of protein microcrystals by use of paramagnetic metal ions for optimizing 1H T1 relaxation. J Magn Reson 184:350–356
182. Wickramasinghe NP, Parthasarathy S, Jones CR et al (2009) Nanomole-scale protein solid-state NMR by breaking intrinsic 1HT1 boundaries. Nat Methods 6:215–218
183. Cavalli A, Salvatella X, Dobson CM et al (2007) Protein structure determination from NMR chemical shifts. Proc Natl Acad Sci USA 104:9615–9620
184. Shen Y, Bax A (2010) Prediction of Xaa-Pro peptide bond conformation from sequence and chemical shifts. J Biomol NMR 46:199–204
185. Wishart DS, Arndt D, Berjanskii M et al (2008) CS23D: a web server for rapid protein structure generation using NMR chemical shifts and sequence data. Nucleic Acids Res 36: W496–W502
186. Robustelli P, Cavalli A, Vendruscolo M (2008) Determination of protein structures in the solid state from NMR chemical shifts. Structure 16:1764–1769
187. Shen Y, Vernon R, Baker D et al (2009) De novo protein structure generation from incomplete chemical shift assignments. J Biomol NMR 43:63–78
188. Li J, Edwards PC, Burghammer M et al (2004) Structure of bovine rhodopsin in a trigonal crystal form. J Mol Biol 343:1409–1438
189. Levitt MH, Oas TG, Griffin RG (1988) Rotary resonance recoupling in heteronuclear spin pair systems. Isr J Chem 28:271–282
190. Oas TG, Griffin RG, Levitt MH (1988) Rotary resonance recoupling of dipolar interactions in solid-state nuclear magnetic resonance spectroscopy. J Chem Phys 89:692–695
191. Szeverenyi NM, Sullivan MJ, Maciel GE (1982) Observation of spin exchange by two-dimensional fourier transform 13C cross polarization-magic-angle spinning. J Magn Reson 47:462–475
192. Suter D, Ernst RR (1982) Spectral spin diffusion in the presence of an extraneous dipolar reservoir. Phys Rev B 25:6038–6041
193. Cornilescu G, Delaglio F, and Bax A (1999) Protein backbone angle restraints from searching a database for chemical shift and sequence homology. J Biomol NMR 13:289–302

Top Curr Chem (2012) 326: 215–242
DOI: 10.1007/128_2011_297

Published online: 5 November 2011

Dynamic Nuclear Polarization: New Methodology and Applications

Kong Hung Sze, Qinglin Wu, Ho Sum Tse, and Guang Zhu

Abstract One way to overcome the intrinsically low sensitivity of Nuclear Magnetic Resonance spectroscopy is to enhance the signal by dynamic nuclear polarization (DNP), where the polarization of high-gyromagnetic ratio (γ) electrons is transferred to the surrounding nuclei using microwave (MW) irradiation. Recent developments in DNP instrumentations and applications have shown that DNP is one of the most effective methods to increase the nuclear spin polarization in inorganic, organic, and biological materials. It is possible to obtain a solution of molecules containing hyperpolarized nuclei in combination with methods to dissolve rapidly the polarized solid sample. In this chapter, a brief introduction on a theoretical basis and some of new DNP applications in NMR spectroscopy as well as medical applications in Magnetic Resonance Imaging (MRI) are described.

Keywords DNP · NMR · MRIh

Contents

1 Introduction 216
2 Theories of Dynamic Nuclear Polarization 217
2.1 Continuous-Wave DNP Mechanism 218
2.2 Pulsed DNP 221
2.3 Polarizing Agents 224

K.H. Sze (✉) and H.S. Tse
Department of Microbiology, The University of Hong Kong, Pokfulam Road, Hong Kong, China
e-mail: khsze@hku.hk

Q. Wu and G. Zhu (✉)
Division of Life Science, The Hong Kong University of Science and Technology, Clear Water Bay, Kowloon, Hong Kong, China
e-mail: gzhu@ust.hk

3 Instrumentation 226
3.1 Microwave Sources 226
3.2 Microwave Waveguides 227
3.3 Probes Used for DNP Experiments 228
4 Applications 228
4.1 Applications to Small Molecules 228
4.2 Applications to Biomolecules 234
4.3 Magnetic Resonance Imaging 235
4.4 Multidimensional Time-Domain Experiment for DNP 237
5 Future Perspectives 239
References 239

1 Introduction

Nuclear Magnetic Resonance (NMR) is an important spectroscopic tool for the identification and structural characterization of molecules in chemistry and biochemistry. The most significant limitation of NMR spectroscopy compared to other spectroscopic techniques is its relatively low sensitivity, which thus often requires long measurement times or large amounts of sample, typically half a milliliter (mL) of sample at rather high concentrations. The origin of low sensitivity in NMR is well known to be due to the small magnetic moment of nuclear spins, which yields small Boltzmann polarizations and weak absorption signals. One of the ways to overcome this low signal-to-noise ratio is to enhance the signal by the creation of hyperpolarized transitions. This can be achieved by a process which was named dynamic nuclear polarization (DNP), where the polarization of high-gyromagnetic ratio (γ) electrons is transferred to the surrounding nuclei using microwave (MW) irradiation.

In 1953, Albert Overhasuser [1] first proposed that it was possible to transfer polarization to nuclei from electrons in metals by saturating the electron transition. This idea was not widely accepted until experimentally verified by Carver and Slichter with low field (3 mT) experiments performed on lithium metal and other materials with mobile electrons [2, 3]. This was soon expanded to be applied to solid dielectrics by Abragam and Proctor [4]. Extension of electron-nuclear and other high polarization transfer experiments involving noble gases, parahydrogen, semiconductors, or photosynthetic reaction centers [5–10] to contemporary solid-state and solution experiments is very appealing, since it could significantly enhance the sensitivity in a variety of NMR experiments. In particular, the theoretical enhancement for electronuclear polarization transfers is approximately (γ_e/γ_H), where the ratio is 660, making the gains in sensitivity ideally very large. Accordingly, during the 1960s and 1970s, there were extensive efforts to perform electron nuclear polarization transfer in liquids and solids. In 1980s, work has been carried out to couple DNP to magic-angle-spinning solid-state NMR (MAS-ssNMR).

This concept of nuclear polarization enhancement, originally proposed by Overhauser in 1953, was first experimentally demonstrated in metals and subsequently in liquids, which are two distinct types of systems with mobile electrons.

Thus, DNP is not a new area of scientific endeavor, but rather one undergoing a transition from low to high magnetic fields and frequencies. This chapter outlines the theoretical descriptions of DNP mechanisms followed by recent developments in DNP instrumentations and applications. DNP has proven to be one of the most effective methods to increase the nuclear spin polarization in inorganic, organic, and biological materials. In combination with methods to dissolve rapidly the polarized solid sample it is possible to obtain a solution of molecules containing hyperpolarized nuclei. This has enabled new applications in NMR spectroscopy as well as medical applications in Magnetic Resonance Imaging (MRI).

2 Theories of Dynamic Nuclear Polarization

In electron-nuclear based DNP experiments, it is required that the electron paramagnetic resonance (EPR) spectrum be irradiated with microwaves that drive the exchange of polarization between the electrons and the nuclear spins. In the case of liquids, these transitions are based on the Overhauser effect and, in solids, other mechanisms – the solid effect (SE), thermal mixing (TM), or the cross effect (CE) – dominate the polarization transfer process. Since DNP experiments require irradiation of the EPR spectrum, they were confined to relatively low magnetic fields because of the paucity of high frequency microwave sources. In particular, the microwave sources used in both the liquid and solid-state experiments were klystrons that operate at 640 GHz, constraining DNP-MAS experiments to 660 MHz ^{1}H frequencies. Thus, for DNP to be applicable to the higher fields employed in contemporary NMR experiments, new instrumental approaches to produce higher frequency microwaves are necessary.

A theoretical analysis of electron and nuclear system requires the quantum mechanical representation. In a DNP experiment, the general static Hamiltonian is written as

$$\begin{aligned} H &= H_E + H_N + H_{EN} \\ &= \omega_{0E} E_Z - \omega_{0N} N_Z + H_{EN}^{\mathrm{is}} + H_{EN}^{\mathrm{di}} \\ &= \omega_{0E} E_Z - \omega_{0N} N_Z + K_{\mathrm{SE}}(E_Z N_Z + E_Y N_Y + E_X N_X) + K_{\mathrm{PSE}} E_X N_Z \end{aligned} \tag{1}$$

where H_E and H_N are the Hamiltonians for electron and nuclear respectively. H_{EN} is the hyperfine coupling, which can be separated into the isotropic hyperfine interaction H_{EN}^{is} and the anisotropic dipolar coupling H_{EN}^{di} between electrons and nucleus. The H_{EN} can be further expressed in a form where the coefficients K_{SE} and K_{PSE} denote the secular and pseudosecular hyperfine interactions. ω_{0E} and ω_{0N} are the electron and nuclear Larmor frequencies.

DNP experiments can be classified based on the polarizing mechanisms. We will discuss the continuous-wave (CW) and time domain polarization mechanisms in the following section.

2.1 Continuous-Wave DNP Mechanism

Continuous wave DNP polarization-transfer mechanisms can be generally classified into four fields, namely the Overhauser Effect (OE), the Solid Effect (SE), the Cross Effect (CE), and Thermal Mixing (TM). They have all been successfully applied to both solid and liquid samples. It is reported that many DNP applications have been performed at low magnetic fields based on these different polarization effects.

2.1.1 Overhauser Effect

The OE DNP mechanism is a relaxation process which relies on the mixing of an electron and a nuclear spin. More specifically, in liquid cases, these relaxation processes are based on time-dependent dipolar and scalar interactions between electrons and nuclei. These interactions are governed by molecular rotational and translational motion in the dipolar case and by chemical exchange and fast relaxation in the scalar case [11]. In solid cases, the mobile electrons are required for the OE relaxation process. These electrons can be offered by the conduction band of metal or one-dimensional organic conductors. In addition, the condition $\omega_{0E}\tau < 1$ (τ, rotational correlation time of paramagnetic species) is required for OE to be established. The latter condition is difficult to satisfy with large value of ω_{0E}. In addition, ω_{0E} is in proportion with the strength of magnetic field. Therefore, the efficiency of the OE polarizing mechanism sharply decreases at high magnetic fields. However, the OE polarization is extremely useful for liquids, for it is the only practical mechanism for the direct application of liquids. We will give the outlines of the main OE features in a later section. A more detailed description can be found in the recent publication and review by Hofer et al. [12].

Figure 1 shows the energy level diagram of the transitions responsible for the OE. W_E and W_N are the rates for the EPR and NMR transitions, respectively. W^0 is the nuclear relaxation rate in the absence of electrons, while W_0 and W_2 are the rates for the zero and double quantum transitions, respectively. In OE polarization, the allowed transitions of EPR are saturated. The observed NMR signals are enhanced by changing the nuclear spin population, which is induced by the zero and double

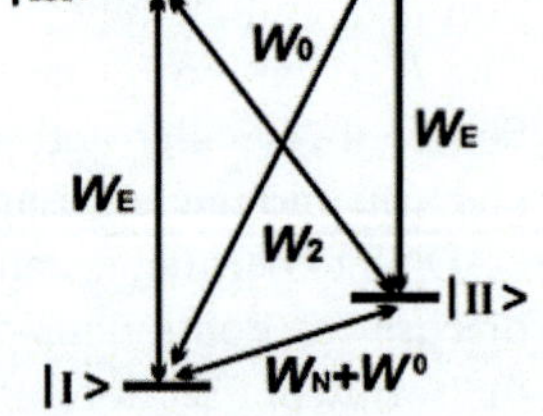

Fig. 1 Energy level diagrams and main transition rates for the Overhauser effect (OE) DNP mechanism

quantum transitions. The enhancement factor is defined as $\xi = N_Z/N_{eq}$, where N_Z and N_{eq} are the nuclei numbers in the polarizing and thermal balance states, respectively. After solving the rate equations for this system, we can obtain the following equation for the signal enhancement:

$$\xi = 1 - \rho\mu\eta\frac{\gamma_E}{\gamma_N} \tag{2}$$

with

$$\rho = \frac{W_2 - W_0}{W_0 + 2W_N + W_2}, \tag{3}$$

$$\mu = \frac{W_0 + 2W_N + W_2}{W_0 + 2W_N + W_2 + W^0}, \tag{4}$$

$$\eta = \frac{\langle E_0\rangle - \langle E_Z\rangle}{\langle E_0\rangle}, \tag{5}$$

where η corresponds to the electron transition saturation factor ranging from 0 (zero saturation, $\langle E_z\rangle = \langle E_0\rangle$) to 1 (complete saturation, $\langle E_z\rangle = 0$). From the above equations, the coupling parameter ρ can vary from -1.0 to 0.5, corresponding to pure scalar coupling and pure dipolar coupling, respectively. In liquid samples, the dipolar coupling is a major interaction leading to a maximum enhancement of 330. On the other hand, ρ also depends on the electron Larmor frequency ω_{0E} and the degree of molecular motion. This relation is shown in Fig. 2 of the Maly et al. review [13]. It is obvious to find that the OE process is inefficient at high magnetic fields. Many scientists (Armstrong and Han, Grucker et al., and Hofer et al.)

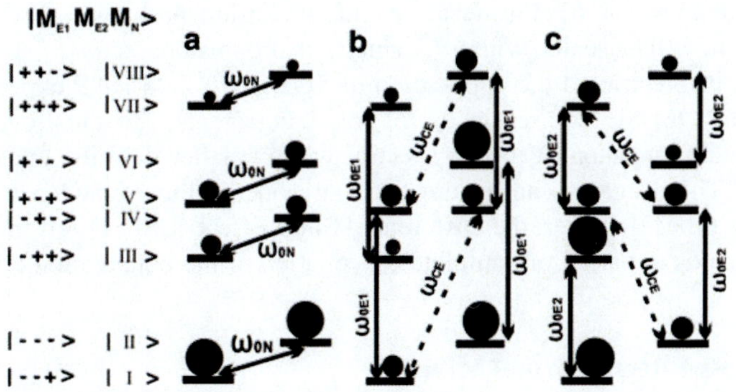

Fig. 2 (**a**) Population distribution at thermal equilibrium for a general three-spin system. (**b**) Saturation of the allowed EPR transitions for one of the dipolar coupled electrons (ω_{0E1}) leads to negative enhancement. (**c**) Saturation of the transition corresponding to the second electron (ω_{0E2}) leads to positive enhancement. M_{E1}, M_{E2}, and M_N are the spin states of electron 1, 2, and nucleus. Reproduced with permission from [13]

discussed this phenomenon and show the determination of the coupling parameter ρ in their articles [12, 14, 15]. The leakage factor μ describes the nuclear spin relaxation by electron spins. The value of μ can be assumed to lie between 0 and 1. The $\mu = 0$ represents no relaxation caused by the electron-nuclear coupling, while $\mu = 1$ means no other relaxation mechanisms ($W^0 = 0$). The saturation factor $\eta = 1$ when the electron transitions are completely saturated. This condition is important in the application of polarizing agents based on nitroxide radicals, although it depends on the number of hyperfine lines in the spectra. For concurrent time-dependent spin exchange, an additional factor σ can be introduced to describe the attenuate contribution on signal enhancement by scalar coupling. Generally, the value of σ can vary between 0 and 1, corresponding to no contribution and large contribution of the scalar coupling, respectively [13].

2.1.2 Solid Effect

The solid effect (SE) is a DNP mechanism which requires states mixing caused by the nonsecular component K_{PSE} of the hyperfine coupling [16]. The pseudosecular term K_{PSE} contains the form $E_Z N^+$ and $E_Z N^-$ (E: Electron spin operator, N: Nucleus spin operator), which leads to a mixing of the states of the system. In SE polarization, the new mix states are generated from the original states with a coefficient p, which can be calculated by the first order perturbation theory and is given by

$$p = -\frac{3\gamma_E \gamma_N}{4\omega_{0N} R^3} \sin\theta \cos\theta e^{-i\phi} \tag{6}$$

where R, θ and ϕ are the polar coordinates describing the electron-nuclear vector. The irradiation probability of zero quantum or double quantum transitions is proportional to $4p^2$ [16]. Furthermore, this transition probability and sensitivity enhancement will be scaled with ω_{0N}^{-2}, since p is proportional to ω_{0N}. Therefore, this correlation has restricted the application of SE in high field DNP experiments.

Practically, the SE requires the use of polarizing agents with a homogeneous EPR line width and an inhomogeneous spectral width smaller than the nuclear Larmor frequency. These agents can ensure that only one of the forbidden transitions is excited at a time. However, the Differential Solid Effect (DSE, [17]) simultaneously exists and leads to partial or complete cancellation of the polarization effect.

2.1.3 Cross Effect/Thermal Mixing

The Cross Effect mechanism is based on allowed transitions and involves the interaction of electron spin packets in an inhomogeneously broadened EPR line. A similar effect, found in a homogeneously broadened EPR line, is called thermal mixing. Wollan proposed a method to process the intermediate case of inhomogeneously and homogeneously broadening at low magnetic field [18]. Since both CE and TM have recently been used at high magnetic fields, the theoretical framework describing these effects needs to be fully established [19–23].

Griffin and co-workers [22] proposed that the CE is defined as a three-spin process, involving the interaction between two dipolar coupled electrons with EPR frequencies ω_{E01} and ω_{E02} that satisfy the relation

$$\omega_{0E2} - \omega_{E01} = \omega_{0N} \tag{7}$$

When the EPR lines are broadened by the inhomogeneous anisotropy, this is the dominant mechanism. The electrons are weakly coupled via electronic cross relaxation. Theoretically, the CE requires that the inhomogeneous breadth Ω is larger than the nuclear Larmor frequency ω_{0N} to make two effective EPR resonance frequencies. Meanwhile, the homogeneous width must satisfy $\Delta < \omega_{0N}$. The biradical chemical polarizing agents can improve the polarization effect of DNP systems to achieve this condition by a dipolar coupling between its two electrons. In contrast, the TM polarization mechanism contains homogeneously broadened EPR lines, where $\Delta > \omega_{0N}$ is satisfied. This condition requires a high concentration of polarizing agent at high magnetic fields, which will restrict the resolution in an MAS NMR experiment.

Figure 2 shows the thermal equilibrium spin population for a three-spin system. Generally there is no degeneracy present. When an appropriate polarizing agent is used, the energy levels $|IV>$ and $|V>$ or $|VI>$ and $|III>$ become degenerated (Fig. 2a). Irradiation of EPR transition and CE transitions leads to positive (Fig. 2b) or negative enhancement (Fig. 2c) of the nuclear polarization.

The signal enhancements of CE/TM are scaled with the magnetic field strength because the line width of the EPR spectra decreases along with the increasing strength of the magnetic field. Despite this drawback, the CE/TM polarizing mechanisms have been widely applied to polarize biological solids at high magnetic fields [24, 25].

The TM effect is less efficient compared to the CE. The TM can be described by a series of interacting systems: the electron Zeeman system (EZS), the electron dipolar system (EDS), and the nuclear Zeeman system (NZS) [19].

2.2 *Pulsed DNP*

S.R. Hartmann and E.L. Hahn reported that pulsed DNP experiments based on coherent polarization transfer such as the Hartmann–Hahn cross polarization (HHCP) are more efficient and do not show a field dependence of the polarization transfer with increasing magnetic field strengths.

HHCP is a fundamental technique widely used in ssNMR spectroscopy. With this mechanism, the polarization of one nucleus can be transferred to another nucleus [26, 27]. This transfer requires the rotating frame Hartmann–Hahn matching condition to be satisfied:

$$\omega_{1N} = \omega_{1E} \tag{8}$$

where ω_{1N} and ω_{1E} are the field strengths of the applied RF fields. In solid-state NMR experiments, the radio frequency fields are strong so their excitation profiles can cover the entire NMR spectrum and they can efficiently spin-lock both spin species. Thus, the Hartman–Hahn condition can be fulfilled.

For DNP experiments, however, there is large inhomogeneous broadening of high field EPR spectra and the spectral breadth Δ can usually exceed several hundred megahertz. In other words, the microwave field magnitude power is one order of magnitude smaller in comparison with the conditions found in ssNMR. The condition of matching electron-nuclear cross polarization (eNCP) cannot be satisfied by modification of the microwave field strength and the RF strength. Therefore, off resonance effects of EPR must be considered in this case.

From thermodynamic theory, the signal enhancement by CP transfer between I (high-γ) and S (low-γ) spins is given by

$$\xi = \left(\frac{\gamma_I}{\gamma_S}\right)\frac{1}{1 + N_S/N_I} \tag{9}$$

The efficient CP transfer between two spins is required so that the concentrations of I is much more than S. In this case, when $N_S \gg N_I$,

$$\xi \approx \frac{\gamma_I}{\gamma_S},$$

and the full ratio γ_I/γ_S can be transferred. NMR-CP experiments with the HHCP schemes are typically performed by transferring polarization from abundant high γ spins I to dilute low γ spins S. The condition of efficient CP transfer can be easily satisfied. However, in DNP experiments, the concentration of unpaired electrons from high γ polarizing agents is about four orders of magnitude lower than the concentration of low γ ^{1}H nucleus. Therefore, the enhancement factor $\xi < 1$ and only a small amount of electrons can be transferred to ^{1}H. The pumping of CP polarization is applied to improve the enhancement because the relaxation time of the electron is much shorter than that of the nucleus.

2.2.1 Integrated Solid Effect

Integrated Solid Effect (ISE) was first introduced by Henstra et al. [17]. It can overcome the low efficiency of SE when the homogeneous width is much larger than the nuclear Larmor frequency ($\Delta \gg \omega_{0N}$), in which the polarization effect could be canceled by simultaneous saturation of the forbidden transitions at $\omega_{0E} \pm \omega_{0N}$. The ISE can preserve the polarization in the case of $\Delta \gg \omega_{0N}$ by inverting a forbidden EPR transition prior to saturation of an allowed transition. This effect can be achieved by using a selective inversion pulse after the irradiation on resonance at $\omega_{0E} \pm \omega_{0N}$, or applying CW microwave irradiation at a fix frequency

ω_{0E} while sweeping the magnetic field through the entire EPR line. In the latter case, if the microwave power used for inversion of the electron spin polarization is sufficient, electron spin will go through an adiabatic fast process. This is considered as adiabatic-ISE. It is reported that ISE experiments were performed at low magnetic fields (9 GHz) using either electrons to polarize ^{29}Si nuclei or transferring the high polarization of a photoexcited triplet to surrounding protons [28–30]. A. Henstra et al. show an ISE build up curve for the adiabatic-ISE. In this experiment, the ^{29}Si signal in p-type Si at 1.2 K is detected and a factor of about 20 larger enhancements is obtained in an adiabatic-ISE rather than DSE [17].

There are certain limitations in the application of both the ISE and the adiabatic-ISE. The ISE requires high-power microwave pulses for the excitation of a large amount of electrons, while the adiabatic-ISE requires an adiabatic magnetic field sweep. On the one hand, the appropriate microwave sources are not commercially available at high microwave frequencies. On the other, the adiabatic magnetic sweep is difficult to perform at high magnetic fields. Although having these limitations, the ISE has a brilliant future in DNP applications because of the large sensitivity enhancement it can achieve.

2.2.2 NOVEL

Nuclear spin orientation via electron spin locking (NOVEL) is a DNP experiment based on coherent pulses. In NOVEL experiments, the electron magnetizations are locked the in electron rotating frame via electron spin-lock sequences. The Hartmann-Hahn condition (rotating frame or lab-frame) can be satisfied if the field strength of the spin-lock pulse meets the condition $\omega_{0E} = \omega_{0N}$; then the polarization can be transferred from the electrons to the nuclei. van den Heuvel et al. [31] show the nuclear polarization buildup curve for a NOVEL experiment in which the electron polarization from pentacene guest molecules in a photoexcited triplet state is transferred to ^{1}H of the naphthalene host crystal. An enhancement of 220 is achieved [31]. To date, NOVEL experiments have only been performed at 9 GHz microwave frequencies [30–33]. Much higher microwave field strengths will be required at higher magnetic fields.

2.2.3 Dressed-State Solid Effect

Dressed-state solid effect (DSSE) is a DNP mechanism based on simultaneous near resonant microwave and RF irradiations, which can be established even in the absence of nonsecular hyperfine coupling terms. This experiment was first introduced by Weis et al. [34]. The mechanism of DSSE is illustrated in Fig. 3. ω_{1E}^{A} and ω_{1E}^{B} are two effective fields which have unequal influence at high frequency DNP polarization, because the microwave power is currently limited. It indicates that the EPR transitions cannot be achieved with the same microwave field strength. The matching condition for DSSE polarization is given by

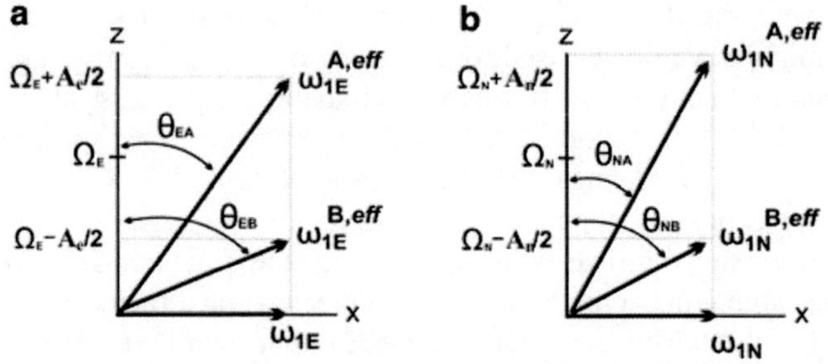

Fig. 3 Illustration of the electron (**a**) and nuclear (**b**) spin effective fields. The effective fields belonging to the EPR and NMR transitions are no longer equal. Ω_E and Ω_N are the resonant offsets of electrons and nuclei, while $A_e$and A_n are the hyperfine coupling constant of electrons and nuclei. $\omega_{1E}^{A,\text{eff}}$ and $\omega_{1E}^{B,\text{eff}}$ ($\omega_{1N}^{A,\text{eff}}$ and $\omega_{1N}^{B,\text{eff}}$) are the unequal effective microwave fields on different resonant offsets of electrons (nuclei). $\theta_{EA/EB/NA/NB}$ are the angles between the effective microwave fields on electrons/nuclei and the z dimension

$$\sqrt{(\delta_E + A_e/2)^2 + \omega_{1E}^2} + \sqrt{(\delta_E - A_e/2)^2 + \omega_{1E}^2}$$
$$= \pm\left[\sqrt{(\delta_N + A_n/2)^2 + \omega_{1N}^2} + \sqrt{(\delta_N - A_n/2)^2 + \omega_{1N}^2}\right] \quad (10)$$

where δ_E and δ_N are the resonant offsets of electron and nuclear spectra, respectively. ω_{1E} and ω_{1N} represent the RF field strengths and A_e and A_n the hyperfine coupling constants. For NMR-CP experiments, the matching condition is given by the equation

$$\delta_N \approx \pm\sqrt{\omega_{1E}^2 + (A_e/2)^2 - \omega_{1n}^2} \approx \pm\sqrt{\omega_{1E}^2 + (A_e/2)^2} \quad (11)$$

in the case that the microwave irradiation is on resonance and the hyperfine coupling is arbitrary. The expected polarizing region is determined by hyperfine couplings.

2.3 Polarizing Agents

Polarizing agents are widely used in DNP experiments. The choices of these agents are strongly dependent on a number of factors, such as the width of the EPR spectrum, the radical solubility and toxicity, the radical reactivity, the temperature dependence of relaxation times, etc. Either exogenous or endogenous radicals can be used as polarizing agents. Figure 4 shows five polarizing agents which

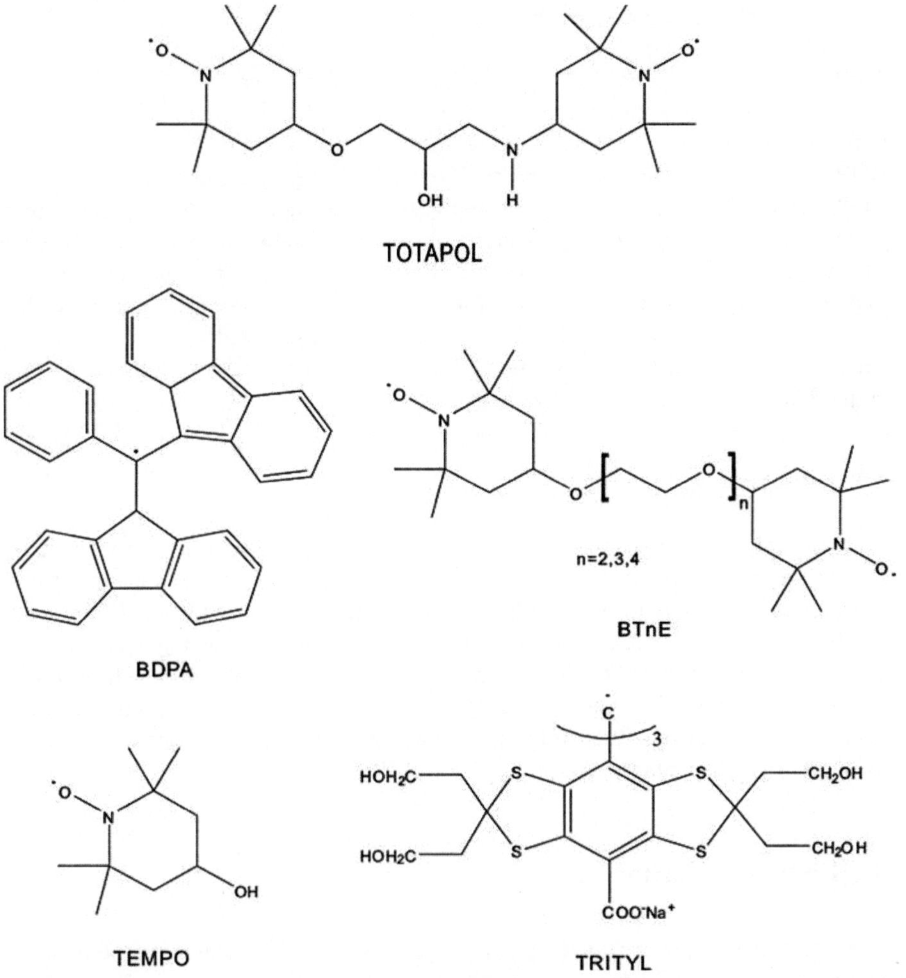

Fig. 4 Chemical structures of five polarizing agents which have been frequently used in high-field DNP experiments

have been frequently used in high-field DNP experiments. The trityl and BDPA are suitable for SE DNP experiments. The TEMPO is used to allow TM polarizing mechanism, while its biradical derivatives, BTnE and TOTAPOL is used to induce the CE polarizing mechanism. These similar nitroxide based radicals and biradicals are stable and soluble in a variety of solvents, allowing them to be applied in DNP experiments involving both liquid- and solid-states as well as biological samples.

2.3.1 Polarizing Agents for the SE

The SE polarizing mechanism requires that the zero and double quantum forbidden transitions are not simultaneously excited and that the enhancement effect is without positive and negative transitions cancellation. To date, two radicals, BDPA and trityl, can satisfy these requirements at high magnetic fields [35, 36]. BDPA is the first polarizing agent applied to investigate the SE mechanism in polystyrene matrix, which reveals an inhomogeneous linewidth of $\Omega \sim 20$ MHz at a 211-MHz machine [37–39]. However, BDPA is insoluble in aqueous solution, making it very difficult to use in biological science. In contrast, the trityl has good solubility in water and a narrow EPR spectrum width; it has therefore been successfully used in aqueous solution DNP applications with an Oxford HyperSense Oxford DNP polarizer [40].

2.3.2 Polarizing Agents for TM and CE

As mention above, under the situation that $\Omega > \omega_{0N}$, the TM and CE mechanisms are the main polarizing mechanisms in DNP experiments. The homogeneously broadened EPR leads to TM while the inhomogeneously broadened EPR results in CE. Nitroxide based radicals and biradicals are suitable polarizing agents for TM and CE because their EPR inhomogeneous line widths are about 600MHz at a ^{1}H Larmor frequency of 211-MHz. Furthermore, these chemical compounds, which have an EPR spectrum consisting of two narrow sharp lines separated by ω_{0N}, are ideal polarizing agents for CE polarization.

3 Instrumentation

The instrumentation for DNP experiments has enjoyed great developments in recent years. Generally speaking, on top of an existing NMR system, a microwave source, a waveguide, and a new probe are extra instrumentation required for performing DNP experiments. The microwave source is used to create the microwave and the waveguide will transmit the microwaves from the source to the NMR probe. The new probe is designed with the additional function that it can irradiate the sample with microwaves. In addition, there are some other requirements for special cases, such as performing MAS below 90 K, irradiating the EPR spectrum with maximum enhancement, and so on.

3.1 Microwave Sources

The power of microwave sources for DNP depends strongly on the nature of the experiment to be performed. The field strength of the microwave ω_{1E} has a relationship with the microwave power P_m and the quality factor Q_m according to the following equation:

$$\omega_{1E} \propto k\sqrt{Q_m \cdot P_m}. \tag{12}$$

Thus a high power microwave source is required for DNP experiments. Currently, the microwave sources can be classified into two different types: solid-state and vacuum electronic devices. Several excellent reviews of the introduction of – and outlook for – microwave source technology are available [41–43].

Among these, gyrotrons and cyclotron resonance masers are high-frequency vacuum electronic devices that have the ability to produce sufficient power in the frequency range of 140–590 GHz for electrons (200–900 MHz for proton nuclei). The electron cyclotron resonance maser can emit the coherent radiation near the relativistic electron cyclotron frequency. The irradiation frequency is given by

$$\omega_g = f\frac{eB_0}{M'm_e c} \tag{13}$$

and

$$M' = \frac{1}{\sqrt{1 - \upsilon^2/c^2}}, \tag{14}$$

where B_0 is the strength of the static magnetic field and e is the charge of the electron. f is the harmonics of the operation mode, m_e the mass of the electron, while M' is a relativistic mass factor given by (14). υ and c are the speeds of the electron and light, respectively. From (13), the coherent radiation frequency ω_g is directly proportional to the strength of the static magnetic field B_0 when the other experimental parameters in (13) remain constant.

3.2 *Microwave Waveguides*

The microwave waveguide is a device which is applied to transmit the microwaves from the source to the NMR probe. The efficient delivery of the microwave irradiation is highly necessary for microwave waveguides. When the microwave frequencies increase, however, the efficiency will decrease. Nowadays, with the help of corrugated waveguides, the loss is almost negligible and the efficiency of the microwave transmission is immensely increased. These devices have already been used in high frequency EPR and DNP applications [24, 44].

3.3 Probes Used for DNP Experiments

The probes for DNP experiment can be designed with or without a resonant structure. For liquid state DNP experiments, the TE_{102} rectangular EPR cavity is used for low microwave frequencies polarization, while the cylindrical TE_{011} cavity is used for high frequencies [45, 46]. In addition, an external RF coil with a slotted cavity is required in a high frequency electron nuclear double resonance (ENDOR) setup to avoid mounting the NMR RF coil inside the resonator. In 1965, Hyde proposed a double resonance probe for ENDOR experiments, which was widely applied later [47]. For MAS experiments, the microwave irradiation is applied vertically to the axis of the rotor in the present design, instead of parallel in the past. Due to space considerations, the microwaves are often placed between the turns of the NMR coil in high frequency situations. The variable tuning circuits are positioned outside the probe and wire transmission is used for the RF power irradiation. This design ensures that the probe also works under low temperature conditions.

4 Applications

NMR is an important spectroscopic tool for the identification and structural characterization of molecules in chemistry and biochemistry. The most significant limitation of NMR compared to other spectroscopic techniques is its intrinsically low sensitivity, which thus often requires long measurement times or large amounts of sample. DNP has been shown to be an effective method to increase the nuclear spin polarization in inorganic, organic, and biological materials; hence DNP has become an attractive technique to boost the sensitivity of NMR signals allowing NMR spectra of small amounts of samples from natural sources or from chemical synthesis to be readily acquired. Perhaps even more interestingly, the availability of the entire hyperpolarized NMR signal in one single scan allows the measurement of transient processes in real time, if applied together with a stopped-flow technique. In combination with the methods to dissolve rapidly the polarized solid sample, it is possible to obtain a solution of molecules containing hyperpolarized nuclei. This has enabled new applications in NMR spectroscopy as well as medical applications in MRI.

4.1 Applications to Small Molecules

By using incipient wetness impregnation of the solid samples with a solution of organic radical species such as TEMPO or TOTAPOL, surface enhanced NMR spectroscopy by DNP can be carried out to allow fast characterization of functionalized solid surfaces. Polarization is transferred from the radical protons of the solvent to the rare NMR active nuclei at natural abundance on the surface. Lesage et al. [48] have applied surface enhanced NMR spectroscopy by DNP to

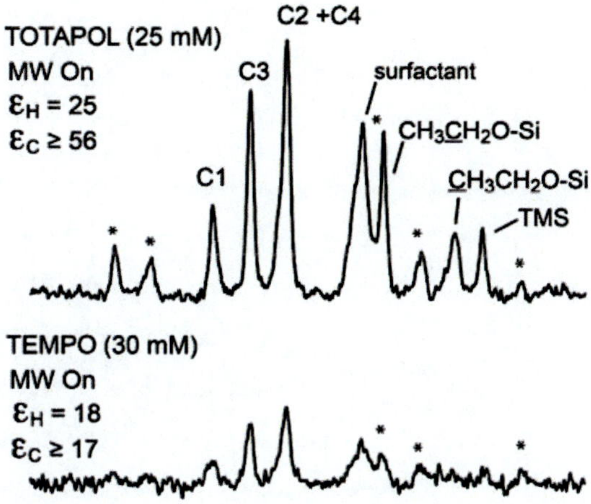

Fig. 5 ^{13}C CP MAS spectra of a silica framework with MW irradiation at 263 GHz to induce DNP. The figures compare the best enhancements observed using TEMPO and TOTAPOL radicals. It should be noted that significant DNP enhancement of the alkyl moiety of the surface ethoxy groups was also observed. Reproduced with permission from [48]

yield a 50-fold signal enhancement on ^{13}C signals for surface species covalently incorporated into a silica framework (Fig. 5). Lelli et al. [49] have applied the technique to study the distribution of surface bonding modes and interaction of functionalized silica materials by observing the ^{29}Si signals directly (Fig. 6). The remarkable gain in sensitivity and time provided by surface-enhanced silicon-29 DNP NMR spectroscopy is on the order of a factor of 400. This has also allowed the acquisition of the previously inconceivable two-dimensional correlation spectra, enabling more detailed characterizations of these functionalized surfaces [48, 49].

The unprecedented sensitivity observed in DNP-NMR means that sufficient signals can be obtained from a single scan, allowing reactions even far from equilibrium to be studied in real time. Of great importance, the DNP-NMR method is compatible with quantitative rate determination experiments where a single spin in the reactant is labeled in its spin state by a selective radio frequency pulse for subsequent tracking through the reaction, allowing the unambiguous identification of its position in the product molecule. Zeng et al. [50] have demonstrated the application of DNP-NMR to the Diels-Alder reaction of 1,4-diphenylbutadiene with 4-phenyl-1,2,4-triazole-3,5-dione, where reaction rates could be obtained accurately and reproducibly. In particular, the high chemical shift specificity afforded by high-resolution NMR permitted the simultaneous determination of reaction rates and mechanistic information in one experiment. Similarly, time-resolved DNP-enhanced NMR has been used to study the enzymatic reaction of the conversion of *N*-benzoyl-L-arginine-ethyl ester into the product *N*-benzoyl-L-arginine by trypsin enzyme (Fig. 7) [51].

The control of circulating drug concentrations by therapeutic drug monitoring (TDM) is vital for drugs when under- or overdosing may lead to loss of therapeutic efficacy or to adverse effects. Effective TDM depends on effective analytical platforms for the fast detection, identification, and quantification of circulating drugs with a narrow therapeutic range. As a result of the low concentrations of drugs and their metabolites in blood plasma, analytical tools are needed to provide high sensitivity and specificity. NMR spectroscopy is quantitative if care is taken

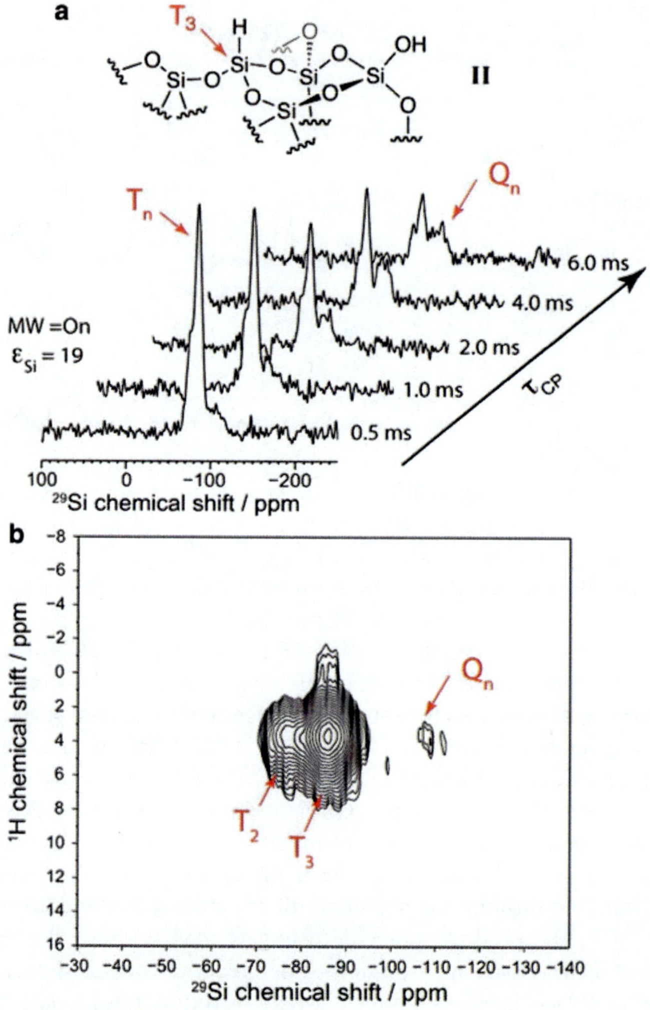

Fig. 6 (**a**) DNP-enhanced silicon-29 CPMAS spectra of compound II as a function of the CP mixing time τ_{CP}. (**b**) Contour plot of a two-dimensional ^{1}H-^{29}Si spectrum of II recorded with DNP. Reproduced with permission from [49]

that signal areas are not affected by different relaxation rates, and the method yields chemical information on the detected molecules in the absence of sample derivatization or sample separation. A principal limitation of conventional ^{1}H NMR is, however, the relatively low signal dispersion over a spectral window of approximately 10 ppm, which restricts the capability to resolve chemical compounds in the complex spectral backgrounds of biofluids. The use of ^{13}C NMR spectroscopy can address poor signal dispersion problems as the ^{13}C chemical shift dispersion is approximately 20-fold larger than that for ^{1}H. Spectral interference of the biofluid

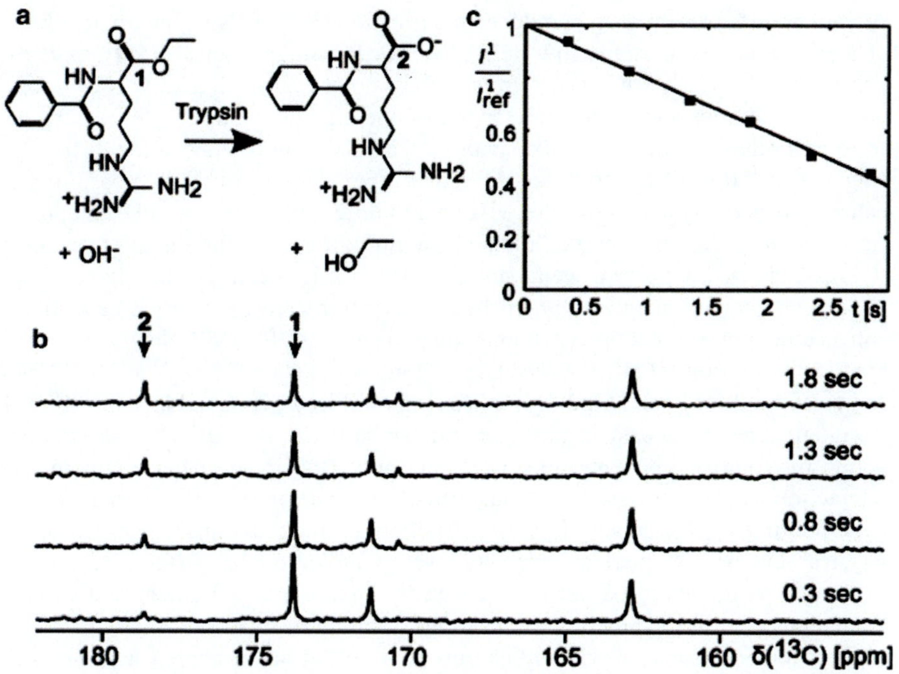

Fig. 7 Illustration of time-resolved DNP enhanced NMR. (**a**) Trypsin catalyzed conversion of BAEE into BA. (**b**) Time-course of the reaction, observed by ^{13}C NMR at natural abundance, using 3.3 mM DNP enhanced BAEE and 54 mM Trypsin. (**c**) Linear fit of normalized intensities from **b**, yielding the rate constant kcat = 12.1 s^{-1}. Reproduced with permission from [51]

background with ^{13}C-enriched drugs is further reduced because of the low ^{13}C abundance (1.1%) in the biofluid background. Therefore ^{13}C DNP-NMR in the form of the hyperpolarization-dissolution method should afford the sensitivity and spectral resolution for the direct detection and quantification of numerous isotopically labeled circulating drugs and their metabolites in single liquid-state NMR transients. Lerche et al. [52] has applied ^{13}C DNP-NMR to perform in vitro quantitative assay of drug and its metabolites in blood plasma. The lower limit of detection for the anti-epileptic drug ^{13}C-carbamazepine and its pharmacologically active metabolite ^{13}C-carbamazepine-10,11-epoxide is 0.08 μg/mL in rabbit blood plasma analyzed by single-scan ^{13}C DNP-NMR. Comparison of quantitative DNP-NMR data with an established analytical method (liquid chromatography–mass spectrometry) yields a Pearson correlation coefficient r of 0.99. These results indicate that ^{13}C DNP-NMR meets the sensitivity and accuracy requirements for quantitative analysis of the drug and its metabolites in blood plasma. It should be noted that all DNP-NMR determinations were performed without analyte derivatization or sample purification other than plasma protein precipitation. Therefore, quantitative DNP-NMR is an emerging methodology which requires little sample preparation and yields quantitative data with high sensitivity for therapeutic drugs.

Advances with para-hydrogen induced polarization (PHIP) also open up new fields of applications for portable low-field DNP NMR. Gloggler et al. [53] report the possibility of tracing drugs down to the micromolar regime. Selectively polarized nicotine quantities similar to those found in one cigarette and morphine extracted from an opium solution were detectable after polarization with para-hydrogen in single-scan ^{1}H NMR experiments. Moreover, they demonstrated the possibility to enhance selectively and detect the ^{1}H-signal of drug molecules with PHIP in proton rich standard solutions that would otherwise mask the ^{1}H NMR signal of the drug.

NMR spectroscopy is a well known and versatile technique for the study of molecular interactions, even when these interactions are relatively weak. Signal enhancement by several orders of magnitude through DNP NMR alleviates several practical limitations of NMR-based interaction studies (Fig. 8) [54]. ^{13}C DNP NMR can be applied for ligand binding studies at natural isotopic abundance of ^{13}C [54]. Resultant screens are easy to interpret and can be performed at ^{13}C concentrations below micromolar concentration. Of importance, the ligand-detected molecular interaction can be assessed and quantified with enzymatic assays that employ hyperpolarized substrates at varying enzyme inhibitor concentrations. Hence, the physical labeling of nuclear spins by hyperpolarization has offered fast novel in vitro experiments to be performed with low material requirement and without the tedious need for synthetic modifications of target or ligand.

This increase in sensitivity by ex situ DNP-NMR has triggered new research avenues, particularly concerning the in vivo monitoring of metabolism and disease by NMR spectroscopy. So far such gains have mainly materialized for experiments that focus on nonprotonated, low-gamma nuclei, targets favored by their relatively long T_1 relaxation times, which enable them to withstand the transfer from the

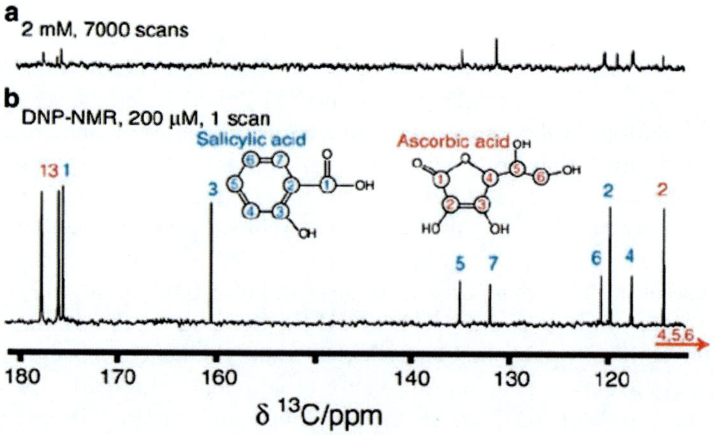

Fig. 8 Comparison of ^{13}C NMR spectra of a ligand mixture of salicylate and ascorbate recorded on a state-of the art NMR spectrometer (18.7 T) with cryogenic probe for signal detection (**a**) and recorded with DNP-NMR on a 9.4 T spectrometer (**b**). Ligands were tenfold concentrated for conventional NMR (**a**) to allow signal detection after 7,000 scans (20 h). Reproduced with permission from [54]

cryogenic hyperpolarizer to the reacting centers of interest. Recent studies have also shown that transferring this hyperpolarization to protons by indirectly detected methods could successfully give rise to ^{1}H NMR spectra of hyperpolarized compounds with a high sensitivity. Harris et al. [55] demonstrates that indirectly detected ^{1}H NMR spectroscopy can also be exploited as time-resolved hyperpolarized spectroscopy by merging with spatially encoded methods. This method can successfully deliver a series of hyperpolarized ^{1}H NMR spectra over a minutes-long timescale. The principles and opportunities presented by this approach were demonstrated by following the in vitro phosphorylation of choline by choline kinase, and by tracking acetylcholine's hydrolysis by acetylcholine esterase, an important enzyme partaking in synaptic transmission and neuronal degradation (Fig. 9) [55].

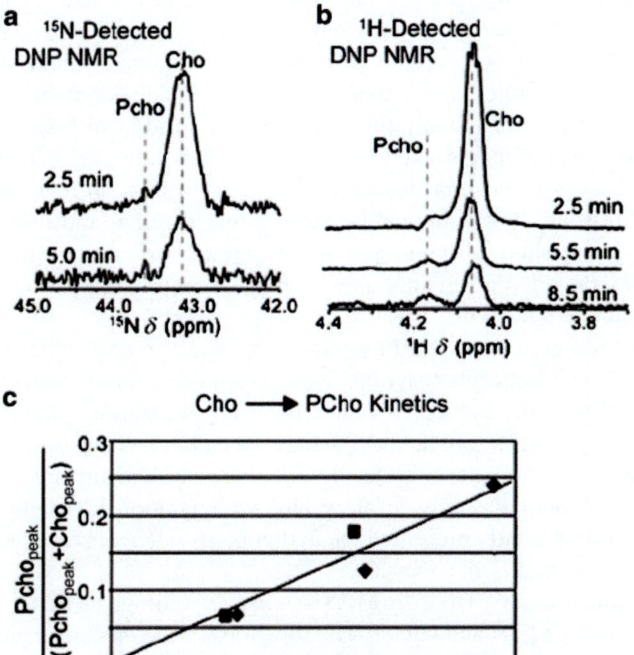

Fig. 9 NMR spectral changes revealed by a 5 mm solution of hyperpolarized choline upon undergoing phosphorylation by 0.5 units of choline kinase. (**a**) Emergence of the new phosphocholine resonance shown by directly detected single-pulse ^{15}N NMR spectroscopy experiments. (**b**) Emergence of the ^{1}H NMR resonance associated with the methylenes in the C2-position of phosphocholine, (**c**) Comparison between the expected enzyme kinetics of kinase with results afforded by the ^{15}N- (&) and ^{1}H-detected (^) hyperpolarized experiments, as derived from the relative peak ratios of the NMR peaks in (**a**) and (**b**). The *straight line* illustrates the best fit of the combined set of data points, and corresponds to an initial phosphorylation rate of 0.3 mM min^{-1} under these conditions. Reproduced with permission from [55]

4.2 Applications to Biomolecules

ssNMR is a powerful technique for the investigation of membrane-associated peptides and proteins as well as their interactions with lipids, and a variety of conceptually different approaches have been developed for their study. The technique is unique in allowing for the high-resolution investigation of liquid disordered lipid bilayers representing well the characteristics of natural membranes. While magic angle solid-state (MAS) NMR spectroscopy follows approaches that are related to those developed for solution NMR spectroscopy, the use of static uniaxially oriented samples results in angular constraints which also provide information for the detailed analysis of polypeptide structures. Solid-state NMR spectroscopy has already in the past provided valuable structural information for biomolecules and major advancement of the technique can be expected when it is possible to overcome the present lack of sensitivity where recording a two-dimensional spectrum can take many days, even for samples consisting of several milligrams of labeled polypeptide. Although high magnetic field spectrometers have already ameliorated the situation, the novel developments of DNP has promised to boost tremendously the sensitivity of solid-state NMR spectroscopy by about two orders of magnitude [56]. Notably, solid-state DNP-NMR has already been applied on oriented bilayer samples [57], as well as on membrane proteins [56, 58]. With such developments the application of three- and four-dimensional NMR experiments should become possible and much extend the ease of structural investigations on oriented membranes as well as improving the conformational details that can be obtained.

Cyanobacteria are widely used as model organisms of oxygenic photosynthesis due to being the simplest photosynthetic organisms containing both photosystem I and II. Photochemically induced dynamic nuclear polarization (photo-CIDNP) ^{13}C MAS NMR is a powerful tool in understanding the photosynthesis machinery down to atomic level. Combined with selective isotope enrichment this technique has now opened the door to study primary charge separation in whole living cells. Janssen et al. [59] recently presented the first photo-CIDNP observed in whole cells of the cyanobacterium Synechocystis.

In the applications of DNP to MAS spectra of biological systems, including studies of lysozyme [20] and bacteriorhodopsin [60], the enhancements have been smaller, with $\varepsilon = 40–50$, which has limited the application of DNP-NMR to biological samples so far. An exception is the amyloidogenic peptide GNNQQNY7 –13 which forms nanocrystals for which the proton T_1 time is long, yielding an ε of ~ 100 [25]. Almost a decade ago in studies of model systems it was observed that deuteration of the solvent resulted in significant increases in ε and subsequently many DNP experiments have employed ^{2}H-labeled solvents such as [D_6]DMSO or [D_8]glycerol/D_2O/H_2O in an approximately 6:3:1 ratio [21, 22, 61]. Akbey et al. [62] recently showed that perdeuteration of a protein has remarkable effects on the observed DNP enhancements. Superior DNP enhancements are obtained for perdeuterated SH3 samples of up to 3.9 and 18.5 times for ^{13}C CP-MAS, and ^{13}C

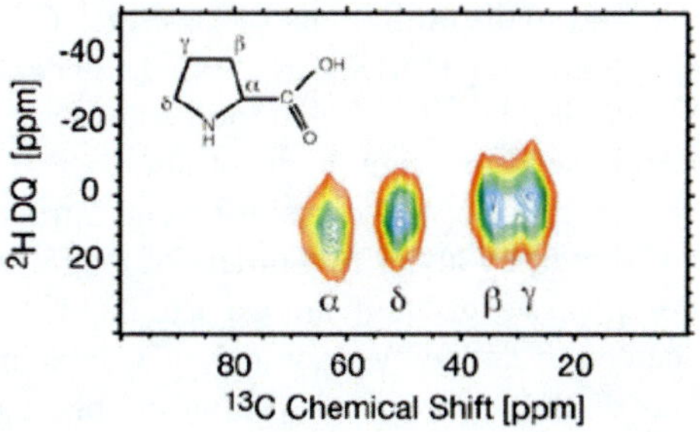

Fig. 10 Two-dimensional DNP-enhanced ^{2}H-DQ-^{13}C correlation spectrum of U-[$^{2}H_{7}$, $^{13}C_{5}$]-proline recorded at 90 K. Reproduced with permission from [63]

MAS experiments, respectively, compared to the same type of experiments in fully protonated SH3. The optimum perdeuteration is found to be approximately 50% which results in the maximum enhancement of $\varepsilon \sim 148$ in a ^{13}C MAS NMR spectrum obtained using a ZrO_2 rotor. Maly et al. [63] also showed that perdeuteration of biological macromolecules for magic angle spinning solid-state NMR spectroscopy can yield high- resolution ^{2}H-^{13}C correlation spectra. They demonstrated that the combination of sample deuteration and DNP yields resolved ^{2}H-^{13}C correlation spectra with a signal enhancement $\varepsilon \geq 700$ compared to a spectrum recorded with microwaves off and otherwise identical conditions. The DNP process is studied using several polarizing agents and the technique is applied to obtain ^{2}H-^{13}C correlation spectra of U-[^{2}H, ^{13}C] proline (Fig. 10). Therefore the use of perdeuterated proteins in DNP-MAS NMR will open new possibilities in the application of these techniques to difficult biological problems.

4.3 Magnetic Resonance Imaging

MRI is one of the most powerful in vivo diagnostic tools due to its superb spatial resolution and the amount of information it allows one to obtain, including anatomical and function information as well as flux, perfusion, and diffusion studies. However, compared with other imaging techniques, MRI's major drawback is its inherent low sensitivity, which has precluded its use as a routine tool in the clinic. Recently, great advances have been achieved on the application of MRI with the enormous sensitivity enhancement afforded by DNP (>10,000-fold) [64]. The strong signal enhancement attainable by DNP has allowed the detection of heteronuclei in MRI, yielding high quality images with very high signal to noise ratios in a few seconds. The applications of hyperpolarized probes to MRI range from vascular imaging, interventional imaging, and perfusion studies to the emerging and challenging field of molecular/metabolic imaging. In fact, the high signal intensities achievable by using hyperpolarized molecules make it possible to detect and image the metabolic products formed upon the administration of the hyperpolarized agent in vivo and allow early diagnosis and assessment of diseases in personalized medicine in a non-invasive manner [64].

The main limitations of current ^{13}C based DNP magnetic resonance spectroscopy imaging (MRSI) are, first, the very short lifetime of the hyperpolarization. For example, the ^{13}C-labeled molecules have a typical life time less than 40 s in vivo. And, second, high concentration of the hyperpolarized molecule is required because, following dissolution and intravenous injection, the ^{13}C-labeled molecule is diluted by about two orders of magnitude before it reaches the tissue of interest [64]. However, there are ways to tackle these limitations. For example, the lifetime can be extended by placing the ^{13}C label in a position in the molecule where there is no or only very weak proton coupling [65]. Accordingly, highly soluble small metabolites, such as [1-^{13}C] pyruvate, [5-^{13}C] glutamine, and [1,4-$^{13}C_2$] fumarate, have been designed and they have been effectively applied to obtain greatly improved MRSI results. Furthermore, these specifically ^{13}C labeled compounds have the advantage of significantly different ^{13}C chemical shifts from the bulk carbon signal and thus can be readily distinguished in the low resolution spectra measured in vivo.

A number of papers dealing with metabolic imaging by ^{13}C-hyperpolarized pyruvate have appeared in the literature since 2006 [66–76]. Pyruvate is a key molecule in major metabolic and catabolic pathways in mammalian cells, as it is converted to alanine, lactate or carbonate to varying extent depending on the status of the cells. Pyruvic acid naturally forms a glass and has been polarized by up to 40%. Golman et al. [66] reported one of the first examples of tumor imaging in vivo by this technique, which showed the maps of distribution of pyruvate, alanine, and lactate after a pyruvate injection into a rat with a P22 tumor. The tumor is clearly revealed by the highest NMR ^{13}C signal from lactate produced within 30 s [66]. Recently, hyperpolarized pyruvate has been utilized as a therapy response marker [72]. Within 24 h of cytotoxic drug treatment the pyruvate–lactate exchange was substantially reduced in lymphoma tumors [72]. It was shown that monitoring changes in hyperpolarized pyruvate-lactate exchange compares favorably with FEG-PET for detecting treatment response [64]. Thus measurement of pyruvate-lactate exchange may be an alternative to FEG-PET for imaging tumor treatment response in the clinic, in particular for tumors that are not FDG-avid, such as the prostate tumor.

Many pathological states are associated with changes in tissue acid–base balance, including inflammation and ischaemia [77–79]. For instance, most tumors have an acidic extracellular pH compared to normal tissue and this can be correlated with prognosis and response to treatment [80–82]. Despite the importance of pH and its relationship to disease, there is currently no clinical tool available to image the spatial distribution of pH in humans. In principle, tissue pH could be determined from ^{13}C-MRSI measurements of endogenous $H^{13}CO_3^-$ and $^{13}CO_2$ using the Henderson-Hasselbalch equation, if there was sufficient signal-to-noise:

$$\mathrm{pH} = \mathrm{p}K_\mathrm{a} + \log([\mathrm{H}^{13}\mathrm{CO}_3^-]/[^{13}\mathrm{CO}_2]) \quad (15)$$

One study has shown that the extracellular pH in tumors can be imaged in vivo from the ratio of the signal intensities of $H^{13}CO_3^-$ and $^{13}CO_2$ following intravenous

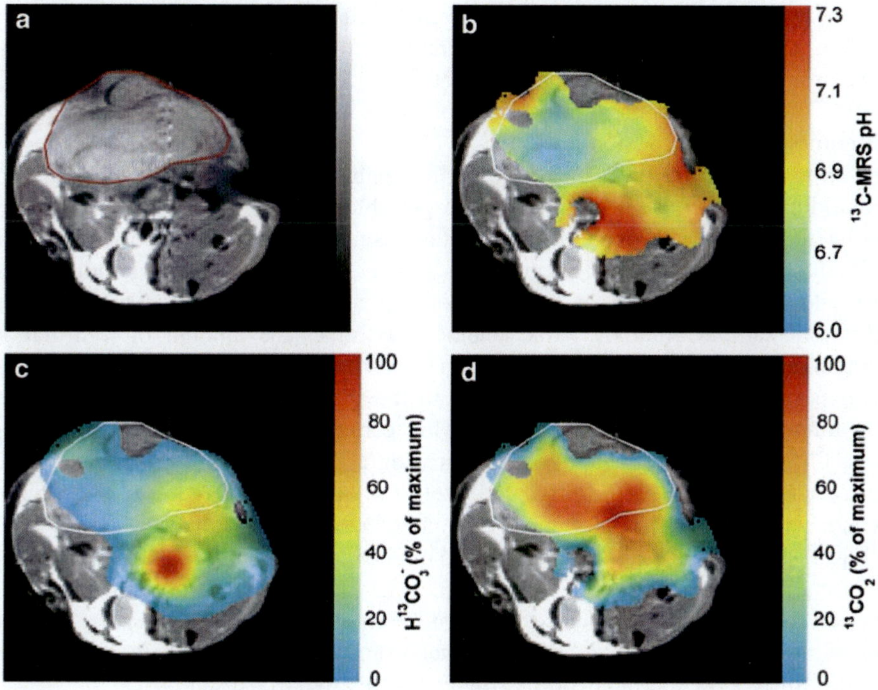

Fig. 11 A pH map of a transverse slice through a mouse implanted with a subcutaneous lymphoma tumor. (**a**) The *black and white* 1H MR image, obtained at 9.4 T, is shown with the tumor outlined in *red*. (**b**) The *false-color* ^{13}C MR pH image is shown separately superimposed over this; the tumor is acidic compared to the surrounding normal tissue. The spatial distribution of $H^{13}CO_3^-$ (**c**) and $^{13}CO_2$ (**d**) are also shown in the lower two images. Reproduced with permission from [64]

injection of hyperpolarized $H^{13}CO_3^-$ [83]. This demonstrated that the extracellular pH in a lymphoma tumor was more acidic than the surrounding tissue (Fig. 11).

4.4 Multidimensional Time-Domain Experiment for DNP

DNP has become an attractive technique to boost the sensitivity of NMR experiments. In the case of ex situ polarizations, two-dimensional (2D) spectra are limited by the short lifetime of the polarization after dissolution and sample transfer to a high field NMR magnet. This limitation can be overcome by various approaches. Ludwig et al. [84] showed how the use of ^{13}C-labeled acetyl tags can help to obtain 2D-HMQC spectra for many small molecules, owing to a nuclear Overhauser enhancement between ^{13}C spins originating from the long-lived carbonyl carbon, which extends the lifetimes of other ^{13}C spins with shorter longitudinal relaxation times. For two-dimensional spectroscopy, a pulse sequence using consecutive small flip angle excitations to read out transverse magnetization while preserving z-magnetization was employed. This procedure enables one to obtain 2D-HMQC spectra over a time frame of the lifetime of the polarization as exemplified for natural abundance nicotinamide. They have also combined this pulse sequence with non-uniform sampling to make the best use of the limited lifetime of the polarization [84].

Short acquisition time and single scan capability of gradient-assisted ultrafast multidimensional spectroscopy make it possible to record 2D spectra of highly polarized spin systems in the liquid state using DNP in conjunction with fast dissolution. Panek et al. [85] presented a slice selective experiment, suitable for back-to-back acquisition of two independent single-scan 2D experiments from different sample volumes. This scheme maximizes the amount of information obtainable from a sample that is prepolarized with a non-repeatable DNP technique. It is particularly suitable for samples with the short longitudinal relaxation times common to proton NMR spectroscopy. This technique is demonstrated by applying two filtered proton 2D COSY experiments on a DNP-polarized mixture of glutamine and glutamate to amplify selectively the correlation pattern of the protons connected to the beta and gamma carbons of either one of the two amino acids (Fig. 12). Particular emphasis was placed on the reproducibility of the experiments, especially the polarization enhancement. It has been demonstrated that the slice selective approach is a useful extension of the single-scan 2D spectroscopy tool set, enabling dynamic studies of DNP-enhanced samples or, alternatively, the acquisition of different experiments on a single one-time polarized sample.

In addition to the DNP methods described above, laser-polarized xenon can be used for the development of NMR and MRI techniques, and it is possible to enhance the nuclear spin polarization of gaseous xenon by four to five orders of magnitude. More details of this technique can be found in [86, 87].

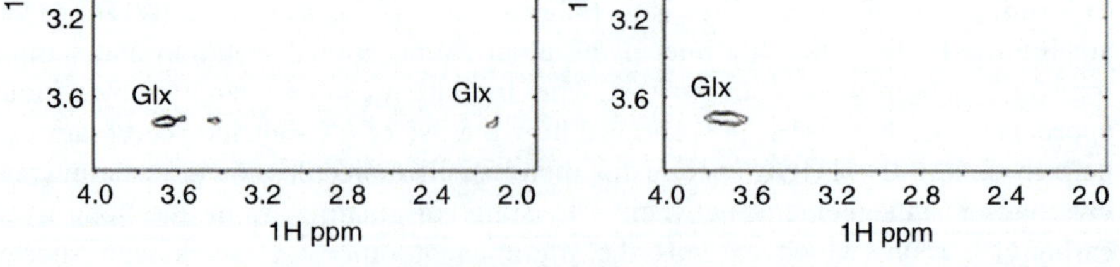

Fig. 12 DNP enhanced proton COSY spectra of glutamine and glutamate mixture acquired for two different evolution delays (Δ) in two separate slices. Reproduced with permission from [85]

5 Future Perspectives

Recent developments of DNP have shown great potential in solid-state and liquid-state NMR spectroscopy and in MRI. As promising as DNP is, the future development of hyperpolarization transition may require great improvement in the following aspects:

1. Development of new instrumentation in DNP. The new design of instruments for DNP should be able to couple conveniently with the current high field NMR spectrometers.
2. New DNP methods should be able to overcome the limitation of the very short lifetime of the hyperpolarization, and should be integrated into the currently available solid-state and liquid-state NMR experiments.
3. Development of new chemical agents to transfer effectively magnetization of electrons to nuclei for the problem orientated applications in NMR spectroscopy and MRI.

We expect a new wave of DNP research in the next few years, which may result in unprecedent breakthroughs of these techniques, consequently allowing us to detect diseases early in a non-invasive way and to study biomolecules in much lower concentrations.

Acknowledgements This research work was supported by grants from Hong Kong Research Grants Council (RGC664109, RGC775508, and RGC776509), AoE/M-06/08, TUYF10SC03, SEG/CUHK09, SEG_HKUST06, and HKUST Research Fund.

References

1. Overhauser AW (1953) Phys Rev 92:411–415
2. Carver TR, Slichter CP (1953) Phys Rev 92:212–213
3. Carver TR, Slichter CP (1956) Phys Rev 102:975–980
4. Abragam A, Proctor WG (1958) Phys Rev 109:1441–1448
5. Duckett SB, Sleigh CJ (1999) Prog Nucl Magn Reson Spectrosc 34:71–92
6. Bargon JNaJ (1997) Prog Nucl Magn Reson Spectrosc 31:293–315
7. Navon G, Song Y-Q, Rõõm T, Appelt S, Taylor RE, Pines A (1996) Science 271:1848–1851
8. Fitzgerald RJ, Sauer KL, Happer W (1998) Chem Phys Lett 284:87–92
9. Cherubini A, Payne GS, Leach MO, Bifone A (2003) Chem Phys Lett 371:640–644
10. Pavlovskaya GE, Cleveland ZI, Stupic KF, Basaraba RJ, Meersmann T (2005) Proc Natl Acad Sci USA 102:18275–18279
11. Loening NM, Rosay M, Weis V, Griffin RG (2002) J Am Chem Soc 124:8808–8809
12. Hofer P, Parigi G, Luchinat C, Carl P, Guthausen G, Reese M, Carlomagno T, Griesinger C, Bennati M (2008) J Am Chem Soc 130:3254–3255
13. Maly T, Debelouchina GT, Bajaj VS, Hu KN, Joo CG, Mak-Jurkauskas ML, Sirigiri JR, van der Wel PCA, Herzfeld J, Temkin RJ, Griffin RG (2008) J Chem Phys 128:052211
14. Armstrong BD, Han S (2007) J Chem Phys 127:104508
15. Grucker D, Guiberteau T, Eclancher B, Chambron J, Chiarelli R, Rassat A, Subra G, Gallez B (1995) J Magn Reson B 106:101–109

16. Abragam A, Goldman M (1978) Rep Prog Phys 41:395
17. Henstra A, Dirksen P, Wenckebach WT (1988) Phys Lett A 134:134–136
18. Wollan DS (1976) Phys Rev B 13:3671–3685
19. Farrar CT, Hall DA, Gerfen GJ, Inati SJ, Griffin RG (2001) J Chem Phys 114:4922–4933
20. Hall DA, Maus DC, Gerfen GJ, Inati SJ, Becerra LR, Dahlquist FW, Griffin RG (1997) Science 276:930–932
21. Hu KN, Yu HH, Swager TM, Griffin RG (2004) J Am Chem Soc 126:10844–10845
22. Hu KN, Bajaj VS, Rosay M, Griffin RG (2007) J Chem Phys 126:044512
23. Song CS, Hu KN, Joo CG, Swager TM, Griffin RG (2006) J Am Chem Soc 128:11385–11390
24. Bajaj VS, Hornstein MK, Kreischer KE, Sirigiri JR, Woskov PP, Mak-Jurkauskas ML, Herzfeld J, Temkin RJ, Griffin RG (2007) J Magn Reson 189:251–279
25. van der Wel PCA, Hu KN, Lewandowski J, Griffin RG (2006) J Am Chem Soc 128: 10840–10846
26. Pines A, Gibby MG, Waugh JS (1972) J Chem Phys 56:1776–1777
27. Conte P, Spaccini R, Piccolo A (2004) Prog Nucl Magn Reson Spectrosc 44:215–223
28. Henstra A, Lin T-S, Schmidt J, Wenckebach WT (1990) Chem Phys Lett 165:6
29. Schmidt J, van den Heuvel DJ, Henstra A, Lin T-S, Wenckebach WT (1992) Isr J Chem 32:165
30. Schmidt J, van den Heuvel DJ, Henstra A, Lin T-S, Wenckebach WT (1992) Pure Appl Chem 64:859
31. van den Heuvel DJ, Henstra A, Lin T-S, Schmidt J, Wenckebach WT (1992) Chem Phys Lett 188:194–200
32. Reynhardt CE, High GL (1998) J Chem Phys 109:4100
33. Takeda K, Takegoshi K, Terao T (2001) Chem Phys Lett 345:166–170
34. Weis V, Griffin RG (2006) Solid State Nucl Magn Reson 29:66–78
35. Ardenkjaer-Larsen JH, Laursen I, Leunbach I, Ehnholm G, Wistrand LG, Petersson JS, Golman K (1998) J Magn Reson 133:1–12
36. Reddy TJ, Iwama T, Halpern HJ, Rawal VH (2002) J Org Chem 67:4635–4639
37. Wind RA, Duijvestijn MJ, Vanderlugt C, Manenschijn A, Vriend J (1985) Prog Nucl Magn Reson Spectrosc 17:33–67
38. Becerra LR, Gerfen GJ, Temkin RJ, Singel DJ, Griffin RG (1993) Phys Rev Lett 71:3561–3564
39. Afeworki M, McKay RA, Schaefer J (1992) Macromolecules 25:4084
40. Ardenkjaer-Larsen JH, Fridlund B, Gram A, Hansson G, Hansson L, Lerche MH, Servin R, Thaning M, Golman K (2003) Proc Natl Acad Sci USA 100:10158–10163
41. Granatstein VL, Parker RK, Armstrong CM (1999) Proc IEEE 87:702
42. Abrams RH, Levush B, Mondelli AA, Parker RK (2001) IEEE Microw Mag 3:61–72
43. Parker RK, Abrams RH Jr, Danly BG, Levush B (2002) Microwave Theory and Technique 50:835–845
44. Woskov PP, Bajaj VS, Hornstein MK, Temkin RJ, Griffin RG (2005) IEEE Trans Microw Theory Tech 53:1863–1869
45. McCarney ER, Armstrong BD, Lingwood MD, Han S (2007) Proc Natl Acad Sci USA 104:1754–1759
46. Dorn HC, Wang J, Allen L, Sweeney D, Glass TE (1988) J Magn Reson 79:404–412
47. Hyde JS (1965) J Chem Phys 43:1806
48. Lesage A, Lelli M, Gajan D, Caporini MA, Vitzthum V, Mieville P, Alauzun J, Roussey A, Thieuleux C, Mehdi A, Bodenhausen G, Coperet C, Emsley L (2010) J Am Chem Soc 132:15459–15461
49. Lelli M, Gajan D, Lesage A, Caporini MA, Vitzthum V, Mieville P, Heroguel F, Rascon F, Roussey A, Thieuleux C, Boualleg M, Veyre L, Bodenhausen G, Coperet C, Emsley L (2011) J Am Chem Soc 133:2104–2107
50. Zeng HF, Lee Y, Hilty C (2010) Anal Chem 82:8897–8902
51. Hilty C, Bowen S (2010) Org Biomol Chem 8:3361–3365
52. Lerche MH, Meier S, Jensen PR, Hustvedt SO, Karlsson M, Duus JO, Ardenkjaer-Larsen JH (2011) NMR Biomed 24:96–103

53. Gloggler S, Emondts M, Colell J, Muller R, Blumich B, Appelt S (2011) Analyst 136: 1566–1568
54. Lerche MH, Meier S, Jensen PR, Baumann H, Petersen BO, Karlsson M, Duus JO, Ardenkjaer-Larsen JH (2010) J Magn Reson 203:52–56
55. Harris T, Giraudeau P, Frydman L (2011) Chem Eur J 17:697–703
56. Bajaj VS, Mak-Jurkauskas ML, Belenky M, Herzfeld J, Griffin RG (2010) J Magn Reson 202:9–13
57. Salnikov E, Rosay M, Pawsey S, Ouari O, Tordo P, Bechinger B (2010) J Am Chem Soc 132:5940
58. Vukojevic V, Bowen AM, Wilhelm K, Ming Y, Ce Z, Schleucher J, Hore PJ, Terenius L, Morozova-Roche LA (2010) Langmuir 26:14782–14787
59. Janssen GJ, Daviso E, van Son M, de Groot HJM, Alia A, Matysik J (2010) Photosynth Res 104:275–282
60. Bajaj VS, Mak-Jurkauskas ML, Belenky M, Herzfeld J, Griffin RG (2009) Proc Natl Acad Sci USA 106:9244–9249
61. Hu KN, Song C, Yu HH, Swager TM, Griffin RG (2008) J Chem Phys 128:052302
62. Akbey U, Franks WT, Linden A, Lange S, Griffin RG, van Rossum BJ, Oschkinat H (2010) Angew Chem Int Ed Engl 49:7803–7806
63. Maly T, Andreas LB, Smith AA, Griffin RG (2010) Phys Chem Chem Phys 12:5872–5878
64. Gallagher FA, Kettunen MI, Brindle KM (2009) Prog Nucl Magn Reson Spectrosc 55:285–295
65. Warren WS, Jenista E, Branca RT, Chen X (2009) Science 323:1711–1714
66. Golman K, Zandt RI, Lerche M, Pehrson R, Ardenkjaer-Larsen JH (2006) Cancer Res 66:10855–10860
67. Merritt ME, Harrison C, Storey C, Jeffrey FM, Sherry AD, Malloy CR (2007) Proc Natl Acad Sci USA 104:19773–19777
68. Chen AP, Albers MJ, Cunningham CH, Kohler SJ, Yen YF, Hurd RE, Tropp J, Bok R, Pauly JM, Nelson SJ, Kurhanewicz J, Vigneron DB (2007) Magn Reson Med 58:1099–1106
69. Kohler SJ, Yen Y, Wolber J, Chen AP, Albers MJ, Bok R, Zhang V, Tropp J, Nelson S, Vigneron DB, Kurhanewicz J, Hurd RE (2007) Magn Reson Med 58:65–69
70. Albers MJ, Bok R, Chen AP, Cunningham CH, Zierhut ML, Zhang VY, Kohler SJ, Tropp J, Hurd RE, Yen YF, Nelson SJ, Vigneron DB, Kurhanewicz J (2008) Cancer Res 68:8607–8615
71. Nelson SJ, Vigneron D, Kurhanewicz J, Chen A, Bok R, Hurd R (2008) Appl Magn Reson 34:533–544
72. Day SE, Kettunen MI, Gallagher FA, Hu DE, Lerche M, Wolber J, Golman K, Ardenkjaer-Larsen JH, Brindle KM (2007) Nat Med 13:1382–1387
73. Merritt ME, Harrison C, Storey C, Sherry AD, Malloy CR (2008) Magn Reson Med 60:1029–1036
74. Schroeder MA, Cochlin LE, Heather LC, Clarke K, Radda GK, Tyler DJ (2008) Proc Natl Acad Sci USA 105:12051–12056
75. Golman K, Petersson JS, Magnusson P, Johansson E, Akeson P, Chai CM, Hansson G, Mansson S (2008) Magn Reson Med 59:1005–1013
76. Schroeder MA, Atherton HJ, Cochlin LE, Clarke K, Radda GK, Tyler DJ (2009) Magn Reson Med 61:1007–1014
77. Adrogue HJ, Madias NE (1998) N Engl J Med 338:26–34
78. Grinstein S, Swallow CJ, Rotstein OD (1991) Clin Biochem 24:241–247
79. Hohn-Berlage M, Okada Y, Kloiber O, Hossmann KA (1989) NMR Biomed 2:240–245
80. Gillies RJ, Raghunand N, Garcia-Martin ML, Gatenby RA (2004) IEEE Eng Med Biol Mag 23:57–64
81. Raghunand N, He X, van Sluis R, Mahoney B, Baggett B, Taylor CW, Paine-Murrieta G, Roe D, Bhujwalla ZM, Gillies RJ (1999) Br J Cancer 80:1005–1011
82. Raghunand N, Mahoney B, van Sluis R, Baggett B, Gillies RJ (2001) Neoplasia (New York, NY) 3:227–235

83. Gallagher FA, Kettunen MI, Day SE, Hu DE, Ardenkjaer-Larsen JH, in't Zandt R, Jensen PR, Karlsson M, Golman K, Lerche MH, Brindle KM (2008) Nature 453:940–U973
84. Ludwig C, Marin-Montesinos I, Saunders MG, Emwas AH, Pikramenou Z, Hammond SP, Gunther UL (2010) Phys Chem Chem Phys 12:5868–5871
85. Panek R, Granwehr J, Leggett J, Kockenberger W (2010) Phys Chem Chem Phys 12:5771–5778
86. Zhou X, Graziani D, Pines A (2009) Proc Natl Acad Sci USA 106:16903–16906
87. Cherubini A, Bifone A (2003) Prog Nucl Magn Reson Spectrosc 42:1–30

Index

A

ABC transporters, 189
Acetylcholinesterase, 22, 233
Activation, structural changes, 199
ADP ribosylation factor (Arf), 61
Alignment, 52
  medium, 47
Amino acids, adjacent, 129
Amphipols, 140, 143
Amyloidogenic peptide GNNQQNY7, 234
Anisotropic electron g-tensor, 42
Antibiotics, 20
AutoDockFilter (ADF), 20
Azurin, 81

B

Bacteriorhodopsin, 127, 192, 234
BDPA, 225
Bicelles, 40, 52, 62, 140, 192
Bloch-McConnell equation, 112
Bound state, 61
Bradykinin (BK), 199

C

Cadmium, ^{113}Cd, 77
Carr–Purcell–Meiboom–Gill (CPMG), 101, 103, 109
^{13}C-carbamazepine, 231
CC-chemokine receptor 5 (CCR5), 143
^{13}C direct detection, 69, 87
Cd_7MT10, 78
Cd_7MT-nc, 74
Cell-free expression, 127
Chalcones, 20
Chemically induced nuclear polarization (CIDNP), 203
Chemical-Shift-ROSETTA, 76, 204
Chemical shifts, 35, 76
  anisotropy (CSA), 153
  perturbations (CSPs), 6, 37, 82
Chemokine CXCL12, 23
Cheshire, 204
Choline kinase, phosphorylation, 233
Collagen gel, 52, 54
Combinatorial isotope labeling, 129
Comparison of Protein Active-Site Structures (CPASS), 19
Conservative mutation, 55
Continuous-wave decoupling, 190
Continuous-wave DNP, 218
Critical micelle concentration (cmc), 129
Cross-correlation, 102
Cross effect/thermal mixing, 220
Cross-polarization, 190
Cryoprobe, 146
CS23D, 204

D

DARR, 200
Detergents, 123, 131
Deuteration, 148
Diacylglycerol kinase (DAGK), 126
Dihexanoylphosphatidylcholine (DHPC), 52, 133
Dimyristoylphosphatidylcholine (DMPC), 52
1,2-Di-O-hexyl-sn-glycero-3-phosphocholine (6-O-PC), 141
Direct Interpretation of Dipolar Couplings (DIDC), 56

Dispersion, 99
Disulfide bond isomerase (DsbB), 126
DNA based media, 53
DNA nanotubes, 53
DNA recognition, 74
DNP, pulsed, 221
Docking, 8
Dodecylphosphocholine (DPC), 130, 133
Dressed-state solid effect (DSSE), DNP, 223
Drug discovery, 1
Dynamic nuclear polarization (DNP), 203, 215, 217
Dynamics, 47, 99

E
Electron nuclear double resonance (ENDOR), 228
Empirical scoring functions, 11
Ensemble, 47
Ensemble minimization, 59
Expression systems, 126

F
FAST-NMR, 1, 18
Field cycling, 108
Flavonoids, 20
Force field-based scoring functions, 11

G
Genetic algorithms, 9
Gfizf35-DNA complex (PDB: 2kmk), 75
Global folds, 123
G-matrix Fourier transform (GFT) NMR, 89
G-protein coupled receptor (GPCR), 61, 127, 196
G-proteins, transducin (Gt), 62
Gramicidin A (gA), 144
G-tetrad DNA, 40

H
HADDOCK, 21
Hadmard NMR, 89
Half-filtered NOESY (intermolecular NOEs), 39
Hartmann–Hahn cross polarization (HHCP), 221
HET-s (218–289) fibril, 197
High-power heteronuclear decoupling, 190
High-throughput screening (HTS), 1
Histidine, 55, 73, 83
Histidine kinases, 189
HypA, 72, 76, 78

I
In silico screening, 1
Integrated solid effect (ISE), 222
Isotope labeling, 123, 148

K
KcsA potassium channel, 145, 195
Knowledge-based scoring functions, 11

L
Labeling, MAS solid-state NMR, 191
Lauryldimethylamine-oxide (LDAO), 137
Lead identification, 23
Ligand affinity screens, 1, 17
Ligand-based NMR screens 4
Ligand conformation, 187
Ligands, conformation/binding, 196
LIGDOCK, 21
Lipids, 19
Lupin, Ap4A hydrolase, 23
Lysolipids, 133
Lyso-myristoyl phosphatidylglycerol (LMPG), 133
Lyso-palmitoyl phosphatidylglycerol (LPPG), 137
Lysozyme, 234

M
Magic-angle spinning (MAS) solid-state NMR, 187, 189
Magnetic resonance imaging (MRI), 215, 235
Membrane mimetic systems, solution NMR, 129
Membrane proteins, activation, 196
  solid-state NMR, 187
Metal coordination, 69, 76
  identification, 76
Metalloproteins, 69
  paramagnetic, 69, 84
Metallothionines, 73, 78
α-Methyl mannose, 40
Methyl protonation, 149, 152
Micelles, 129
  mixed, 137
  reverse, 144

Microwave waveguide, 227
Mitochondrial uncoupling protein 2 (UCP2), 127, 137, 161
Molecular docking, 1, 7
MOLSDOCK, 8
Monte Carlo algorithms, 9
Multidimensional decomposition (MDD), 157

N
Nanodiscs, 140, 142
New molecular entities (NMEs), 2
Nicotinic acid, 5
[NiFe] hydrogenases, 76, 78
Nitroxide, 41, 86, 162, 220, 226
NMR, 1ff
  ligand affinity screens, 4
  relaxometry, 108
  structure restraints, 123
NMRScore, 21
NOE, 35, 38, 154
Non-uniform sampling (NUS), 155
^{15}N relaxation, 99
Nuclear magnetic relaxation dispersion (NMRD), 108
Nuclear spin orientation via electron spin locking (NOVEL), DNP, 223
Nucleic acids, magnetic susceptibility, 53

O
Off-resonance error, 103
Oligomers, 47
Outer membrane proteins (Omps), 127
Overhauser effect, 218
1-Oxyl-tetramethyl-3-pyrroline-3-methyl-methanethiosulfonate (MTSL), 162

P
Palmitoyltransferase PagP, 127
Para-hydrogen induced polarization (PHIP), 232
Paramagnetic NMR, 41
Paramagnetic relaxation-assisted condensed data collection (PACC), 204
Paramagnetic relaxation enhancement (PRE), 41, 48, 71, 162
Paramagnetism-based structural restraints, 84
Peroxiredoxins, 25
2-Phenoxybenzoic acid, 5
Phosphatidylcholines, 130, 133
Plant homeo domain (PHD) finger, 76
Plastocyanin, 81
Polarizing agents, 224
Polyacrylamide gel, charged, 53
Poly(ADP-ribose) (PAR), 75
Polyketides, 20
PRE, 35
Projection reconstruction (PR) method, 89
Protein–protein interactions, weak (wPPIs), 35
Proteins, 99
  backbone dynamics, 99
  dynamics, 201
  expression, 123
  E. coli, 126
  flexibility, 13
  paramagnetic, 86
Pseudocontact shifts (PCS), 35, 41, 42, 85
Purple membranes (PM), 61

R
Receiver operating characteristic (ROC) curve, 16
Recoupling, MAS solid-state NMR, 190
Relaxation, 99
  delay, high magnetic field strength, 104
  dispersion, 108
  optimization, 152
Residual chemical shift anisotropy (RCSA), 55, 158
Residual dipolar coupling (RDC), 35, 40, 47, 158
  accuracy improvement, 55
Retinal, 196, 198, 203
Rhodopsin, 61, 126, 188, 192, 196, 198
Rusticyanin, 81

S
Sampling, non-uniform, 90, 155
Scoring, 10
Sensitivity enhancement, 203
Sensory rhodopsin II, 126
Shape-based scoring functions, 12
Singular-value-decomposition (SVD), 51
Sodium dodecyl sulfate (SDS), 130
SOFAST/BEST NMR, 89
SOFAST-HMQC, 7
Solid effect (SE), DNP, 220
Solomon–Bloembergen (SB) equation, 85
Spin labels, 162
Spin-lock R1ρ experiment, 101
Stereo-array isotope labeling (SAIL), 151
Structural changes/dynamics, 187
Structure determination, 187

T
Tabu search algorithms, 9
Tandem zinc fingers (TZFs), 75
Tanimoto index, 17
Target-based NMR screens, 6
TEMPO, 225, 228
Therapeutic drug monitoring (TDM), 229
Therapeutic targets, 18
TOTAPOL, 225, 228
Transducin, 62
Transferred NOE (trNOESY), 38
Transverse relaxation optimized spectroscopy (TROSY), 37, 153

U
Ubiquitin, 60
UCP2, 127, 137, 161

V
VAMP2, 202
VDAC, 127, 148, 155, 157
Virtual screening/assessment, 1, 14

X
Xplor-NIH, 59

Y
YkuJ, 62

Z
ZINC 310454, 23